Thermodynamics

Thermodynamics

James H. Luscombe

CRC Press
Taylor & Francis Group
Boca Raton London New York

CRC Press is an imprint of the
Taylor & Francis Group, an **informa** business

CRC Press
Taylor & Francis Group
6000 Broken Sound Parkway NW, Suite 300
Boca Raton, FL 33487-2742

First issued in paperback 2020

© 2018 by Taylor & Francis Group, LLC
CRC Press is an imprint of Taylor & Francis Group, an Informa business

No claim to original U.S. Government works

Version Date: 20180219

ISBN-13: 978-0-367-57199-3 (pbk)
ISBN-13: 978-1-138-54298-3 (hbk)

Visit the Taylor & Francis Web site at
http://www.taylorandfrancis.com

and the CRC Press Web site at
http://www.crcpress.com

Contents

Preface

> The peculiar multiplicity of formulation and reformulation of the basic thermodynamic formalism is responsible for the apparent complexity of a subject that in its naked form is quite simple.—H.B. Callen[1, p85]

THERMODYNAMICS and statistical mechanics are core components of physics curricula. Statistical mechanics, together with quantum mechanics, provides a framework for relating the macroscopic properties of large collections of atoms (such as in a solid) to the microscopic properties of their constituents. The term *macroscopic* is difficult to define because samples of matter barely discernible to human senses contain enormous numbers of atoms. Statistical mechanics uses probability to make predictions in the face of incomplete information. Predictions are theoretically possible when the state of every atom in a system is known. The trouble is, such information is not available for the enormous number of atoms comprising macroscopic systems, and we must resort to probabilistic methods. Historically, statistical mechanics was developed to account for the results of thermodynamics, a phenomenological theory that presupposes no knowledge of the microscopic components of matter, and instead makes use of concepts such as temperature and entropy. The relation between the two subjects exemplifies a common theme in physics: The role of microscopic theory is to explain the results of macroscopic theory.

In years past, thermodynamics and statistical mechanics were taught separately. With the invariable compression of curricula (squeeze more content into time-constrained curricula), there has been a movement to combine the subjects under the rubric of *thermal physics*. Students can find themselves on shaky ground when thermodynamic ideas are introduced as an outgrowth of the statistical approach, because they're not familiar with thermodynamics. Thermodynamics is a difficult subject that takes time to learn. It's difficult because of its generality. The application of general theories to specific instances taxes a student's understanding of the entire theory; it takes time to become proficient in thermodynamics, to learn its scope and methods.

I have undertaken to write a relatively brief, yet in-depth review of thermodynamics, a *précis*, with emphasis on the structure of the theory. The heart of thermodynamics is entropy—students must learn from the outset that thermodynamics mainly *is* about entropy.[1] To that end, Chapter 1, Concepts of Thermodynamics, has been written with the goal of introducing everything that can be said using only the first law of thermodynamics. The rest of the book therefore is about entropy in some way or form. Entropy, which was discovered in the analysis of steam engines, pertains to all forms of matter (solid, liquid, gas) but also to such disparate systems as information and black holes: Entropy is a universal feature of the physical world. Like anything genuinely new, it cannot be reduced to concepts gained through prior experience. Entropy in my opinion should be learned first from the phenomenological perspective. When later it's stated that the statistical theory of entropy is in accord with the results of thermodynamics, consistency requires that the latter stand in its own right. While thermodynamics is firmly rooted in classical physics, there are a number of instances where thermodynamics uncannily anticipates the existence of quantum mechanics (noted throughout the book). This book is offered against the backdrop of the corpus of physical theory to which the student is unabashedly assumed to have been exposed. There *are* works on thermodynamics that

---

[1] Maxwell wrote, "The touchstone of a treatise on thermodynamics is what is called the second law."[2, p667]

try to keep the subject in a "walled-off garden"; this isn't one of them. I don't pretend that other parts of physics don't exist and will readily point out parallels with other branches of physics.

Still, thermodynamics *is* a "dead" subject, right? That might have been a valid observation at one time, but it's not the case today. Several developments at the forefront of science require that one hone one's understanding of entropy: information theory, the physics of computation, and black hole physics. The book consists of two parts. Part I, Thermodynamics Basics, covers the essential background in thermodynamics required for a study of statistical mechanics. Part II, Additional Topics in Thermodynamics, covers modern yet more specialized applications.

We develop the subject from the beginning, but at a level and pace of advanced undergraduate/beginning graduate students. Sprinkled throughout Part I are sections on mathematical topics (indicated with asterisks); these are "just-in-time" reviews of selected areas of mathematics of particular utility to thermodynamics. I note here the sign convention adopted in this book: Heat and work are treated as positive if they represent energy transfers *to* the system: $\Delta U = Q + W$.

Throughout the book I've reproduced passages from the writings of Gibbs, Clausius, Maxwell, Planck, and others. In my opinion, it's instructive for students to see how the founders of thermodynamics grappled with very subject they are encountering. No attempt has been made to offer a history of thermodynamics.

I thank my colleagues Brett Borden for being my LATEX guru and David Ford for loaning me his laptop that I took on a backpacking trip to Spain. That laptop went into the backpack and followed me around; I would work on this book sipping *una caña*. I thank my colleague Andrès Larraza who, without knowing it, would catalyze sudden insights into thermodynamics with his offhand remarks. I thank Ted Jacobson for useful discussions on black hole thermodynamics. I thank Blake McCracken for comments on the manuscript, and Evelyn Helminen for making the figures. I thank the editorial staff at CRC Press, in particular Francesca McGowan and Rebecca Davies. I thank my family, for they have seen me too often buried in a computer. To their queries, "How's it coming?" came the usual reply: slow, glacially slow. My wife Lisa I thank for her encouragement and consummate advice on how not to mangle the English language. Finally, to the students of the Naval Postgraduate School, I have learned from you, more than you know. Keep in mind that science is a "work in progress"; more is unknown than known.

James Luscombe

Monterey, California

I

Thermodynamics Basics
(for Advanced Students)

CHAPTER 1

Concepts of thermodynamics

Equilibrium, energy, and irreversibility

THERMODYNAMICS is a science of matter presupposing no knowledge of the constitution of matter! As far as thermodynamics is concerned, matter could equally well be a sack of hammers as a box of atoms. For that reason, a high level of confidence is accorded to the results of thermodynamics: Our concepts of matter may change, but conclusions reached by means of thermodynamics will not. While nothing prevents us from using our knowledge of the microscopic nature of matter to augment the understanding achieved by thermodynamics, the central concepts of thermodynamics—energy and entropy—were formulated without a picture of the internal constitution of matter. That, in a nutshell, is the *strength* of thermodynamics: independence of the details of physical systems. Entropy, discovered in the analysis of heat engines, has nothing *specifically* to do with steam engines and is a universal feature of the physical world that scientists and engineers must get to know. The purpose of this chapter is to introduce the concepts of thermodynamics, its basic vocabulary, and the first law of thermodynamics—conservation of energy. Entropy is a consequence of the second law of thermodynamics, the direction of heat flow, which is introduced in Chapter 2.

1.1   THE MANY AND THE FEW

The language of thermodynamics is couched in the measurable properties of macroscopic systems. A definition of *macroscopic* is elusive, as well as its correlative *microscopic*. A representative figure for the number of atoms in macroscopic samples of matter is provided by *Avogadro's number*, $N_A \equiv 6.022 \times 10^{23}$. What Avogadro's number counts is discussed below; the point is that *even minute amounts of matter contain enormous numbers of atoms* and can be considered macroscopic.[1] That's the many.

What we can say about macroscopic systems is what we can *measure*. That's the few. The number of quantities that can be measured is minuscule in comparison to the number of atoms in macroscopic matter. Thermodynamics is a *phenomenological* theory: It *describes* interrelations among measurable quantities, but in a strict sense does not *explain* anything.[2] The subject of statistical mechanics, which builds on thermodynamics, is an explanatory framework for relating the physical properties of macroscopic collections of atoms to the microscopic properties of their parts.

---

[1] A cube of solid matter 1 μm on a side, barely visible to the human eye, contains on the order of 10^{10} atoms.

[2] Class assignment: Discuss the difference between description and explanation.

Thermodynamics deals with a special yet ubiquitous state of matter, that of *thermodynamic equilibrium* (defined in Section 1.2). Remarkably, systems having enormous numbers of microscopic degrees of freedom can, in equilibrium, be described by just a few macroscopic quantities called *state variables*, those that characterize the *state* of equilibrium.[3] Quantum mechanics pertains at the microscopic level, yet the methods of quantum mechanics can practically be applied only to systems with just a few degrees of freedom, say[4] $N \sim 10$. Whatever the size of systems that can be treated quantum mechanically, a microscopic description is impossible[5] for $N \sim N_A$. Not only is it impossible, it's *pointless*. Even if one could solve for the wave function of N_A particles, what would you do with that information? Wouldn't you immediately seek to reduce the complexity of the data through some type of averaging procedure? *Thermodynamics is concerned with the average properties of macroscopic systems in thermal equilibrium.*

Sidebar discussion: What does Avogadro's number count? N_A is defined such that the mass of a *mole* of an element equals the element's atomic weight, in grams. That such a number is possible is because 1) the mass of an atom is almost entirely that of its nucleus and 2) nuclei contain integer numbers of essentially identical units, *nucleons* (protons or neutrons). In free space protons have mass $m_p = 1.6726 \times 10^{-24}$ g and neutrons $m_n = 1.6749 \times 10^{-24}$ g. In nuclei their mass is smaller because of the nuclear binding energy. By the equivalence of mass and energy ($E = mc^2$), the mass of a bound object is less than the mass it has as a free object. The *atomic mass unit*, u, is defined as one twelfth the mass of a ^{12}C atom, $1\,\mathrm{u} \equiv \frac{1}{12}\mathrm{m}(^{12}\mathrm{C}) = 1.6605 \times 10^{-24}$g (which is *less* than m_p or m_n). Avogadro's number is defined so that the mass of one mole of ^{12}C is exactly 12 g,

$$N_A \cdot \mathrm{m}(^{12}\mathrm{C}) = N_A \cdot 12 \cdot \frac{1}{12}\mathrm{m}(^{12}\mathrm{C}) = 12 N_A \cdot 1.6605 \times 10^{-24}\ \mathrm{g} \equiv 12\ \mathrm{g}\,,$$

implying that $N_A = (1.6605 \times 10^{-24})^{-1} = 6.022 \times 10^{23}$. How many molecules are in 1 cm^3 of water? The mass density of water is about 1 gm/cm^3. The molar mass of H_2O is 18.015 g; call it 18 g. One gram of H_2O is then $\frac{1}{18}$ of a mole, implying $\frac{1}{18}N_A \approx 3.3 \times 10^{22}$ molecules.

1.2 EQUILIBRIUM AND TIMELESSNESS

A macroscopic system is said to be in equilibrium *when none of its properties appear to be changing in time*. The *state* of equilibrium is specified by the values of its state variables, which are what we can measure. In contrast to the enormous number of microscopic degrees of freedom possessed by macroscopic systems, there are only a *handful* of state variables. Thermodynamics therefore represents a huge reduction in the complexity of description. How is such a simplification achieved? The signals produced by measuring instruments represent *time averages* of microscopic processes occurring on time scales far shorter than that over which instruments can typically report data. A pressure gauge, for example, reports the average force per area produced by numerous molecular impacts. Thermal equilibrium is the state where, to quote Richard Feynman, "all the 'fast' things have happened and all the 'slow' things not."[6, p1] We know for example that in 5 billion years the sun will run out of hydrogen and possibly explode, disastrously affecting thermal equilibrium on Earth. *All systems change in the course of time*; it's a matter of time scales.[6] Equilibrium is the

---

[3]The term *state variable* is due to J. Willard Gibbs:[3, p2] "The quantities V, P, T, U, and S are determined when the state of the body is given, and it may be permitted to call them *functions of the state of the body*."

[4]Microscopic descriptions are not possible with the usual methods of quantum mechanics when the number of particles exceeds a small number. Non-classical correlations arise called *entanglement*, making the wave function a complicated object to deal with analytically. Computers can simulate the quantum physics of approximately 40 particles.[4] Walter Kohn argued in his Nobel lecture that the many-particle wave function $\psi(r_1, \cdots, r_N)$ is *not a legitimate scientific concept for* $N > N_0$, where $N_0 \approx 10^3$.[5] For $N \sim N_A$, one seeks averages provided by thermodynamics and statistical mechanics.

[5]Even if one could conceive a computer large enough to solve for the dynamical evolution of N_A particles, where would one get the initial data with sufficient precision to permit such calculations?

[6]If it seems we're being needlessly fussy over the definition of equilibrium, the question of what constitutes an equilibrium state has an impact on the third law of thermodynamics, Chapter 8.

macroscopically quiescent state where the system ceases changing over the time it takes to complete experiments; it's *effectively* a timeless state described by *variables that are independent of time.*[7] While time does not appear explicitly in the state variables of thermodynamics, it occurs *implicitly* in the *direction* of time! As we'll see, *irreversibility* prescribes the *order* of states as they occur in natural processes, giving an interpretation to the direction of time (Section 1.7). Time is implicit in thermodynamics with its emphasis on *processes*, which occur in time.[8]

A thermodynamic description would not be possible if the equilibrium properties of systems were not *reproducible*. It's a fact of experience that systems in equilibrium can be reproducibly specified by just a few conditions.[9] Equilibrium is characterized by the *values* of state variables, not by *how* they have come to have their values. In equilibrium, any memory is lost of how the system came to be in equilibrium. In saying that a sample of water has a certain temperature, one has no knowledge of how many times in the past week it has frozen and thawed out. There *are* systems where this feature does not hold: Imagine a paperclip you've been bending back and forth so many times that it's just about to break. The present state of that system *is* dependent on its history. *History-dependent substances are outside the scope of classical thermodynamics*, although it could be said that such systems are not in equilibrium. Examples of state variables are pressure P, volume V, and temperature T; these three show up in most thermodynamic descriptions.[10] Other examples are magnetization M, chemical potential μ, and entropy S.

State variables can be classified as *intensive* and *extensive*. Intensive quantities *have the same values at different spatial locations of the system* (P, T, and μ) and are independent of the size of the system. Extensive quantities *are proportional to the number of particles in the system* and are characteristic of the system *as a whole* (S, V, and M); extensive quantities are *additive* over independent subsystems.[11] Mass is extensive: Doubling the mass while maintaining the density fixed, doubles the volume. We'll see that intensive quantities occur in the theory as *partial derivatives* of one extensive variable with respect to another; intensive quantities are proportionality factors between extensive quantities under precisely defined conditions.

Thermodynamic state space (or simply *state space*) is a mathematical space of the values of state variables (which by definition are their equilibrium values). A state of equilibrium is therefore represented by a point in this space.[12] Thermodynamics, however, is not concerned solely with systems in equilibrium, but with *transitions* between equilibrium states. Between any two points of state space, a *possible* change of state is implied. There are an *unlimited* number of possible "paths"

---

[7]We say *effectively* timeless because systems remain in equilibrium *until the environment changes*. The distinction between a system and its environment is discussed in Section 1.4.

[8]It's sometimes said that thermodynamics should be called *thermostatics* because of the timeless nature of the equilibrium state. If all there was to thermodynamics was the first law (defined shortly), there would be merit to this idea. The second law, however (Chapter 2), a recognition of the existence of irreversibility, prescribes a *time order* to the states of systems as they occur in natural processes. The term "thermostatics" would completely hide the past-and-future relation of states in irreversible processes, arguably the central result of the theory of thermodynamics.

[9]One can ask just *how* reproducible equilibrium states are. *Fluctuations* in the values of state variables due to the randomness of molecular motions are treated in statistical mechanics. Fluctuations are difficult to motivate in thermodynamics because state variables are presumed to have fixed values in equilibrium. See the discussions in Sections 3.10 and 12.3.

[10]We could "go Dirac" and label an equilibrium state as a list of the values of its state variables, such as $|PVT\rangle$.

[11]Intensivity and extensivity hold strictly for systems that are *infinite* in size; the concept would not apply for systems that are so small as to be dominated by surface effects.[7, p10]

[12]To quote Gibbs, "Now the relation between the volume, entropy, and energy may be represented by a surface, most simply if the rectangular coordinates of the various points of the surface are made equal to the volume, entropy, and energy of the body in its various states."[3, p33] Gibbs added a footnote: "Professor J. Thomson has proposed and used a surface in which the coordinates are proportional to the volume, pressure, and temperature of the body. It is evident, however, that the relation between the pressure, volume, and temperature affords a less complete knowledge of the properties of the body than the relation between volume, entropy, and energy. For, while the former relation is entirely determined by the latter, and can be derived from it by differentiation, the latter relation is by no means determined by the former." Gibbs is saying that it's "better" to represent the equilibrium state in terms of *extensive* variables rather than intensive because intensive variables can be uniquely expressed as derivatives between extensive variables. We'll see in Chapter 4 that through the use of Legendre transformations, intensive variables can be used as independent variables.

between the same endpoints in state space. Because the values of state variables are independent of the history of the system, *changes* in state variables must be described independently of the manner by which change is brought about. This type of path independence is ensured mathematically by requiring differentials of state variables to be *exact differentials*.

1.3 EXACT DIFFERENTIALS*

We begin our series on the mathematics of thermodynamics (indicated with an asterisk) in which we pause to review mathematical topics of particular utility to thermodynamics. The differential of a function $f(x, y)$ is, from calculus,

$$\mathrm{d}f = \left(\frac{\partial f}{\partial x}\right)_y \mathrm{d}x + \left(\frac{\partial f}{\partial y}\right)_x \mathrm{d}y \,.$$

In thermodynamics it pays to be fastidious about indicating which variables are held fixed in partial derivatives, and we'll adhere to that practice. Functions usually (but not always) have continuous mixed second partial derivatives, $\partial^2 f / \partial y \partial x = \partial^2 f / \partial x \partial y$. In the notation where we indicate what we're holding fixed, equality of mixed partials would be written $(\partial/\partial y \, (\partial f/\partial x)_y)_x = (\partial/\partial x \, (\partial f/\partial y)_x)_y$.

Suppose you're given an expression involving the differentials of two independent variables, $P(x, y)\mathrm{d}x + Q(x, y)\mathrm{d}y$, what's known as a *Pfaffian differential form*.[13] A question we must answer is, *is a Pfaffian the differential of a function*? That is, is there a function, call it $g(x, y)$, such that $\mathrm{d}g = P(x, y)\mathrm{d}x + Q(x, y)\mathrm{d}y$, so that $\mathrm{d}g$ can be integrated? For a Pfaffian to be the differential of an integrable function, the *integrability condition* must be satisfied:

$$\left(\frac{\partial P}{\partial y}\right)_x = \left(\frac{\partial Q}{\partial x}\right)_y \,. \tag{1.1}$$

If $P(x, y)$ and $Q(x, y)$ satisfy Eq. (1.1), $\mathrm{d}g$ is said to be an *exact differential*. The issue of integrability is fundamental to thermodynamics; we consider it again in Chapter 10.

Example. Is $(x^2 + y)\mathrm{d}x + x\mathrm{d}y$ an exact differential? Applying Eq. (1.1), $P(x, y) = x^2 + y$ and $Q(x, y) = x$. Thus, $\partial P/\partial y = 1 = \partial Q/\partial x$. Yes, it's an exact differential.

Exact differentials are well suited to serve as the differentials of state variables. It is necessary and sufficient for the integral of a Pfaffian to be independent of the path of integration that it be exact.[9, p353] In such a case,

$$\int_A^B (P(x, y)\mathrm{d}x + Q(x, y)\mathrm{d}y) = \int_A^B \mathrm{d}g = g(B) - g(A) \,. \tag{1.2}$$

If a Pfaffian $P(x, y)\mathrm{d}x + Q(x, y)\mathrm{d}y$ is exact, the function $g(x, y)$ (that it represents the differential of) can be found by integration (up to a constant), $g(x, y) = g(x_0, y_0) + \int_{x_0, y_0}^{x, y} (P(x', y')\mathrm{d}x' + Q(x', y')\mathrm{d}y')$. For example, $(x^2 + y)\mathrm{d}x + x\mathrm{d}y$ is the differential of $g(x, y) = g_0 + \frac{1}{3}x^3 + xy$, where g_0 is a constant. It follows from Eq. (1.2) that the integral of an exact differential around a *closed path* is zero,

$$\oint (P(x, y)\mathrm{d}x + Q(x, y)\mathrm{d}y) = \oint \mathrm{d}g = 0 \,. \tag{1.3}$$

---

[13]See for example [8, p18]. The term "Pfaffian" is old fashioned; today a linear differential form would be referred to as a "1-form." For this book, Pffafian will suffice.

Equation (1.3) is an important, oft-used property of state variables.

Notice how knowledge of $P(x,y)\mathrm{d}x + Q(x,y)\mathrm{d}y$ differs from the usual problem in calculus where for a given function you're asked to find its differential. Here you are given a differential expression and are asked, first, is it the differential of a function, and, second, what is that function? Only when you have answered the first question can the second be answered. Don't all differentials represent the differential of a function? No. Take the above example and modify it slightly: Is $(x^2 + 2y)\mathrm{d}x + x\mathrm{d}y$ exact? No, because $\partial(x^2+2y)/\partial y \neq \partial(x)/\partial x$. *There is no function* in this case, call it $h(x,y)$, such that $\mathrm{d}h = (x^2 + 2y)\mathrm{d}x + x\mathrm{d}y$. Differential forms that are not exact are called *inexact*. Are inexact differentials "useless"? The subject of thermodynamics is loaded with them! A special notation is used to indicate inexact differentials: đ. Thus we can write $đh = (x^2 + 2y)\mathrm{d}x + x\mathrm{d}y$; $đh$ is *notation*, shorthand for the inexact Pfaffian $(x^2 + 2y)\mathrm{d}x + x\mathrm{d}y$. Can inexact differentials be integrated? Certainly, but here's the crucial difference: The value of the integral depends on the path of integration. That's why the integral of an inexact differential does not represent a function in the usual sense. What meaning can be given to an expression like $h(x,y) = \int^{x,y} đh$ when the value assigned to the integral is different for every path of integration? The "function" defined this way depends not just on (x,y) but on "how you get there," or in the language of thermodynamics, on the *process*.

An *integrating factor* is a function that, when multiplied by an inexact differential, becomes exact. Example: Let $đh = (3xy + y^2)\mathrm{d}x + (x^2 + xy)\mathrm{d}y$. Is it exact? (No—Check it!) But if we multiply by x, $x đh = (3x^2y + xy^2)\mathrm{d}x + (x^3 + x^2y)\mathrm{d}y$, which *is* exact. (Check it!) In this case the variable x is an integrating factor and we can write $x đh = \mathrm{d}f$ where $\mathrm{d}f = (3x^2y + xy^2)\mathrm{d}x + (x^3 + x^2y)\mathrm{d}y$. In general, an integrating factor $\lambda(x,y)$ is that which produces an exact differential when multiplied by an inexact differential: $\mathrm{d}f = \lambda(x,y)đh$. The existence of integrating factors is a topic of *fundamental* importance to thermodynamics, something we return to in Chapter 10.

1.4   INTERNAL ENERGY: WORK, HEAT, AND BOUNDARIES

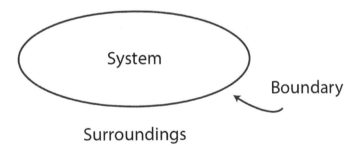

Figure 1.1   The boundary separates a system from its surroundings

We've been somewhat cavalier in our use of the word "system." Obviously, to apply a theory one must know *what* the theory is being applied to! Yet, a frequent source of difficulty in applying thermodynamics comes from confusion over what comprises the system. As shown in Fig. 1.1, it should be possible to draw a *boundary* around that little part of the universe we consider to be of interest—the system. There *must* be a demarcation between the system and its *environment* or *surroundings*: One can't model the entire universe.[14] The boundary is not simply a mental construct,

---

[14]Or can we? Cosmology is the study of the universe, and thermodynamical ideas feature in that subject. While we're on the topic, expositions on thermodynamics have a tendency to invoke the universe ("entropy change of the universe"). It's true that what's not the system is everything else, and thus logically not-system = universe, but such an inference cannot be taken literally. Only bodies external to a system reasonably within its vicinity can be expected to be in relation with

it's an actual physical boundary, *not part of the system*. In thermodynamics, what something *is* is what it *does*, and the behavior of systems is *selected* by the nature of the boundaries.[15]

At this point we distinguish two kinds of boundaries (or *walls* or *partitions*): *diathermic* and *adiabatic*, those that do and do not conduct heat.[16] Adiabatic boundaries allow a system to interact with its surroundings *through mechanical means only* (no flow of heat or matter). Adiabatic walls are an idealization, but they can be so well approximated in practice, as in a vacuum flask, that it's not much of a stretch to posit the existence of perfectly heat-insulating walls. Systems enclosed by adiabatic walls are said to be *adiabatically isolated*. Diathermic walls allow heat transport; systems separated by diathermic walls are said to be in *thermal contact*. Walls can be moveable. *The two types of boundary allow us to distinguish two types of energy transfer between the system and its environment*: *work* and *heat*.

Through much experimentation (starting in the 1840s with the independent work of James Joule and Robert Mayer), it's been established that in the performance of *adiabatic work*, W_{ad} (work on adiabatically isolated systems), the transition $i \to f$ it produces between reproducible equilibrium states (i, f), depends only on the *amount* of work and not on *how* it's performed. For a fluid system, part of the work could come from compression and part could come from stirring. Regardless of how the proportions of the *types* of work are varied, the same *total amount* of adiabatic work results in the same transition $i \to f$. *This discovery is of fundamental importance*. If the transition $i \to f$ is independent of the means by which it's brought about, it can only depend on the initial and final states. It implies the existence of a physical quantity associated with equilibrium states that "couples" to adiabatic work, the *internal energy* U that depends only on the *state* of the system, such that[17]

$$\Delta U = U_f - U_i = W_{ad} . \tag{1.4}$$

Internal energy is a state variable. Adiabatic work done *on* the system *increases* its internal energy, and is taken as a positive quantity. Adiabatic work done *by* the system (somewhere in the environment a weight is higher, a spring is compressed) is accompanied by a *decrease* in internal energy, and is taken as a negative quantity. *Changes in internal energy come at the expense of adiabatic work done on or by the system*. Equation (1.4) expresses *conservation of energy*: If we don't let heat escape, work performed on a system is *stored* in its internal energy, energy that can be recovered by letting the system do adiabatic work on the environment. Internal energy is the *storehouse* of adiabatic work.

Now let work W be performed under *nonadiabatic* conditions. It's found, for the same transition $i \to f$ produced by W_{ad}, that $W \neq \Delta U$. *The energy of mechanical work is not conserved in systems with diathermic boundaries*. Conservation of energy, however, is one of the sacred principles of physics and we don't want to let go of it. The principle can be restored by recognizing different *forms of energy*.[18] The *heat absorbed by the system*, Q, is defined as the difference in work

$$Q \equiv \Delta U - W \tag{1.5}$$

---

the original system. What goes on elsewhere is immaterial. One can't generalize physical laws based on local phenomena (thermodynamics) to cosmological distances. The large-scale structure of the universe is described by the general theory of relativity, wherein the concept of energy is conceived differently than in thermodynamics (scalar vs. tensor).

[15] As with any boundary value problem (studied in mathematical physics), what systems do is determined by *boundary conditions*, the specification of how the system interacts with its environment.

[16] We'll introduce a third kind of boundary in Section 3.5 that allows the passage of particles into the system: permeable boundaries. We introduce in Section 3.3 the concept of an isolated system—no interactions of any kind.

[17] So, we have the almighty equal sign, our first equality between physical quantities. An equation $A = B$ should not be read simply as A "equals" B, but rather A is *the same thing as* B. In writing Eq. (1.4), $\Delta U = W_{ad}$, we're saying that the change in a physical quantity, the internal energy U, is the same as the amount of adiabatic work done on the system. If that's all there was to the internal energy, Eq. (1.4) would simply be a change of variables. But there *is* more to internal energy than W_{ad}. There are different *forms* of energy, with ΔU the sum of the work done on the system and the heat transferred to it, Eq. (1.5). The equal sign in Eq. (1.4) is a *physical* equality that defines the internal energy.

[18] Neutrinos were invented for the purpose of preserving energy conservation in nuclear beta decay.

that effects the same change in state on systems with the two types of boundaries, $Q = W_{ad} - W$. If it takes more work W to produce the same change of state as that under adiabatic conditions, $Q < 0$: heat *leaves the system* by flowing through the boundary; $Q > 0$ corresponds to heat entering the system.[19] Equation (1.5) is the *first law of thermodynamics*. In Section 3.5 we allow a third type of wall (a *permeable boundary*) that permits the flow of particles. There we introduce another kind of work (*chemical work*), the energy required to change the amount of matter in the system. The point here is that *the nature of the boundaries allows us to classify different types of energy*.

Three types of systems can be distinguished by the nature of their boundaries and the interactions they permit with the environment (summarized in Table 1.1). Adiabatically isolated systems are enclosed by adiabatic walls; they interact with the environment through mechanical means only. *Closed systems* interact with the environment thermally and have a fixed amount of matter. *Open systems* allow the flow of energy and matter.

Table 1.1   Systems, boundaries, and interactions

System	Boundaries	Interaction with environment
Adiabatically isolated	Adiabatic	Mechanical (adiabatic work only)
Closed	Diathermic	Thermal (flow of heat, mass fixed)
Open	Permeable	Permits flow of energy and matter

Sidebar discussion: Internal vs. mechanical energy
Internal energy is analogous to, but not the same as mechanical energy. The mechanical work done

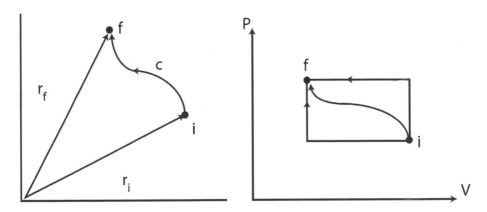

Figure 1.2   Mechanical work is independent of path for conservative forces (left); adiabatic work is independent of process (right)

by a force F on a particle of mass m in producing an infinitesimal displacement dr is $F \cdot dr = m(dv/dt) \cdot dr = m(dv/dt) \cdot v dt = \frac{1}{2}m d(v^2)$. Over a finite displacement $W_{if} = \int_C F \cdot dr = \frac{1}{2}m(v_f^2 - v_i^2) \equiv \Delta K$, where C is a path connecting r_i with r_f (see Fig. 1.2). Thus we have the

---

[19]We have adopted a consistent *sign convention*: Q and W are positive if they represent energy transfers *to* the system. In other books, positive work is work done *on* the environment. That sign convention (what we don't use) derives from what we want in an engineering sense from heat engines: work output as a result of heat input. I prefer to treat the sign conventions for Q and W on equal footing—energy transfers to the system (in the form of work or heat) are considered positive.

work-energy theorem: The change in kinetic energy ΔK equals the work done on the particle, $\Delta K = W_{if}$. The theorem requires that we know the trajectory of the particle: We need to know everything about the motion before we can apply the theorem! For *conservative forces*, W_{if} is independent of path and depends only on the endpoints. A force is conservative if the vector field $\boldsymbol{F}(\boldsymbol{r})$ is derivable from the gradient of a scalar field $V(\boldsymbol{r})$ (the potential energy function), $\boldsymbol{F}(\boldsymbol{r}) = -\boldsymbol{\nabla}V(\boldsymbol{r})$. In that case, $W_{if} = \int_C \boldsymbol{F} \cdot \mathrm{d}\boldsymbol{r} = -\int_C \boldsymbol{\nabla}V \cdot \mathrm{d}\boldsymbol{r} = -\int_C \mathrm{d}V = -(V_f - V_i) \equiv -\Delta V$; W_{if} is the same for any path between the endpoints in physical space. For conservative forces, therefore, $K_f + V_f = K_i + V_i$: *Mechanical energy $E \equiv K + V$ is invariant*; it's *conserved* along the trajectory. Work can appear as kinetic or potential energy, interconvertible forms of energy with $\Delta E = 0$. In thermodynamics, adiabatic work W_{ad} performed on the system induces the same transition $i \to f$ between equilibrium states (i, f) *regardless of how the work is done* (the "path"). There is a quantity, internal energy, that depends only on the endpoints *in state space* such that $\Delta U_{if} = W_{\mathrm{ad}}$. Internal energy is analogous to mechanical energy, except that U is on the "potential" side of the ledger (internal energy excludes the kinetic energy of the system as a whole). Internal energy is recoverable (conserved) by letting the system perform adiabatic work on the environment. For systems with diathermic boundaries, *mechanical energy is not conserved*: The energy content of the system cannot be accounted for solely in terms of adiabatic work; energy is passed to the system in another form, *heat*. Heat can be quantified as the *difference* in work $Q \equiv W_{\mathrm{ad}} - W$ that brings about the same change in state $i \to f$ when the system has the two types of boundaries. Internal energy thus pertains to mechanically non-conservative systems with $\Delta U = W + Q$.

The division of internal energy into work and heat lines up with the distinction between macroscopic and non-macroscopic, i.e., microscopic. Work is defined in thermodynamics more broadly than in mechanics as a *generalized displacement* $\mathrm{d}X$ of an extensive quantity X multiplied by its conjugate *generalized force* Y, an intensive quantity such that the product YX has the dimension of energy (Section 1.9). *Work is associated with changes in observable macroscopic quantities. What's not work is heat*, which is *energy transferred to microscopic degrees of freedom*. One of the consequences of the second law of thermodynamics is the existence of new extensive state variable, entropy S, such that a small quantity of heat $đQ$ transferred to a system is associated with a generalized displacement, $đQ = T\mathrm{d}S$ (Chapter 3).[20] We'll see (Chapter 7) that entropy is a quantitative measure of the number of microscopic configurations of the system ("microstates") that are consistent with an equilibrium state specified by given values of the state variables.

We can write the first law in differential form (compare with Eq. (1.5)),

$$\mathrm{d}U = đQ + đW \, . \tag{1.6}$$

A small change in internal energy $\mathrm{d}U$ can be produced by transferring to the system a small quantity of heat $đQ$ and/or performing on the system a small amount of work $đW$. The Joule-Mayer experiments established that a definite, reproducible amount of work would raise the temperature of a sample of water the same as a known quantity of heat, and thus heat and work are *interchangeable* forms of energy.[21] The number of combinations of $đQ$ and $đW$ that affect the same change $\mathrm{d}U$ is obviously unlimited. The notation of Eq. (1.6) tells the story. The left side indicates that $\mathrm{d}U$ depends only on the initial and final equilibrium states (because it's exact), and that there *is* a physical quantity U, internal energy, of which $\mathrm{d}U$ represents a small change. The right side of Eq. (1.6) indicates that there do *not* exist substances Q and W of which $đQ$ and $đW$ represent small changes; $đQ$ and $đW$ represent *modes* of infinitesimal energy transfer to the system. Work is not a substance; neither is heat. Both are forms of energy. An early (obsolete) idea was that there's a substance called *caloric*

---

[20]Thus, $\mathrm{d}S = đQ/T$; T^{-1} is the integrating factor for $đQ$.

[21]The *mechanical equivalent of heat*, 4.184 J/cal (a *calorie* is a non-SI unit of energy, that is required to raise the temperature of one gram of water by one degree Celcius), is recognized today as the specific heat of water. Specific heat is defined in Section 1.11.

in every body that cannot be *created*, only transferred.[22] Today our view is that heat is something that *happens* as changes are made to a system. Heat is transferred between objects in processes, but is not something pre-existing in an object. When you "fry" electronic components in the laboratory, the joke is that you "let the smoke out." Was the smoke already contained in that diode, or did it occur as a result of what you did to it?[23] The same is true of heat: a process-related quantity, not something pre-existing; *heat has no independent existence*. We have in the first law that the sum of two inexact differentials is exact, which is the mathematical expression of the physical content of Eq. (1.6), that energy can be transformed from one form to another. For *finite* changes, Eq. (1.6) is written as in Eq. (1.5), $\Delta U = W + Q$.

The laws of thermodynamics

We will continually be referring to the laws of thermodynamics, and it's useful to gather them in one location together with succinct summaries, even though we can't do them justice here.

- Zeroth law (Section 1.5): Temperature exists as a property of equilibrium systems as does an equation of state, a functional relation among state variables.

- First law (Section 1.4): Internal energy U is established as a state variable. Internal energy is conserved; work and heat are interconvertible forms of energy.

- Second law (Chapter 2): Heat flows spontaneously from hot to cold, never the reverse. The existence of a new state variable, entropy S is established (Chapter 3), as well as absolute temperature T (Section 2.4).

- Third law (Chapter 8): Changes in entropy vanish as T approaches absolute zero. It's not possible to achieve absolute zero temperature.

1.5 EMPIRICAL TEMPERATURE AND EQUATION OF STATE

Figure 1.3 System A in equilibrium with B and C, implies equilibrium between B and C

Equilibrium is *transitive*: If system A is in equilibrium with B and B is in equilibrium with C, *then A is in equilibrium with C* (see Fig. 1.3). This fact of experience is known as the *zeroth law of thermodynamics* which, as we now show, implies the existence of an intensive state variable known as temperature.[24]

---

[22]Joule discovered in 1840 that the flow of current I in a resistance R is accompanied by the development of heat proportional to $I^2 R$ (*Joule heating*). Joule concluded from his experiments that caloric (heat) is *created*, which cannot be if caloric is a conserved substance transmitted between bodies. Objections to Joule's findings centered on the fact because current is caused by chemical reactions (in a battery), such reactions are the source of caloric, which is somehow redistributed to the resistor. Joule then produced currents using electromagnetic induction, which also led to the production of heat in a resistor. Critics countered that caloric somehow leaked out of the magnets used to produce induced currents. Joule concluded that because the method of producing currents turns out to be irrelevant, caloric is created rather than displaced between objects. Established theories shown to be wrong often die slow deaths.

[23]Analogously, neutrinos are created in the process of beta decay; they're not pre-existing in the nucleus.

[24]The name "zeroth law" (to reflect its logically prior status to the first law) entered the physics lexicon in 1939[10, p56]: "This postulate of the '*Existence of temperature*' could with advantage be known as *the zeroth law of thermodynamics*." The principle, however, had long before been recognized. It appeared in Planck's work as early as 1897.[11, p2]

A *simple fluid* (composed of a single chemical species) is characterized by two state variables, P and V. Two simple fluids, A and B, are thus described by four variables, P_A, V_A, P_B, V_B. Placing the fluids in thermal contact, there will be a relation between the four equilibrium values of these variables $F(P_A, V_A, P_B, V_B) = 0$, where the function F depends on the fluids (water, alcohol, etc.). We show that $F = 0$ occurs in the form $\phi_A(P_A, V_A) - \phi_B(P_B, V_B) = 0$, i.e., F (a function of four variables) separates into functions ϕ_A, ϕ_B, each a function of the individual system variables.

Consider three simple fluids, A, B, and C. For A and B in equilibrium, there's a functional relation between the variables

$$F_1(P_A, V_A, P_B, V_B) = 0 . \tag{1.7}$$

Similarly, for B in equilibrium with C and C in equilibrium with A we have the functional relations

$$F_2(P_B, V_B, P_C, V_C) = 0 \qquad F_3(P_C, V_C, P_A, V_A) = 0 . \tag{1.8}$$

The two equations in Eq. (1.8) can be solved for P_C: $P_C = f_2(P_B, V_B, V_C)$ and $P_C = f_3(P_A, V_A, V_C)$. Thus,

$$f_2(P_B, V_B, V_C) = f_3(P_A, V_A, V_C) . \tag{1.9}$$

By the zeroth law, for A and B separately in equilibrium with C, A and B are in equilibrium. Equation (1.9) must therefore be equivalent to Eq. (1.7). But Eq. (1.7) is independent of V_C. The functions f_2, f_3 must be such that V_C cancels out of Eq. (1.9). The most general form that would permit this is (for functions α and β)

$$f_2(P_B, V_B, V_C) = \phi_2(P_B, V_B)\alpha(V_C) + \beta(V_C)$$
$$f_3(P_A, V_A, V_C) = \phi_1(P_A, V_A)\alpha(V_C) + \beta(V_C) .$$

Substituting these equations into Eq. (1.9), we have that $\phi_1(P_A, V_A) = \phi_2(P_B, V_B)$. QED.

Repeating the analysis for (B, C) in equilibrium with A and (C, A) in equilibrium with B, we arrive at the equalities $\phi_1(P_A, V_A) = \phi_2(P_B, V_B) = \phi_3(P_C, V_C)$. *There's something the same about systems in equilibrium*, what's called the temperature. If we consider a single system to be composed of a network of sub-systems, all in thermal contact, then the state of equilibrium is characterized by a single temperature, the same at all points of the system (an intensive quantity). The functions ϕ_1, ϕ_2, and ϕ_3 of the state variables of systems A, B, and C are *equations of state*, with their common value (in equilibrium) the *empirical temperature*, θ:

$$\phi(P, V) = \theta . \tag{1.10}$$

The temperature θ is called empirical because it's determined by a *thermometric property* of particular substances (a property that changes with temperature, such as the height of a column of mercury). The θ-temperature of an object can be measured by placing it in contact with a thermometer (literally thermo-meter) constructed from the particular substance.

In establishing Eq. (1.10) we considered the simplest system, with equilibrium described by two state variables. The argument can be repeated for substances having equilibrium characterized by any set of variables $\{X_i\}_{i=1}^n$, and we'd arrive at an equation of state and empirical temperature,[25] $\phi(X_1, \cdots, X_n) = \theta$. The set of all states having the property $\phi(X_1, \cdots, X_n) = \theta_0$ for a fixed value of θ_0 is referred to as an *isotherm*.[26]

---

[25] A thermodynamic description must involve at least two state variables. Equilibrium is characterized by temperature (that's one). In adiabatic processes, non-thermal energy is imparted to the system by changes in at least one other variable (that's two), what in Section 1.9 we term a *deformation coordinate*. There must be a *non-deformation* or *thermal* coordinate, together with at least one deformation coordinate.

[26] An isotherm is technically a *hypersurface* in state space. Our familiar notion of surface (such as the surface of an orange) is a two-dimensional manifold embedded in three dimensional space; a hypersurface is an $(n-1)$-dimensional manifold embedded in n-dimensional space. For simplicity we'll refer to hypersurfaces as surfaces.

Temperature correlates with our physiological sensations of hot and cold,[27] but should not be confused with *heat*. Heat is a *mode* of energy transfer (Section 1.4); it has no independent existence. Temperature is a state variable. We show in Chapter 2 that an *absolute temperature* T can be defined that's independent of the thermometric properties of substances. In what follows we'll use T exclusively to denote absolute temperature, reserving θ for empirical temperature. Almost all formulas in thermodynamics involve the absolute temperature T.

Note what's happened here: We've discovered a state variable (temperature) associated with *equilibrium itself*; *temperature is not a property of the microscopic constituents of the system*, it's a property possessed by a system in equilibrium. We'll see that entropy has the same quality of not being a microscopic property of matter, and is a property of the equilibrium state.

1.6 EQUATION OF STATE FOR GASES

State variables do not have to be independent of each other, and they usually aren't. Much of the mathematical "apparatus" of thermodynamics is devoted to exposing interrelations among state variables. We're aided in this task through the existence of equations of state, functional relations among state variables. In this section we focus on the equation of state for gases.

The quantity PV has the dimension of energy.[28] (Check it!) Given the emphasis on internal energy as a state variable, let's guess an equation of state in the form $PV = \psi$, where ψ is a characteristic of equilibrium having the dimension of energy. Indeed, the *constant-volume gas thermometer* uses P as the thermometric property.[29] It's found experimentally that for n moles of *any gas* the quantity $PV/(nT)$ approaches the same value $R = 8.314$ J (K mole)$^{-1}$, the *gas constant*, at low pressures.[30] The gas constant is one of the fundamental constants of nature.

The *ideal gas* is a fictional substance *defined* as having the equation of state $PV = nRT$ for all pressures—the *ideal gas law*. The ideal gas law can be derived assuming non-interacting atoms, and hence is an idealization. *All gases become ideal at sufficiently low pressure*. The ideal gas law is also written $PV = Nk_BT$, where N is the number of particles and $k_B \equiv R/N_A = 1.381 \times 10^{-23}$ J K^{-1} is *Boltzmann's constant*. The gas constant (or Boltzmann's constant) is one of the few places where experimental data enters the theory of thermodynamics.

For a mixture of ideal gases, each with its own *mole number* n_i ($n = \sum_i n_i$), *Dalton's law* is that the pressure P of the mixture is equal to the sum of the *partial pressures* $P_i \equiv n_iRT/V$ that each component would exert were it the only gas present, with $P = \sum_i P_i$. Ideal gases contained in the same volume are not "aware" of each other; they exert their own pressure independent of whether other gases are present. The ratio $P_i/P = n_i/n \equiv x_i$ is called the *mole fraction*.

One way to characterize deviations from the ideal gas law is through the *virial expansion*, an expansion in powers of the density of the gas,

$$P = \frac{nRT}{V}\left[1 + \frac{n}{V}B_2(T) + \left(\frac{n}{V}\right)^2 B_3(T) + \cdots\right]. \tag{1.11}$$

The quantities $B_n(T)$, the *virial coefficients*, can be measured and they can be calculated using the methods of statistical mechanics. The virial coefficients depend on the strength of inter-particle interactions.

---

[27]That is, increasing empirical temperature—derived from the thermal behavior of materials—generally correlates with our experience of increasing hotness.

[28]Pressure therefore is an *energy density*.

[29]See any book on thermodynamics (except this one) for a discussion of thermometers and temperature scales, for example Zemansky.[12]

[30]Absolute temperature naturally has the dimension of energy (see Section 2.4). By measuring T in Kelvin (a non-dimensional unit), we need a conversion factor between energy and degrees Kelvin—what's supplied by the gas constant, i.e. RT has the dimension of energy. The ratio $Z \equiv PV/(nRT)$ is termed the *compressibility factor*. It's found for all gases that $Z \rightarrow 1$ at sufficiently low pressures.[13].

The *van der Waals equation of state* attempts to take into account deviations from the ideal gas by introducing the effects of 1) the finite size of atoms and 2) their interactions. Johannes van der Waals reasoned that the volume available to the atoms of a gas is reduced from the volume of the container V by the volume occupied by atoms. Thus, he modified the ideal gas law, purely on phenomenological grounds, to

$$P = \frac{nRT}{V - nb} ,$$

where $b > 0$ is an experimentally determined quantity for each type of gas. Clearly, the more moles of the gas, the greater the *excluded volume*. Van der Waals further reasoned that the pressure would be reduced by attractive interactions between atoms. The decrease in pressure is proportional to the probability that two atoms interact; this, in turn is proportional to the square of the particle density, $(n/V)^2$. In this way, van der Waals wrote down

$$P = \frac{nRT}{V - nb} - a\left(\frac{n}{V}\right)^2 ,$$

where $a > 0$ is another material-specific parameter to be determined from experiment. The van der Waals equation of state is usually written in the form

$$\left(P + a\left(\frac{n}{V}\right)^2\right)(V - nb) = nRT . \tag{1.12}$$

One can show that Eq. (1.12) yields an expression for the second virial coefficient in terms of the van der Waals parameters,

$$B_2^{VW} = b - \frac{a}{RT} . \tag{1.13}$$

This expression predicts that $B_2(T)$ should be negative at low temperatures and positive at high temperatures, what's observed for many gases. The van der Waals equation of state is a reasonably successful model of the thermodynamic properties of gases. While it does not predict all properties of real gases, it predicts enough of them for us to take the model seriously.[31]

1.7   IRREVERSIBILITY: TIME REARS ITS HEAD

We introduce the distinction between *reversible* and *irreversible* processes. A process is reversible if it can be *exactly reversed through infinitesimal changes in the environment*. In a reversible process the system *and* the environment are restored to their original conditions, a tall order! By this definition, in an irreversible process *the system cannot be restored to its original state without leaving changes in the environment.*[32]

A change of state can occur reversibly only if the forces between the system and the surroundings are essentially balanced at all steps of the process, and heat exchanges occur with essentially no temperature differences between system and surroundings. A process is termed *quasistatic* if it's performed slowly (so the system does not appreciably deviate from equilibrium) and work is performed *only* by forces conjugate to generalized displacements, i.e., forces of friction have been

---

[31] Van der Waals received the 1910 Nobel Prize in Physics.

[32] It's instructive to quote from Maxwell,[2, p642] as the notion of reversibility extends back to the beginnings of the subject: "A physical process is said to be reversible when the material system can be made to return from the final state to the original state under conditions which at every stage of the reverse process differ only infinitesimally from the conditions at the corresponding stage of the direct process. All other processes are called irreversible. Thus the passage of heat from one body to another is a reversible process if the temperature of the first body exceeds that of the second only by an infinitesimal quantity, because by changing the temperature of either of the bodies by an infinitesimal quantity, the heat may be made to flow back again from the second body to the first. But if the temperature of the first body is higher than that of the second by a finite quantity, the passage of heat from the first body to the second is not a reversible process, for the temperature of one or both of the bodies must be altered by a finite quantity before the heat can be made to flow back again."

eliminated. *Reversible processes are quasistatic*, but not all quasistatic processes are reversible (as we show below). Because the points of state space represent equilibrium states, *quasistatic processes can be plotted in state space*. Non-quasistatic processes (what can't be plotted in state space) are termed *non-static*.[33]

The concept of reversibility is problematic because it conceives of the *reversal* of processes in *time*. Strictly speaking, reversible processes are idealizations that don't exist: Equilibrium states do not involve change (time), yet reversible processes are taken to proceed through a sequence of equilibrium states. Real changes in state occur at a finite rate, so the intermediate states cannot strictly be equilibrium states. Practically speaking, there will be some small yet finite rate at which processes occur such that *disequilibrium* has no observable consequences within experimental uncertainties. All processes are irreversible therefore; *some* processes, however, can be idealized as reversible in limiting cases.[34] Reversible processes are highly useful idealizations, akin to adiabatic boundaries, and we'll refer to them frequently.

Example. Consider a gas confined to a cylinder with a movable piston. If the volume increases by dV against an external pressure P', the work done *by* the system is $P'dV$, and thus the work done *on* the system is $-P'dV$. If the expansion is done *slowly and without friction*—a quasistatic process—the gas pressure P will equal the external pressure, $P = P'$. The work done in a quasistatic isothermal expansion of an ideal gas from volume V_i to volume V_f is thus $W = -\int_{V_i}^{V_f} PdV = -nRT \int_{V_i}^{V_f} dV/V = -nRT \ln (V_f/V_i)$, which depends on the initial and final states. The work is still process dependent, however—it applies for an isothermal process.

Example. At normal pressure ($P = 1$ atm) ice is in equilibrium with water at $T = 273.15$ K. Ice can be melted reversibly at $T = 273.15$ K by applying heat from a reservoir[35] for which the temperature only slightly exceeds $T = 273.15$ K. A slight lowering of the temperature is then sufficient to reverse the direction of the process. If ice is placed in water at room temperature ($T \approx 293$ K), the melting process is irreversible.

Example. Chemical reactions can be controlled with electrochemical cells (Chapter 6). The reaction $CuSO_4 + Zn \leftrightarrows ZnSO_4 + Cu$ can be reversibly controlled by applying an opposing voltage to the cell slightly less than the electromotive force (emf) $\mathcal{E}$ of the cell. A small increase in the voltage is sufficient to reverse the direction of the reaction.

While some real processes can be idealized as reversible, other processes are *intrinsically irreversible*. That there *are* "un-reversible," i.e., irreversible processes lies at the heart of thermodynamics. The *free expansion* (described in Section 3.3 and analyzed in Section 4.9) is a non-static, irreversible process. In a free expansion, the gas in an isolated system is allowed to spontaneously expand into an evacuated chamber; such a system cannot be restored to its initial state leaving the environment unchanged. A more subtle example is as follows. Consider a gas enclosed by adiabatic boundaries at pressure and volume (P_0, V), and perform work on the system in the form of stirring,

---

[33]Non-static processes $i \rightarrow f$ cannot be represented as a continuous set of points (or curve) in state space connecting the equilibrium states (i, f). The process "leaves" state space at the initial state i and returns at the final state f. In that sense, the concept of a non-static process evokes the Heisenberg formulation of quantum mechanics, where all we can know is what we can observe.

[34]Planck wrote:[11, p85] "Whether reversible processes exist in nature or not, is not *a priori* evident or demonstrable. There is, however, no purely logical objection to imagining that a means may some day be found of completely reversing some process hitherto considered irreversible: one, for example, in which friction or heat conduction plays a part."

[35]Heat reservoirs are defined in Section 2.2.

holding V fixed. As a result P will increase to a value $P_1 > P_0$. Can this process be reversed, can P be reduced from $P_1 \to P_0$? Stirring the "other way" will only increase P further. With V fixed, there's no way to *lower* P for this adiabatically isolated system; the process is irreversible. Irreversibility reveals the existence of an important attribute of equilibrium states: *inaccessibility.* In this example, the state of a system, call it K_0, characterized by the values of its state variables (P_0, V) cannot be attained by means of adiabatic work from any of its states (P, V), denoted K, with $P > P_0$. The states K and K_0 *do not stand in an equivalent relation to each other.* All states K are attainable (*accessible*) from K_0, symbolized $K_0 \to K$, but K_0 is inaccessible from any state K, $K \nrightarrow K_0$. Processes un-reversible in time therefore imply a *direction in time* for the occurrence of states linked by irreversible processes. For K_0 inaccessible from K, if the system is in K_0 at time t and K at t', then $t' > t$. We can obviously label the states *time ordered* by irreversible processes using the times at which such states occur. Besides time, however, another way to characterize the succession of states in irreversible processes is with the values of a state function, entropy, which stand in correspondence with the time order established by irreversibility (Chapter 3). (We'll show that the entropy of isolated systems only increases, never decreases.) While the first law of thermodynamics is a statement about energy conservation (regardless of the issue of reversibility), the second law (from which entropy follows) codifies our experience of irreversibility. Irreversibility is an intrinsic aspect of the macroscopic world.

1.8   CONSTRAINTS AND STATE VARIABLES

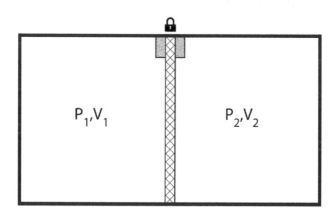

Figure 1.4   Movable partition separating compartments constrained by locking mechanism

An important concept is that of *constraints.* A constraint is the *restriction* of a state variable to certain values. Every state of equilibrium is, by definition, one of *constrained equilibrium,* in which state variables have their equilibrium values. Processes occur in which state variables are constrained, such as T in an isothermal expansion. It's a subtle concept, but constraints are associated with the *existence* of state variables. Basically, if something can be constrained, there's something *to be* constrained. There's a *one-to-one correspondence between constraints and state variables.* If a macroscopic quantity can be constrained, it implies the existence of a state variable as a *descriptor* of a macroscopic system. Consider gases in two compartments, separated by a movable adiabatic boundary. The system as a whole is held at a fixed temperature. Initially the partition is held in place by some locking mechanism (see Fig. 1.4). The state of this system is described by three variables, which could be chosen as T and the pressures P_1 and P_2, where $P_1 \neq P_2$. Alternatively, the state variables could be chosen as T, P_1, and the volume of one of the compartments, say V_1. The constraints on this system are the fixed temperature, the fixed volume $V = V_1 + V_2$, and the location of the partition. Now remove the locking mechanism on the partition, which will move until the pressures equalize,[36] $P_2 = P_1$. At this point, the system is described by *two* independent variables, say T and P_1. *The removal of a constraint results in the loss of an independent state variable.* Conversely, the addition of a constraint implies the *creation* of a new independent state variable.[37] To restore the partition to its previous location work must be done on the system. Thus, *eliminating constraints is an irreversible process.* We discuss in Section 3.3 how removal of constraints, irreversibility, and the increase in entropy are connected.

1.9 THE MANY FACES OF WORK

The work done on a system in an infinitesimal reversible expansion is $-P\mathrm{d}V$. While $-P\mathrm{d}V$ is the most well-known expression for $\mathrm{d}W$, there are others. Consider a length L of wire under tension J. The work done in stretching the wire reversibly by $\mathrm{d}L$ is $J\mathrm{d}L$. The work done in reversibly expanding the surface area of a film by an amount $\mathrm{d}A$ is $\sigma\mathrm{d}A$, where σ is the surface tension. These examples involve the *extension* or *deformation* of the boundaries of the system: volume, length, area. More generally, work involves any change of macroscopically observable properties, *not necessarily involving the boundaries of the system.* The latter entails system interactions with externally imposed *fields*. The work done in reversibly changing the magnetization $\boldsymbol{M}$ of a sample by $\mathrm{d}\boldsymbol{M}$ in a field $\boldsymbol{H}$ is $\mu_0\boldsymbol{H}\cdot\mathrm{d}\boldsymbol{M}$ (see below). The work done in reversibly changing the electric polarization $\boldsymbol{P}$ by $\mathrm{d}\boldsymbol{P}$ in electric field $\boldsymbol{E}$ is $\boldsymbol{E}\cdot\mathrm{d}\boldsymbol{P}$.

Expressions for work occur in the form $\mathrm{d}W = Y\mathrm{d}X$, *generalized work*, the product of intensive quantities Y (generalized forces: $-P, J, \sigma, \mu_0\boldsymbol{H}, \boldsymbol{E}$), with the differentials of *extensible quantities*, $\mathrm{d}X$ (generalized displacements: $\mathrm{d}V, \mathrm{d}L, \mathrm{d}A, \mathrm{d}\boldsymbol{M}, \mathrm{d}\boldsymbol{P}$). An extensible quantity is (as the name suggests) one that can be extended, the change in an extensive variable. The infinitesimal work done on a system is the sum of the individual differentials, $\mathrm{d}W = \sum_i Y_i\mathrm{d}X_i$. The first law can therefore be written $\mathrm{d}U = \mathrm{d}Q + \sum_i Y_i\mathrm{d}X_i$. The quantities X_i are known as the *deformation coordinates* of the system. Energy transfers effected through changes in deformation coordinates are referred to as *extensible work*. Systems, however, must possess one *non-deformation* (or *thermal*) coordinate because work can be done on a system keeping deformation coordinates fixed. In stirring the fluid in an adiabatic enclosure, work is done on the system, but the deformation coordinates don't change.

Magnetic work

Because of the maxim that "the B-field does no work," we calculate the work done in reversibly changing the magnetization of an object. Energy is conveyed in the process of magnetization, and unless that energy is in the form of heat, *work is done by a magnetic field.*

Consider a solenoid of length L with N turns and cross-sectional area A carrying a current I (supplied by a battery) that contains a material of magnetization M. The magnetic intensity $H = NI/L$ couples to the current, independent of the material in the solenoid. The customary expression from electromagnetic theory for the magnetic field is $B = \mu_0(H + M)$. In this familiar formula M is the magnetization *density*; $B = \mu_0(H + M)$ is a *relation among intensive quantities.* The paradigm in thermodynamics, however, is that an infinitesimal amount of work is the product of an intensive quantity (a parameter set by the environment, H), and the differential of an extensive quantity. For our use here we take M to be the *total* magnetic moment of the system. In terms of the

---

[36]We show in Section 3.10 that equilibrium is characterized by the equality of intensive variables between a system and its surroundings, those associated with conserved quantities that can be exchanged between system and environment.

[37]That adding or removing constraints increases or decreases the number of state variables is reminiscent of creation and annihilation operators in the quantum theory of many particle systems. Adding or removing constraints is formally a mapping between thermodynamic state spaces of different dimensions, just as creation and annihilation operators are mappings between Hilbert spaces of differing dimensions corresponding to the number of identical particles.

"thermodynamic M," B is related to M by

$$B = \mu_0 \left(\frac{NI}{L} + \frac{M}{AL} \right) .$$

In this expression N and L are extensive quantities; the current is intensive. If for fixed current the magnetization changes, $dB = (\mu_0/AL)\, dM$. A change in B produces a change in flux $\Phi = NAB$, inducing a back emf,

$$\mathscr{E} = -\frac{d\Phi}{dt} = -NA\frac{dB}{dt} = -\mu_0 \frac{N}{L} \frac{dM}{dt} .$$

To maintain constant current, the battery must do work to overcome the induced emf. In passing a charge dq through the circuit, the battery does work (on the system)

$$dW = |\mathscr{E}|\, dq = \mu_0 \frac{N}{L} \frac{dM}{dt} dq = \mu_0 \frac{N}{L} \frac{dq}{dt} dM = \mu_0 \frac{NI}{L} dM = \mu_0 H dM .$$

We have assumed that $\boldsymbol{M}$ is aligned with $\boldsymbol{H}$. If not, the work generalizes to

$$dW = \mu_0 \boldsymbol{H} \cdot d\boldsymbol{M} . \tag{1.14}$$

The *system* in this case is the magnetized sample; the H field is a constant parameter characterizing the environment, just as temperature is a parameter set by the environment. Should we enlarge the scope of the system, we would need to include the energy of the field ($\frac{1}{2}\int \boldsymbol{B} \cdot \boldsymbol{H} dV$) in our definition of internal energy.

1.10   CYCLIC RELATION*

Consider three variables connected through a functional relation, $f(x, y, z) = 0$. Any two can be taken as independent, and each can be considered a function of the other two: $x = x(y, z)$, $z = z(x, y)$, or $y = y(x, z)$. We can form the differential of x in terms of the differentials of y and z,

$$dx = \left(\frac{\partial x}{\partial y} \right)_z dy + \left(\frac{\partial x}{\partial z} \right)_y dz , \tag{1.15}$$

or we can take the differential of z in terms of x and y,

$$dz = \left(\frac{\partial z}{\partial x} \right)_y dx + \left(\frac{\partial z}{\partial y} \right)_x dy . \tag{1.16}$$

Combine Eq. (1.16) with Eq. (1.15):

$$0 = \left[\left(\frac{\partial x}{\partial z} \right)_y \left(\frac{\partial z}{\partial x} \right)_y - 1 \right] dx + \left[\left(\frac{\partial x}{\partial y} \right)_z + \left(\frac{\partial x}{\partial z} \right)_y \left(\frac{\partial z}{\partial y} \right)_x \right] dy . \tag{1.17}$$

For Eq. (1.17) to be valid for independent dx and dy, we conclude that

$$1 = \left(\frac{\partial x}{\partial z} \right)_y \left(\frac{\partial z}{\partial x} \right)_y \tag{1.18}$$

$$-1 = \left(\frac{\partial x}{\partial y} \right)_z \left(\frac{\partial y}{\partial z} \right)_x \left(\frac{\partial z}{\partial x} \right)_y . \tag{1.19}$$

Equation (1.18) is called the *reciprocity relation* and Eq. (1.19) the *cyclic relation*. These relations are used frequently in the theory of thermodynamics. (See also Eq. (3.48).)

1.11 RESPONSE FUNCTIONS

The seemingly quiescent equilibrium state (specified by the values of state variables) is associated (at the microscopic level) with incessant and rapid transitions among the microscopic configurations of a system—its microstates. In the time over which measurements are made, macroscopic systems "pass" through a large number of microstates. Measurements represent time averages over the microstates that are consistent with system constraints (fixed volume, fixed energy, etc.). In statistical mechanics averages are *calculated* from the microscopic energies of such states.[38] In thermodynamics, however, relatively few quantities are actually calculated. What thermodynamics does (spectacularly) is to reveal *interrelations* between measurable quantities. The mathematical theory of thermodynamics is a "consistency machine" that relates known information about a system to other quantities that might be difficult to measure. To make the machine *useful* it must be fed information. *Response functions*, what we introduce now, are quantities readily accessible to experimental measurement. These are *derivatives* that measure, under controlled conditions, how measurable thermodynamic quantities vary with respect to each other.

Expansivity and compressibility

The *thermal expansivity* and *isothermal compressibility* are by definition

$$\alpha \equiv \frac{1}{V}\left(\frac{\partial V}{\partial T}\right)_P \qquad \beta \equiv -\frac{1}{V}\left(\frac{\partial V}{\partial P}\right)_T . \tag{1.20}$$

These measure the fractional change in volume that occur with changes in T and P. We'll show in a later chapter that β is *always positive*, Eq. (3.51); whereas α has no definite sign. What about the third derivative from this trio of variables, $(\partial P/\partial T)_V$? Is it given its own definition as a response function? Nope, no need. Using Eq. (1.19),

$$\left(\frac{\partial P}{\partial T}\right)_V = -\frac{(\partial V/\partial T)_P}{(\partial V/\partial P)_T} = \frac{\alpha}{\beta} . \tag{1.21}$$

Knowing α and β, we know the derivative on the left of Eq. (1.21). We're free to assume that V, T, and P are connected by an equation of state $V = V(T, P)$. Thus,

$$dV = \left(\frac{\partial V}{\partial T}\right)_P dT + \left(\frac{\partial V}{\partial P}\right)_T dP = \alpha V dT - \beta V dP . \tag{1.22}$$

With measurements of α and β, Eq. (1.22) can be integrated to experimentally determine the equation of state. Equation (1.22) demonstrates the "differentiate-then-integrate" strategy: The laws of thermodynamics typically give us information about the *differentials* of various quantities, which can then be integrated to find the equation of state or the internal energy function.

Heat capacity

The *heat capacity*, $C_X \equiv (dQ/dT)_X$, measures the heat absorbed per change in temperature, where X signifies what's held fixed in the process. Typically measured are C_P and C_V. Heat capacities are tabulated as *specific heats*, the heat capacity per mole or per gram of material.[39] What does a large heat capacity signify? For a small temperature change ΔT, say one degree, $\Delta Q = C_X \Delta T$, i.e., *the heat capacity is the heat required to change the temperature of an object by one degree* (from $T \to T + \Delta T$). A system characterized by a large heat capacity requires a relatively large amount

---

[38]In statistical mechanics, time averages are replaced with ensemble averages.

[39]A useful mnemonic is "heat capacity of a penny, specific heat of copper."

of heat to change its temperature, larger than for systems with a small heat capacity, for which the same temperature change can be accomplished with a smaller amount of heat transferred.[40]

There's an important connection between C_P and C_V, Eq. (1.29), that we now derive. Assume that[41] $U = U(T, V)$. Then,

$$dU = \left(\frac{\partial U}{\partial T}\right)_V dT + \left(\frac{\partial U}{\partial V}\right)_T dV = đQ - PdV , \tag{1.23}$$

where the second equality is the first law of thermodynamics. Thus

$$\frac{đQ}{dT} = \left(\frac{\partial U}{\partial T}\right)_V + \left[P + \left(\frac{\partial U}{\partial V}\right)_T\right]\frac{dV}{dT} . \tag{1.24}$$

By holding V fixed, we have from Eq. (1.24)

$$C_V = \left(\frac{đQ}{dT}\right)_V = \left(\frac{\partial U}{\partial T}\right)_V . \tag{1.25}$$

Equation (1.25) is very useful, and is an example of a *thermodynamic identity*, one of many that we'll derive: A measurement of C_V is a measurement of $(\partial U/\partial T)_V$. From Eq. (1.23) therefore, $[dU]_V = C_V dT$, where the notation indicates that V is held fixed. Holding P fixed in Eq. (1.24),

$$C_P = \left(\frac{đQ}{dT}\right)_P = C_V + \left[P + \left(\frac{\partial U}{\partial V}\right)_T\right]\left(\frac{\partial V}{\partial T}\right)_P , \tag{1.26}$$

where we've used Eq. (1.25). Turning Eq. (1.26) around, and using Eq. (1.20),

$$\left(\frac{\partial U}{\partial V}\right)_T = \frac{C_P - C_V}{\alpha V} - P . \tag{1.27}$$

Equation (1.27) relates the derivative $(\partial U/\partial V)_T$ (which is not easily measured) to quantities that are more readily measured. Once entropy is introduced as a state variable, together with the theoretical device of *Maxwell relations* (Section 4.5), we'll discover considerable leeway in evaluating derivatives like $(\partial U/\partial V)_T$. It's shown in Exercise 4.6 that

$$\left(\frac{\partial U}{\partial V}\right)_T = T\left(\frac{\partial P}{\partial T}\right)_V - P = T\frac{\alpha}{\beta} - P , \tag{1.28}$$

another connection between a derivative and measurable quantities.[42] With measurements of C_V, α, and β, we have the derivatives $(\partial U/\partial T)_V$ and $(\partial U/\partial V)_T$ (Eq. (1.25) and Eq. (1.28)), from which $U(T, V)$ can be obtained through integration. Equating Eqs. (1.28) and (1.27), we have the desired relation:

$$C_P - C_V = \frac{\alpha^2}{\beta}TV . \tag{1.29}$$

Because $\beta > 0$ (Section 3.10), Eq. (1.29) implies that $C_P > C_V$, *one of the key predictions of thermodynamics*.[43] It's shown in Section 3.10 that $C_V > 0$, Eq. (3.51).

---

[40]The definition of heat capacity is analogous to the definition of electrical capacitance, $Q = CV$. A large capacitor requires a large amount of charge to raise the voltage by one volt, $\Delta Q = C\Delta V$.

[41]We're free to take U to be a function of whatever independent variables we choose.

[42]It turns out that Eq. (1.28) is one of the most important equations of thermodynamics from a theoretical perspective; see Chapter 10. The group of terms $\left[P + (\partial U/\partial V)_T\right](\partial T/\partial P)_V$ serves as a proxy for temperature, what systems in mutual equilibrium have in common.

[43]Heat added at constant volume causes the temperature to rise ($C_V > 0$). Heat added keeping the pressure fixed (for the same amount of a substance) could be done by allowing the volume to increase, and thus the system would do work on the environment. More heat at constant P would have to be supplied for the same temperature increase as for constant V, $C_P > C_V$.

For the *ideal gas*, we have the special results (see Exercise 1.6),

$$C_P - C_V = nR \,, \tag{1.30}$$

and

$$\left(\frac{\partial U}{\partial V}\right)_T = 0 \,. \tag{1.31}$$

Equation (1.31), known as *Joule's law*, can be taken as an alternate *definition* of the ideal gas: A gas whose internal energy is a function of temperature only, $U = U(T)$. The ideal gas ignores interparticle interactions; its internal energy is independent of the volume. The heat capacity C_V for the noble gases (He, Ne, Ar, Kr, and Xe) has the measured value $1.50nR$ for moderate temperatures (temperatures not so low that the gas liquefies). For the *ideal* gas, C_V is exactly $\frac{3}{2}nR$ for all temperatures (see Eq. (7.34)); noble gases thus approximate ideal gases for moderate temperatures. The internal energy function of the ideal gas thus has the simple form (using $C_V = (\partial U/\partial T)_V = \frac{3}{2}nR$ and Eq. (1.31)),

$$U = \frac{3}{2}nRT = \frac{3}{2}Nk_BT \,. \tag{1.32}$$

Equation (1.32) shows that the gas constant (or Boltzmann's constant) is related to the *average* kinetic energy of the molecules of an ideal gas.[44] Because all gases become ideal at sufficiently low pressure, the gas constant measures a universal property of gaseous matter at temperature T.

The temperature dependence of $C_V(T)$ is a probe of how matter absorbs energy at the microscopic level; the full power of statistical mechanics is required to understand its behavior. At low temperatures gases condense into liquids and eventually solidify. The heat capacity of many solids is $3nR$, the *law of Dulong and Petit*.[45] At very low temperatures, the law of Dulong and Petit breaks down and heat capacities *vanish*, as the third law of thermodynamics indicates must happen (Chapter 8). It's found that the heat capacity of many solids vanishes like T^3 as $T \to 0$.

CHAPTER SUMMARY

This (lengthier than most) chapter has presented the foundations of thermodynamics organized around three concepts: equilibrium, internal energy, and irreversibility. Many definitions have been introduced which form the basic vocabulary of the subject. If there wasn't something called entropy out there, this chapter would be just about all one needs to know about thermodynamics. As it is, however, and as we'll see, thermodynamics is all about entropy.

- Thermodynamics is the study of macroscopic systems in equilibrium, together with the processes that drive transitions between them.[46] Equilibrium is the quiescent state where nothing appears to be happening, where all the fast things have happened and the slow things not.

- States of equilibrium are specified by the values of state variables, measurable properties of equilibrium systems. State variables are independent of time and of the history by which equilibrium has been established. Differentials of state variables are exact differentials.

- Thermodynamics achieves a huge reduction in the complexity of description, from an enormous number of microscopic degrees of freedom to just a few state variables. In the time required to make measurements, macroscopic systems pass through a large number of microstates; macroscopically measurable quantities represent averages over microstates compatible with system constraints.

---

[44]The average in this case is established by dividing the total energy U by N, $U/N = \frac{3}{2}k_BT$.

[45]In statistical mechanics, a useful result is derived, the *equipartition theorem* that accounts for the classical heat capacities.

[46]A more learned definition of thermodynamics is offered on page 146.

- Thermodynamic state space is the mathematical space of all values of state variables. Every point in state space represents a possible equilibrium state.

- State variables are classified as extensive or intensive, those that do and do not scale with the size of the system. Intensive variables have the same values at different spatial locations of the system. The values of state variables are connected either through equations of state or by the laws of thermodynamics.

- The thermal behavior of macroscopic systems is determined by the nature of the boundaries separating the system from the environment. Boundaries control the types of interactions a system is allowed to have with its environment.

- Thermodynamics distinguishes three types of boundaries: adiabatic, diathermic, and permeable. Adiabatic boundaries allow a system to interact with its environment through mechanical means only (no interchange of matter or heat with the surroundings). Diathermic boundaries allow thermal interactions with the environment. Permeable walls allow the system to exchange matter and energy with the environment.

- The different types of boundary enable the formulation of the zeroth and first laws of thermodynamics. Diathermic boundaries allow thermal energy exchange between systems, implying the existence of temperature as a state variable, what systems in equilibrium have in common, as well as the existence of equations of state, functional relations among state variables. The Joule-Mayer experiments demonstrate, on adiabatically isolated systems, the existence of internal energy as a state variable, the storehouse of adiabatic work.

- Adiabatically isolated systems conserve energy in that the energy of mechanical work (an interaction of the system with an external agency) is stored in the internal energy of the system, $\Delta U = W_{\mathrm{ad}}$, energy that can be recovered by letting the system perform adiabatic work on the environment. The same total amount of adiabatic work induces the same transition between reproducible equilibrium states (i, f), independent of the kind of work performed. This type of path independence indicates that U is a state variable with $\Delta U = U_f - U_i$.

- Mechanical energy is not conserved, however, for systems enclosed by diathermic boundaries. Heat is the difference in work required to induce the same transition $i \to f$ for the system enclosed by the two types of boundary: $Q \equiv W_{\mathrm{ad}} - W$.

- The first law of thermodynamics $\mathrm{d}U = \mathrm{d}W + \mathrm{d}Q$ has that a small change $\mathrm{d}U$ in internal energy can be affected by the transference of a small quantity of heat $\mathrm{d}Q$ and/or the performance of a small amount of work $\mathrm{d}W$. Work and heat are not substances; they represent the effects of thermodynamic processes. The notation $\mathrm{d}W$ and $\mathrm{d}Q$ indicates that these are process-dependent (inexact) differentials, whereas $\mathrm{d}U$ (exact differential) indicates a small difference in a physical quantity, energy. Energy is stored in a system, but not work or heat, which are modes of affecting the transfer of energy in and out of a system. For finite changes, the first law is written $\Delta U = Q + W$.

- Extensible work is the energy $\sum_i Y_i \mathrm{d}X_i$ required to change the values of the deformation coordinates of the system X_i, those that represent changes in observable macroscopic properties, where each generalized displacement $\mathrm{d}X_i$ is multiplied by a conjugate intensive quantity, the generalized force Y_i. Systems must have one non-deformation (thermal) coordinate because work can be done on systems keeping deformation coordinates fixed.

- Reversible processes can be exactly reversed, restoring system and environment to their original states. In irreversible processes, the original state of the system cannot be restored without leaving changes in the environment. Reversible processes are carried out slowly so that state variables have well defined values at every step of the process. Irreversible processes (literally un-reversible) imply an order in time by which the states in such processes occur. The time ordering implied by irreversibility intimates the existence of a state function (entropy) as a way to characterize the equilibrium states linked by irreversible processes.

- Response functions are derivatives of thermodynamic variables measured under controlled conditions. Thermodynamic derivatives that cannot be easily measured can invariably be expressed in terms of response functions. Much of the "art" of thermodynamics consists of deriving thermodynamic identities.

- The heat capacity at constant volume $C_V = (\text{d}Q/\text{d}T)_V$ of the ideal gas has the value $\frac{3}{2}nR$ where the gas constant R is a fundamental constant of nature. The ideal gas is defined by the equation of state $PV = nRT$ for all temperatures; real gases become ideal at sufficiently low pressures. For an ideal gas $(\partial U/\partial V)_T = 0$, i.e., U is a function only of the temperature, $U = U(T) = \frac{3}{2}nRT$. The ideal gas ignores inter-particle interactions and the internal energy is independent of the volume.

EXERCISES

1.1 How many atoms of silicon are in a cube 1 μm on a side? The mass of a mole of silicon is approximately 28 grams, and the density of silicon is approximately 2.3 g cm^{-3}.

1.2 Integrate the differential $(x^2+y)\text{d}x+x\text{d}y$ from (0,0) to (1,1) along two paths: first from (0,0) to (1,0) to (1,1), and then from (0,0) to (0,1) to (1,1). Is the value of the integral the same? Now repeat this exercise with the differential $(x^2 + 2y)\text{d}x + x\text{d}y$.

1.3 Is $x^2y^4\text{d}x + x^3y^3\text{d}y$ an exact differential? Can you find an integrating factor? Integrating factors are not unique.

1.4 Derive Eq. (1.13), the second virial coefficient obtained from the van der Waals equation of state.

1.5 Verify the cyclic relation Eq. (1.19) using (P, V, T) for (x, y, z) and the ideal gas equation of state.

1.6 Show that for the ideal gas, $\alpha = T^{-1}$ and $\beta = P^{-1}$.

1.7 Take $U = U(T, P)$ and $V = V(T, P)$.

a. Show that
$$\frac{\text{d}Q}{\text{d}T} = \left(\frac{\partial U}{\partial T}\right)_P + P\left(\frac{\partial V}{\partial T}\right)_P + \left[\left(\frac{\partial U}{\partial P}\right)_T + P\left(\frac{\partial V}{\partial P}\right)_T\right]\frac{\text{d}P}{\text{d}T}.$$

b. Show that this result implies another set of thermodynamic identities
$$C_P = \left(\frac{\partial U}{\partial T}\right)_P + P\left(\frac{\partial V}{\partial T}\right)_P \qquad \left(\frac{\partial U}{\partial P}\right)_T = -\frac{\beta}{\alpha}(C_P - C_V) + P\beta V.$$

1.8 From the results of Exercise 1.7, and Eqs. (1.28), (1.29), and (1.20) show that the first law can be written in either of the two ways
$$\text{d}Q = C_V\text{d}T + \frac{T\alpha}{\beta}\text{d}V \qquad \text{d}Q = C_P\text{d}T - \alpha VT\text{d}P.$$

1.9 Using the results of Exercise 1.8,

    a. Show that under adiabatic conditions

$$\left(\frac{\mathrm{d}P}{\mathrm{d}V}\right)_{\text{adiabatic}} = -\frac{\gamma}{\beta V},$$

where $\gamma \equiv C_P/C_V$ is the *ratio of the heat capacities*. Equation (1.29) implies that $\gamma > 1$. The locus of points in state space reachable by a process in which $\mathrm{d}Q = 0$ is known as an *adiabat*.

    b. Show that under isothermal conditions

$$\left(\frac{\mathrm{d}P}{\mathrm{d}V}\right)_{\text{isothermal}} = -\frac{1}{\beta V}.$$

    Hence we have the result

$$\left(\frac{\mathrm{d}P}{\mathrm{d}V}\right)_{\text{adiabatic}} = \gamma \left(\frac{\mathrm{d}P}{\mathrm{d}V}\right)_{\text{isothermal}}.$$

    Because $\gamma > 1$, $\mathrm{d}P/\mathrm{d}V$ on adiabats is always larger in magnitude than $\mathrm{d}P/\mathrm{d}V$ along isotherms.

1.10 For an adiabatic process involving the ideal gas, show that either $PV^\gamma = $ constant or $TV^{\gamma-1} = $ constant. Use the result of Exercise 1.9 applied to the ideal gas.

1.11 Show for the ideal gas that $(\partial U/\partial P)_T = 0$. Hint: One could simply apply the chain rule to Eq. (1.31), or one could use the result derived in Exercise 4.6. The internal energy of the ideal gas is a function only of the temperature—Joule's law. Joule discovered this result by allowing gases to expand into a vacuum under adiabatic conditions (Section 4.9). An ideal is such that $PV = nRT$ with the proviso that $U = U(T)$.

1.12 The isothermal bulk modulus is defined as

$$B_T \equiv -V \left(\frac{\partial P}{\partial V}\right)_T.$$

Show that $\left(\dfrac{\partial P}{\partial T}\right)_V = \alpha B_T$, where α is the thermal expansivity.

1.13 a. Show for an adiabatic transformation of the ideal gas that the first law can be written

$$\frac{C_V}{T}\mathrm{d}T + \frac{nR}{V}\mathrm{d}V = 0.$$

    Does this expression represent an exact differential? If so, is there is a quantity $f(T,V)$ that's constant in an adiabatic process? Hint: Use the value of C_V for the ideal gas.

    b. Show that for a general transformation of an ideal gas

$$\mathrm{d}Q = C_V \mathrm{d}T + \frac{nRT}{V}\mathrm{d}V.$$

    Is this an exact differential? Is there a quantity $Q(T,V)$ stored in the system?

    c. Is there an integrating factor that would turn $\mathrm{d}Q$ into an exact differential?

Second law of thermodynamics

Direction of heat flow

THE first law of thermodynamics places restrictions on the possible changes of state a system may undergo, namely those that conserve energy. There are processes, however, that while *conceivable* as energy conserving, are never observed. An ice cube melts into a puddle of water, but a puddle never gathers itself into an ice cube. What distinguishes naturally-occurring processes is the *direction of heat flow*. The *second law of thermodynamics* codifies our experience that *heat does not spontaneously flow from cold to hot*. The *directionality* of spontaneous heat flows has numerous implications in accounting for which processes are observed to occur in nature.

2.1 THERMODYNAMICS OF CYCLES: SYSTEM AS A BLACK BOX

A *cyclic process* returns a system to its initial state. After one cycle, $\Delta U = 0$ (U is a state variable), and thus $\Delta U = 0 = Q + W$. The work done *by* a system in a cycle, $-W$, is therefore equal to the net heat absorbed by the system, $-W = Q$, "energy out equals energy in," the *conversion of heat into work*. A *heat engine* converts heat into work. A heat engine run backwards can act as a *refrigerator* or as a *heat pump*, depending on whether the primary goal is to remove heat from a cold object or deliver heat to a hot object.

At various points of a cycle, heat is either absorbed from, or expelled to, the environment, so that the *net* heat transferred to the system $Q \equiv Q_{in} + Q_{out}$, where by our sign convention $Q_{out} < 0$. The *efficiency* η of a heat engine is defined as the work produced per heat absorbed,

$$\eta \equiv \frac{-W}{Q_{in}} = \frac{Q_{in} + Q_{out}}{Q_{in}} = 1 - \frac{|Q_{out}|}{Q_{in}}, \tag{2.1}$$

Can $\eta = 1$? Nothing in the first law would preclude that, yet we know from experience that 100% efficiency doesn't sound right. We'll show as a consequence of the second law that there's a maximum efficiency, η_{max}, with $\eta \le \eta_{max} < 1$. *There must be heat expelled*, or *waste heat*, $|Q_{out}| \ne 0$. The second law teaches us that it *takes* energy to *transform* energy.[1]

---

[1]The second law is sometimes called the *law of the transmutability of energy*. If energy *did* spontaneously flow from cold to hot, inequalities in temperature would be created that could be used to perform useful work with a heat engine. Free energy! Real refrigerators *require* work to create temperature inequalities.

Aside from their practicality as heat engines or refrigerators, cyclic processes are an ingenious *theoretical* device that allow us to formulate concepts without reference to the properties of matter. Through the use of cycles we can treat systems as "black boxes" where their properties are eliminated in favor of what's observed external to the system.

2.2   CLAUSIUS AND KELVIN STATEMENTS OF THE SECOND LAW

There are two equivalent ways of stating the second law—the Clausius and the Kelvin versions—and you need to be aware of both.[2] The *Clausius form of the second law* states that:

> It is impossible to devise a process that produces no effect other than the transfer of heat from a colder to a hotter object.

Note the emphasis on producing *no other effect*, what we'll refer to as the "sole-result clause." In other wordings of the Clausius and Kelvin statements, emphasis is placed on *working in a cycle*. Cyclical processes are *useful* in ensuring a system is returned to its starting point, allowing the sole-result clause to be more readily tested, but they're not strictly necessary.[3] Work performed on the system (as in a refrigerator) violates the sole-result clause. Only if *all* we have effected is a transfer of heat from cold to hot, can a violation of the second law be claimed.

Example. Let a cylinder of gas be in thermal contact with a cold object. Expand the gas so that it absorbs heat from the cold object. Now isolate the gas—surround it with adiabatic walls. Allow the gas to come to equilibrium with a hot object. Compress the gas so that heat is delivered to the hot object. We've delivered heat from cold to hot, yet there's no violation of the second law, which forbids processes whose *sole result* is the transfer of heat from cold to hot.

The *Kelvin form of the second law* states that:

> It is impossible to devise a process that produces no effect other than the extraction of heat from a single reservoir and the performance of an equal amount of work.

A *heat reservoir* is an object with a large heat capacity, so large that it's capable of absorbing or rejecting heat without undergoing appreciable changes in temperature.[4] The Kelvin statement implies there *must* be waste heat, $Q_{out} \neq 0$; a 100% efficient heat engine is impossible. One cannot extract heat from a *single* reservoir (read single temperature) and convert it entirely into work; there must be at least two reservoirs (temperatures) involved. Note the emphasis on *converting heat into work*; there are no restrictions on converting work into heat.[5]

The two forms of the second law are equivalent: *The falsity of one implies the falsity of the other*. To show that, it helps to have a symbolic diagram as a representation of heat engines (a *heat engine diagram*). The left portion of Fig. 2.1 shows a circle (heat engine working in a cycle—"black box") that absorbs heat Q_1 from the reservoir at temperature T_1 and expels Q_2 to the reservoir at temperature T_2, where $T_1 > T_2$. In the process, work is done on the environment with $Q_1 = |Q_2| + |W|$; energy in equals energy out.

Assume, contrary to the Clausius statement, that we *can* devise an engine whose sole effect is to deliver heat from cold to hot. By having this hypothetical engine work in conjunction with a normal heat engine, the combined engine violates the Kelvin statement. The middle portion of Fig. 2.1 shows an engine that (in violation of the Clausius statement) absorbs Q_0 from a cold reservoir

---

[2]Knowing the Clausius and Kelvin statements of the second law is highly useful in eliminating spurious arguments; they should be considered part of the "intellectual toolbox" of physical scientists.

[3]Look ahead to Chapter 12. Maxwell's demon was invented in order to effect a non-cyclical violation of the second law.

[4]Throw an ice cube in the ocean; does the temperature of the ocean change?

[5]You can rub your hands together all day, converting work into heat.

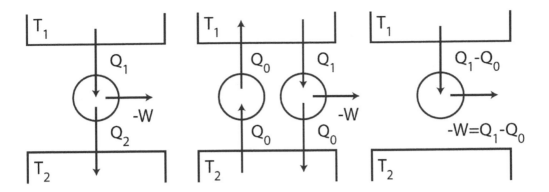

Figure 2.1   A violation of the Clausius statement implies a violation of the Kelvin statement

and delivers Q_0 to a hot reservoir, with no work involved. The other engine absorbs Q_1 from the hot reservoir and delivers Q_0 to the cold reservoir. Either the *size* of the normal engine can be adjusted to deliver Q_0 in one cycle, or the *rate* at which it operates can be adjusted to deliver Q_0 in the same time that the first engine absorbs Q_0. The net effect of these engines working in combination is shown in the right part of the figure—a composite engine that absorbs $Q_1 - Q_0$ from a single reservoir and delivers the same amount of work to the environment, in violation of the Kelvin statement.

The left part of Fig. 2.2 shows a refrigeration cycle that absorbs heat Q_C at the cold reservoir and delivers Q_H to the hot reservoir, as a result of the work W performed on the system. Energy conservation requires that $|Q_H| = W + Q_C$. The middle portion of Fig. 2.2 shows an engine that violates the Kelvin statement by absorbing heat Q_1 from a single reservoir and turning it completely into work. The work produced by this engine is used to drive a refrigerator. The net effect is an engine that delivers heat from a cold object to a hot object with no other effect, in violation of the Clausius statement.

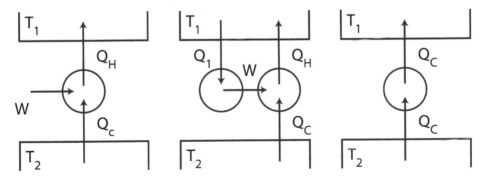

Figure 2.2   A violation of the Kelvin statement implies a violation of the Clausius statement

The negation of one statement implies the negation of the other, what we can indicate symbolically as $\overline{C} \Rightarrow \overline{K}$ and $\overline{K} \Rightarrow \overline{C}$. By the rules of logic, the two statements are equivalent, $C \Leftrightarrow K$. Note that we have not *proved* either statement, merely that they are equivalent; one cannot prove the second law any more than one can prove the first law—both are codifications of experience. One can neither prove nor disprove phenomenological laws; one can only report violations of them to your local physics authority.

2.3 CARNOT THEOREM: UNIQUENESS OF ADIABATS

Carnot cycle

A *Carnot cycle* (shown in Fig. 2.3) is one that operates reversibly along two isotherms and two adiabats (or *isentropes*). The slope dP/dV is *steeper* for adiabats than for isotherms (Exercise 1.9). A reversible cycle specified as exchanging heat only along isotherms is automatically a Carnot cycle because those parts of the cycle not involving heat transfer must necessarily be (reversible) adiabats.

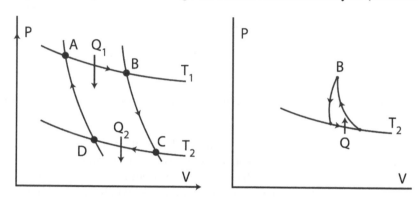

Figure 2.3 Carnot cycle (left). Adiabats must be unique (right).

A key point is the *uniqueness* of adiabats. Suppose *two* adiabats connect point B in Fig. 2.3 with the isotherm at temperature T_2. One could then extract heat from a single reservoir and, operating in a cycle, convert it entirely into work, in violation of the Kelvin statement. The uniqueness of adiabats can readily be demonstrated for a system with two independent variables. With $dQ = 0$ in the first law, and with $U = U(P, V)$,

$$\left(\frac{dP}{dV}\right)_{\text{adiabat}} = -\left(\frac{\partial U}{\partial P}\right)_V^{-1}\left[P + \left(\frac{\partial U}{\partial V}\right)_P\right]. \tag{2.2}$$

The gradient of an adiabat is thus specified by Eq. (2.2) at every point in the P-V plane. Through integration, Eq. (2.2) will produce a unique adiabat starting from a given point.[6] For systems having three or more variables, the uniqueness of adiabats (required by the second law) needs further analysis and is the subject of Chapter 10.

Carnot theorem: $\eta \leq \eta_R$

Carnot's theorem states:

> All reversible engines operating between the same reservoirs have the same efficiency, which exceeds that of any non-reversible engine.

To prove the assertion, envision a Carnot cycle operating in conjunction with an assumed more-efficient engine, and show that we're led to a violation of the second law. In the left portion of Fig. 2.4, Carnot engine C operates between reservoirs at temperatures T_1 and T_2 with $T_1 > T_2$. Operating between the same reservoirs is a hypothetical engine[7] H having greater efficiency than C. By assumption therefore

$$\eta_H = \frac{|W_H|}{Q_{H_1}} > \eta_C = \frac{|W_C|}{Q_{C_1}}. \tag{2.3}$$

---

[6]The solutions of first-order differential equations are unique under very general conditions.[14, p22]

[7]H and C in Fig. 2.4 do *not* mean hot and cold!

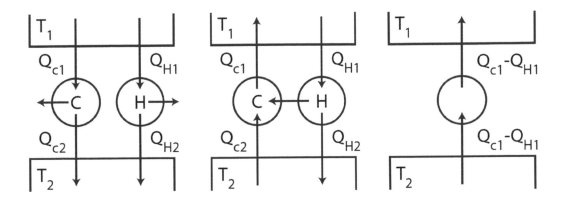

Figure 2.4   Proof of the Carnot theorem

Use the work output of H to run C backwards (C is reversible). Adjust C so that it absorbs the work produced by[8] H, $|W_C| = |W_H|$. From the inequality (2.3) therefore

$$Q_{H_1} < Q_{C_1}, \qquad (2.4)$$

i.e., *the more-efficient engine absorbs less heat for the same work output.* Apply the first law to the middle diagram in Fig. 2.4. Because for the composite engine $W = 0$, we have for the net heats $|Q_{\text{out}}| = Q_{\text{in}}$, or that $Q_{C_2} + Q_{H_1} = |Q_{C_1}| + |Q_{H_2}|$. Thus, $Q_{C_2} - |Q_{H_2}| = |Q_{C_1}| - Q_{H_1} > 0$, where from (2.4) $Q_{H_1} < |Q_{C_1}|$. The composite engine therefore works as drawn in the right portion of Fig. 2.4, *in violation of the Clausius statement.*

The assumption $\eta_H > \eta_C$ thus leads to a violation of the second law, and we conclude that $\eta_H \le \eta_C$. Proof seemingly over. You could ask, however, what properties of C were used in the proof? The only feature is that it's *reversible.* Because no properties of H were invoked either, what if H is reversible, call it R? By what we've just shown, $\eta_C \ge \eta_R$. But now interchange the roles of the engines: Assume that C is *more* efficient than R, $\eta_C > \eta_R$, and use it to run R backwards. Repeating the argument we would be led to a violation of the second law, implying $\eta_R \ge \eta_C$. Consistency requires that $\eta_R = \eta_C$. *All reversible engines operating between the same reservoirs are equally efficient.* Because C is reversible, we conclude that

$$\eta \le \eta_R. \qquad (2.5)$$

This inequality is Carnot's theorem: *The efficiency cannot exceed that of a reversible engine.* We show in Chapter 3 that the inequality (2.5) leads directly to the existence of entropy.

2.4   ABSOLUTE TEMPERATURE

Carnot's theorem implies that η_R *cannot depend on any specific property of an engine* (the *working substance* of the engine), because it's the same for all reversible engines. The quantity η_R can only depend on the reservoirs—it's all that remains—and the equilibrium state of reservoirs is characterized by its temperature (zeroth law). Empirical temperature, however, is measured using the thermometric properties of substances which we are free to choose (Section 1.5). How can η_R be universal when the reservoir temperatures depend on arbitrary substances? *The universality of Carnot's theorem implies that temperature can be formulated in a universal manner.*

---

[8]From Fig. 2.3 the location of the adiabats can be adjusted so that the area enclosed by the cycle (work performed) has any prescribed value.

For the efficiency η_R to depend only on the reservoirs (Carnot's theorem), let's posit an "efficiency function" f of the reservoir temperatures,

$$f(\theta_1, \theta_2) = Q_2/Q_1 \,, \tag{2.6}$$

where we revert to the empirical temperature notation θ (Section 1.5), and we've made use of Eq. (2.1). For convenience all occurrences of the symbol Q in this section represent absolute magnitudes, minus signs having been relegated[9] to Eq. (2.1). We first show that $f(\theta_1, \theta_2)$ occurs as a ratio of functions of a single variable, $\phi(\theta_2)/\phi(\theta_1)$ (see Eq. (2.9)).

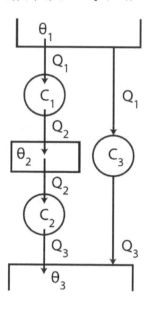

Figure 2.5   Derivation of the absolute temperature

Figure 2.5 shows a Carnot engine C_1 operating between reservoirs at temperatures θ_1 and θ_2, and a second Carnot engine[10] C_2 operating between θ_2 and θ_3. Engine C_1 absorbs heat Q_1 at θ_1 and expels Q_2 at θ_2. Adjust the size of C_2 so that it absorbs Q_2 at θ_2 and expels Q_3 at θ_3. The net effect is a single engine C_3 that absorbs Q_1 at θ_1 and expels Q_3 at θ_3. Apply Eq. (2.6) to each engine:

$$\frac{Q_2}{Q_1} = f(\theta_1, \theta_2) \qquad \frac{Q_3}{Q_2} = f(\theta_2, \theta_3) \qquad \frac{Q_3}{Q_1} = f(\theta_1, \theta_3) \,. \tag{2.7}$$

Equation (2.7) implies the *functional relation* satisfied by f

$$f(\theta_1, \theta_2) f(\theta_2, \theta_3) = f(\theta_1, \theta_3) \,. \tag{2.8}$$

To satisfy Eq. (2.8), $f(\theta_1, \theta_2)$ must have the form

$$f(\theta_1, \theta_2) = \phi(\theta_2)/\phi(\theta_1) \,. \tag{2.9}$$

With $f(\theta_1, \theta_2)$ in the form of Eq. (2.9), we have "passed the buck" onto an unknown function $\phi(\theta)$ that depends on the temperature of a *single* reservoir.

---

[9]The heats Q in equation Eq. (2.6) refer to reversible heat transfers.

[10]These engines do work on the environment; we're just not showing it in Fig. 2.5.

Combining Eq. (2.9) with Eq. (2.6), we have that the ratio of the heats reversibly expelled to and absorbed from reservoirs has the property that it equals the ratio of a function of θ evaluated at the reservoir temperatures *no matter what empirical temperature is used for each reservoir*,

$$\frac{Q_2}{Q_1} = \frac{\phi(\theta_2)}{\phi(\theta_1)} \,. \tag{2.10}$$

That's a tall order. We appear to require a "universal decoder function" ϕ that enforces Eq. (2.10) for any θ. The only way Eq. (2.10) can apply for *any* empirical temperature is if ϕ is a property of heat reservoirs that's independent of material composition. We assert the existence of a universal property of reversible heat flows—*absolute temperature T*—defined such that the ratio of absolute temperatures is the same as the ratio of heats reversibly drawn from or expelled to the reservoirs,

$$\frac{T_2}{T_1} \equiv \left(\frac{Q_2}{Q_1}\right)_{\text{reversible}} . \tag{2.11}$$

Equation (2.11) implies an *invariant* for reversible cycles, which we'll see is the entropy:

$$\left(\frac{Q_1}{T_1}\right)_{\text{rev}} = \left(\frac{Q_2}{T_2}\right)_{\text{rev}} . \tag{2.12}$$

The efficiency of a reversible engine can thus be given in terms of absolute temperatures. Combining Eq. (2.11) with Eq. (2.1), we have that the maximum efficiency of a cycle operating between hot and cold reservoirs with absolute temperatures T_h and T_c is

$$\eta_R = 1 - \frac{T_c}{T_h} \,. \tag{2.13}$$

Note that Eq. (2.13) has been obtained without relying on a particular equation of state (such as the ideal gas). It's shown in Exercise 2.4 that the efficiency of a Carnot cycle using the ideal gas as the working substance is the same as Eq. (2.13). While such a result is demanded by the Carnot theorem, it's "nice" to show it explicitly.

Carnot's theorem thus implies the *existence* of absolute temperature. To make it *useful*, we must adopt a unit of absolute temperature. The *Kelvin scale* conventionally assigns the value 273.16 K to the *triple point* of water, a reproducible state. The triple point is the *unique* combination of (T, P) at which the solid, liquid, and gas phases of a single substance coexist in equilibrium (Chapter 6). When heat Q_{tp} is reversibly transferred from a reservoir held at the triple point of H_2O, the temperature T (in Kelvin) of another reservoir at which heat Q is reversibly transferred, is

$$T = 273.16 \, (Q/Q_{tp}) \text{ K} \,. \tag{2.14}$$

Equation (2.14) gives an operational sense of what it means to say that one temperature is twice as large as another. Conversely, the smaller Q is, the smaller is the value of T, indicating that $T \to 0$ is possible.[11] Whether $T = 0$ is experimentally achievable is another matter (the province of the third law of thermodynamics, Chapter 8); the larger point is that absolute temperature provides a framework in which zero temperature can be discussed.[12] As practically necessary as the Kelvin scale is, it's something of a red herring. There's no *dimension* embodied in the Kelvin or any other temperature scale. *There is nothing fundamental about the Kelvin temperature scale.* We see from Eq. (2.11) that *absolute temperature naturally has the dimension of energy.* We need a conversion factor between the unit of energy, Joule, and the unit of temperature, Kelvin; this is supplied by Boltzmann's constant, $k_B = 1.38 \times 10^{-23}$ J K^{-1}.

---

[11]Thermometers based on empirical temperatures cannot be used for arbitrarily low temperatures; substances undergo phase transitions, altering their thermometric properties.

[12]Which is a colder day: 0 °C or 0 °F? Using empirical temperature scales one cannot unambiguously compare the coldness or hotness of objects.

CHAPTER SUMMARY

This chapter introduced the second law of thermodynamics, that heat does not spontaneously flow from cold to hot. The Carnot theorem, consistent with the second law, states that the efficiency of heat engines cannot exceed that of reversible engines operating between the same reservoirs. The universality of Carnot's theorem implies that temperature can be formulated in an equally universal manner, independent of the thermometric properties of materials, the absolute temperature T.

- The Clausius form of the second law is that heat does not flow from cold to hot with no other effect. The Kelvin form is that heat cannot be entirely converted to work at a single temperature. The two are equivalent and imply each other.

- The ratio of absolute temperatures T_1 and T_2 is the ratio of the absolute values of the heats reversibly transferred with two reservoirs at those temperatures, $Q_1/Q_2 = T_1/T_2$.

- There is an upper bound on the efficiency of a heat engine, $\eta = 1 - |Q_{\text{out}}|/Q_{\text{in}} \le 1 - T_c/T_h$, where T is the absolute temperature (Carnot's theorem).

EXERCISES

2.1 Referring to the right portion of Fig. 2.2, show that heat Q_C is delivered to the hot reservoir.

2.2 Derive Eq. (2.2). Assume that the internal energy is given in the form $U = U(P, V)$.

2.3 Referring to the left portion of Fig. 2.2, the *coefficient of performance* ω of a refrigerator is defined as the heat removed from the cold object per work input, $\omega \equiv Q_C/W$. Show that

$$\omega = \frac{Q_C}{|Q_H| - Q_C}.$$

The coefficient of performance of a heat pump ω_{HP} is defined as the heat delivered to a hot object per work input, $\omega_{HP} \equiv Q_H/W$. Show that $\omega_{HP} = 1 + \omega$.

2.4 Show explicitly that the efficiency of a Carnot cycle using the ideal gas as the working substance is given by $\eta = 1 - T_c/T_h$. Use the results of Exercise 1.10.

2.5 You've been brought in to consult on the development of a new engine, with the possibility that your "sweat equity" would count towards acquiring stock in the company. Should you get involved with this project? The design requirements are:

  a. It must deliver a power of 21 kW at the operating frequency of 600 rpm;

  b. It must, operating in a cycle between heat reservoirs at 500 K and 350 K, extract 6000 J of heat from the high-temperature reservoir while expelling 3900 J to the low-temperature reservoir.

2.6 The *Stirling cycle* consists of four steps: an isothermal expansion at a temperature T_h, a step where heat is removed at constant volume (*isochoric*), an isothermal compression at temperature T_c, and a step where heat is added at constant volume. Assume the ideal gas as the working substance.

  a. On a P-V diagram, indicate where heat enters and leaves the system.

  b. Calculate the heat delivered to and expelled from the system in the isochoric steps. Assume that the heat capacity is constant over the range of temperatures involve. What is the net heat transferred to the system in these steps?

  c. Derive an expression for the efficiency of the Stirling cycle. Compare with η_R.

CHAPTER 3

Entropy

On the other hand, a method involving the notion of *entropy*, the very existence of which depends on the second law of thermodynamics, will doubtless seem to many far-fetched, and may repel beginners as obscure and difficult of comprehension.
—J.W. Gibbs, 1873[3, p11]

3.1 CLAUSIUS INEQUALITY

S TARTING from Carnot's theorem $\eta \leq \eta_R$, we have from Eqs. (2.1) and (2.13),

$$1 - \frac{|Q_2|}{Q_1} \leq 1 - \frac{T_2}{T_1} ,$$

where $T_1 > T_2$, and thus we have the inequality

$$|Q_2| \geq Q_1 \frac{T_2}{T_1} ,$$

where equality holds for reversible cycles. For fixed values of Q_1, T_1, and T_2, the greater the *inefficiency* of the engine (and hence the less the work done), *the greater is the heat expelled at the lower temperature*, $|Q_2|$. The inequality can be written (because $Q_2 < 0$):

$$0 \geq -\frac{|Q_2|}{T_2} + \frac{Q_1}{T_1} = \frac{Q_2}{T_2} + \frac{Q_1}{T_1} .$$

This inequality (which applies to a cycle operating between two reservoirs) can be generalized by noting that an *arbitrary* cycle can be realized out of multiple *infinitesimal* cycles each operating between two reservoirs (see Fig. 3.1).[1] Applying the basic inequality to each of them,

$$\sum_i đQ_i/T_i \leq 0 , \tag{3.1}$$

where $đQ_i$ is the infinitesimal heat added to, or expelled from, the system *at the reservoir temperature T_i*. The sum in (3.1) is non-positive because Q/T for heats expelled at the lower temperatures (negative quantities) exceed in magnitude Q/T for heats absorbed at the higher temperatures. Passing to the limit, we have the *Clausius inequality*, that for any cycle[15, p133]

$$\oint đQ/T \leq 0 . \tag{3.2}$$

---

[1]The P-V plane can be covered by a non-orthogonal coordinate system of adiabats and isotherms where a given point sits at the intersection of an isotherm and an adiabat. The *uniqueness* of adiabats is therefore a key issue in the theory of thermodynamics, what we touched on in Section 2.3. It's a central issue in the Carathéodory formulation of the second law, Chapter 10.

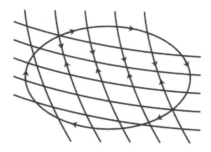

Figure 3.1   An arbitrary cycle is the resultant of multiple infinitesimal cycles

3.2   THE BIRTH OF ENTROPY

> Although the necessity of this theorem admits of strict mathematical proof ... it nevertheless retains an abstract form, in which it is with difficulty embraced by the mind, and we feel compelled to seek for the precise physical cause, of which this theorem is a consequence.—R. Clausius, 1862[15, p219]

For reversible cycles (the case of equality in (3.2)), we have the fundamental result:

$$\oint (\mathrm{d}Q)_{\text{rev}} / T = 0 \, . \tag{3.3}$$

Because the closed-loop integral vanishes, the quantity $(\mathrm{d}Q)_{\text{rev}} / T$ is an exact differential, and *hence represents the differential of a state variable*. Voilà! We have a new state variable called *entropy*, S, defined such that[2]

$$\mathrm{d}S \equiv (\mathrm{d}Q)_{\text{rev}} / T \, . \tag{3.4}$$

The word *entropy* was intentionally chosen by Rudolf Clausius to be close to the word *energy*.[3] As a state variable, *entropy is defined only in equilibrium*. The difference in entropy between equilibrium states is obtained by integrating its differential

$$S(B) - S(A) = \int_{A}^{B} (\mathrm{d}Q)_{\text{rev}} / T \tag{3.5}$$

for any process connecting A and B such that *heat is transferred reversibly to the system at all steps of the process*.[4] The quantity T appearing in Eq. (3.5) is the absolute temperature at which heat is transferred into or out of the system. The inverse temperature T^{-1} is therefore an *integrating factor* for the inexact differential $(\mathrm{d}Q)_{\text{rev}}$. Note that, with Eq. (3.5), *we are no longer referring to cycles, but with processes connecting discrete points in state space*.

---

[2]Entropy makes its entrance onto the stage of physics in the form of a differential. We know what $\mathrm{d}S$ is, but do we know what S is? Stay tuned.

[3]"But as I hold it to be better to borrow terms for important magnitudes from the ancient languages, so that they may be adopted unchanged in all modern languages, I propose to call the magnitude S the *entropy* of the body, from the Greek word τροπή, *transformation*. I have intentionally formed the word *entropy* so as to be as similar as possible to the word *energy*; for the two magnitudes to be denoted by these words are so nearly allied in their physical meanings, that a certain similarity in designation appears to be desirable."[15, p357]

[4]The quantity of heat $\int_{A}^{B} \mathrm{d}Q$ required to bring a system from state A to state B is not uniquely defined; it depends on the path between A and B (Section 1.4). The question of how much heat is contained in the system is therefore meaningless. We now have a quantity closely related to heat transfer which *is* defined by the state of the system. For a system brought from A to B, the quantity $\int_{A}^{B} (\mathrm{d}Q)_{\text{rev}} / T$ has a value independent of path, so long as the path is reversible.

We have been led mathematically to the *existence* of entropy, yet we have no idea what it *is*.[5] Insight can be had by examining its dimension, energy divided by absolute temperature. As noted in Section 2.4, absolute temperature naturally has the dimension of *energy*; the Kelvin is a non-dimensional unit. *Entropy is naturally a dimensionless quantity*, a fact obscured by quoting its value[6] in SI units, J/K. That's why formulas for the entropy always contain as a prefactor the conversion between Joules and Kelvin, the Boltzmann constant or the gas constant. Knowing that S/k_B is a dimensionless characterization of equilibrium suggests that entropy has a *qualitative* aspect to it, in addition to quantitative. We discuss in this chapter some of the properties of entropy and its consequences. We return in Chapter 7 to its meaning.

3.3 ENTROPY, IRREVERSIBILITY, AND DISORGANIZATION

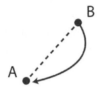

Figure 3.2   Irreversible process from $A \to B$, reversible process from $B \to A$

Consider, as in Fig. 3.2, a cycle consisting of an irreversible process from point A in state space to point B, followed by a reversible process from B to A. The irreversible process is indicated as a dashed line because it cannot be represented in state space (Section 1.7). From the Clausius inequality, therefore, $\int_A^B \mathrm{d}Q/T + \int_B^A (\mathrm{d}Q)_{\mathrm{rev}}/T \leq 0$. Because the process from B to A is reversible, however, we have $\int_A^B \mathrm{d}Q/T - \int_A^B (\mathrm{d}Q)_{\mathrm{rev}}/T \leq 0$, or

$$\int_A^B (\mathrm{d}Q)_{\mathrm{rev}}/T \equiv \int_A^B \mathrm{d}S \geq \int_A^B \mathrm{d}Q/T \,,$$

from which we arrive at the *Clausius inequality in differential form* (because A and B are arbitrarily chosen states):

$$\mathrm{d}S \geq \mathrm{d}Q/T \,. \tag{3.6}$$

The quantity $\mathrm{d}Q$ in (3.6) is not restricted to a reversible heat transfer (which applies in the case of equality). The closed-loop integral of the inequality (3.6) reproduces the inequality (3.2).

The inequality (3.6) informs us that $\mathrm{d}Q/T$ *does not account for all contributions to* $\mathrm{d}S$. There must be another *type* of entropy to close the gap between $\mathrm{d}Q/T$ and $\mathrm{d}S$. Define the difference[7]

$$\mathrm{d}S_i \equiv \mathrm{d}S - \mathrm{d}Q/T \,. \tag{3.7}$$

Combining Eq. (3.7) with (3.6), we have an equivalent form of the Clausius inequality

$$\mathrm{d}S_i \geq 0 \,. \tag{3.8}$$

---

[5]It's a natural question to ask, but try not to become obsessed with it. The question presumes that it can be answered in terms of familiar concepts. What if entropy is something *new*, not reducible to concepts gained through prior experience? Get to know its properties; then you'll know what *it* is. How does it change with temperature under constant pressure? How does it change with volume at constant temperature? If we know how it behaves, we know a great deal about what it *is*.

[6]Entropy is often quoted in the non-SI unit e.u. for "entropy unit." One e.u. is one calorie per Kelvin per mole, or 4.184 Joules per Kelvin per mole.

[7]Clausius termed $\mathrm{d}S_i$ the *uncompensated transformation*, the transformation of a system not "compensated" (caused by) by heat transfers from the environment.[15, p363]

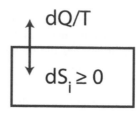

Figure 3.3   Entropy transport to and from environment, entropy creation from irreversibility

Heat transfers occur through system boundaries, which can be *positive, negative, or zero*. Only for reversible heat transfers is $\Delta S = \int (\text{d}Q)_{\text{rev}}/T$. *There is an additional, positive contribution to the entropy produced by irreversible processes internal to the system*, $\Delta S_i > 0$. Entropy can vary for two reasons, and two reasons only: either from entropy transport[8] to or from the surroundings by means of heat exchanges, or by the *creation of entropy* inside the system from irreversible changes in state, which is always positive (see Fig. 3.3.) *Irreversible processes create entropy.*

   Equation (3.7) exhibits the same logic as the first law, Eq. (1.5), in that whereas the heat transferred to the system is the difference between the change in internal energy and the work done, $Q = \Delta U - W$, the entropy created through irreversibility is the difference between the change in entropy and that due to heat transfers, $\Delta S_i = \Delta S - \int \text{d}Q/T$.

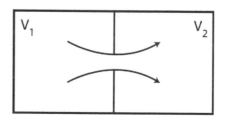

Figure 3.4   Free expansion through a ruptured membrane

   Thus, there are processes between equilibrium states that produce entropy changes *not due to heat exchanges with external reservoirs*. The classic example is the *free expansion* (see Fig. 3.4), the flow of gas through a ruptured membrane that separates a gas from a vacuum chamber in an otherwise isolated system: Even for no heat transfer with the environment, there's an increase in entropy. (We return to the free expansion in Section 4.9.) An *isolated system* is one that cannot exchange energy or matter with the environment. With $\text{d}Q = 0$, we have from Eq. (3.6),

$$dS \geq 0 . \tag{3.9}$$

This is a remarkable state of affairs. For systems not interacting with the environment (isolated), *one may still affect changes between equilibrium states!* The "interaction" in a free expansion consists of rupturing a membrane or opening a valve, acts that can be accomplished without transferring energy to the system, if only because the valve and the walls are not part of the system. That transitions between equilibrium states can be induced without transferring energy indicates *there must be another state variable to describe macroscopic systems.*[9] Entropy is that new state variable.

---

[8]Entropy can flow, just like energy. See Section 3.11.

[9]That is, there's more to the description of equilibrium states than internal energy and temperature, what one would infer from the first law of thermodynamics.

The inequality (3.9) is one of the most far-reaching results of physics: *The entropy of an isolated system can never decrease.* As an isolated system comes to equilibrium, therefore, *entropy achieves a maximum value.*[10] It's instructive to trace the steps leading to (3.9): It's a consequence of the Clausius inequality (3.6), which derives from the inequality established by Carnot's theorem, (2.5), which in turn obeys the second law of thermodynamics. *The directionality of spontaneous heat flows implies the inequality* (3.9), which prescribes the *time order* of spontaneously occurring events, those for which entropy increases. In Section 1.7 we discussed the ordering in time of states linked by irreversible processes. Now we have a state variable S, *the values of which can be put into correspondence with the time order established by irreversibility.* The laws of physics are usually equalities ($F = ma$, $E = mc^2$, $\Delta U = Q + W$), whereas (3.9) is an *inequality.* Therein lies its universality! The second law is *not* an equality, $A = B$. It specifies the time progression of spontaneous processes *regardless of the system.*

In the free expansion, the rupturing of the membrane is the *removal of a constraint.* With the system initially constrained to the volume V_1, by rupturing the membrane the constraint is removed and the system evolves to occupy a new volume, $V_1 + V_2$. Macroscopic systems naturally "explore" all available possibilities, which is associated with an increase in entropy. The removal of constraints is irreversible (Section 1.8). Once removed, *constraints cannot be put back*, at least not without the performance of work, implying irreversibility. Entropy can thus be seen as a measure of the *degree of constraint*: The second law captures the tendency of systems to spontaneously evolve from states of more to less constraint. Because the entropy of isolated systems can only increase, states can be classified according to the *order* in which they occur in a process of progressively removing constraints. Constraints *restrict* the possibilities available to systems, and in that sense they *organize* them. A system undergoing a sequence of changes brought about by the removal of constraints becomes increasingly less organized. *There must be a connection between entropy and disorganization.* We return to that idea in Chapter 7.

The time ordering of states linked by irreversible processes is a distinctive feature of thermodynamics, *unlike any other in physics.* The equations of Newtonian dynamics are *time-reversal invariant*, in which the role played by past and future occur in a symmetric manner;[11] processes proceed the same if the sense of time is reversed, $t \to -t$. That symmetry is broken at the macroscopic level where there's a *privileged direction of time*, an inevitable order to the sequence of spontaneous events, the *arrow of time.*[12] *Why* microscopic theories are time-reversal invariant while the macroscopic world is not, we cannot say. One answer is that macroscopic systems involve so many microscopic constituents, the application of Newton's laws must involve approximations that break time-reversal symmetry, yet what we observe in the macro-world knows nothing of our approximations. And Newtonian dynamics has been superseded by quantum mechanics and relativity, theories that are also time-reversal symmetric. The arrow of time reflects there *not* being equilibrium: As long as *disequilibrium* exists, entropy will increase. The second law cannot be derived from microscopic laws of motion; *it's a separate principle.* Irreversibility is an undeniable feature of the macroscopic world, *our experience of which is codified in the second law.*[13]

Let's summarize the laws of thermodynamics developed so far. The zeroth law establishes temperature as a state variable, what systems in equilibrium have in common. The first law establishes internal energy as a state variable and places restrictions on the possible states that can be connected by transitions, namely those that conserve energy. The second law establishes entropy as a state variable that determines the *order* of states as they occur in spontaneous processes. Whether the transition $A \to B$ is possible in an isolated system is determined by whether $S(A) < S(B)$.

---

[10]Because processes occur in time, with $dS \geq 0$ for an isolated system, time implicitly sneaks into the theory. The entropy of an isolated system keeps increasing until it attains the maximum value it can attain subject to constraints on the system.

[11]Maxwell's equations of electrodynamics are also time-reversal symmetric.

[12]The term arrow of time (actually *time's arrow*) was introduced by Eddington.[16, p69]

[13]Said differently, the second law is the explicit recognition in the pantheon of physical laws of the existence of irreversibility.

3.4 OPENING THE BLACK BOX: GIBBSIAN THERMODYNAMICS

> It is an inference naturally suggested by the general increase of entropy which accompanies the changes occurring in any isolated material system that when the entropy of the system has reached maximum, the system will be in a state of equilibrium. Although this principle has by no means escaped the attention of physicists, its importance does not appear to have been duly appreciated. Little has been done to develop the principle as a foundation for the general theory of thermodynamic equilibrium.
> —J.W. Gibbs, 1878[3, p354]

How does our newly-found state variable fit in with the structure of thermodynamics? It was the great contribution of Gibbs to recognize that entropy and energy allow one to *analytically characterize the thermodynamic properties of materials*. Focusing on the *state of equilibrium* rather than on processes is called *Gibbsian thermodynamics*. In the Clausius-Kelvin form of thermodynamics, the system is considered a "black box" with all relevant information derived from observable properties of the system's interaction with the environment. Gibbs in essence opened the black box.[14] Much of what we consider today to be thermodynamics can be traced to the work of Gibbs.

With $(dQ)_{rev} = TdS$ and $dW = -PdV$ we have from Eq. (1.6)

$$dU = TdS - PdV \qquad (3.10)$$

when it's understood that all infinitesimal changes are reversible. With Eq. (3.10) we have the first law expressed in terms of exact differentials. Equation (3.10) can be integrated to find ΔU for *any reversible process connecting the initial and final states*. A little reflection, however, shows that Eq. (3.10) is *holds for any change of state, however accomplished*, as long as there is a *conceivable* reversible path connecting the initial and final equilibrium states. From now on, we'll refer to Eq. (3.10) (and its generalization in Eq. (3.16)) as the first law of thermodynamics.[15]

As an example, consider an ideal gas that escapes into an evacuated chamber, doubling its volume in the process. The change in entropy can be calculated if we can conceive a hypothetical reversible process that connects the initial and final states. The system is adiabatically isolated; no heat flows to the system. The energy of an ideal gas is independent of volume, Eq. (1.31), and thus the temperature does not change. Imagine that the volume is slowly expanded, keeping the temperature fixed. With $dU = 0$ in Eq. (3.10), we have $dS = (P/T)dV = Nk_BdV/V$. Thus, $\Delta S = Nk_B \ln 2$. Note that $\Delta S > 0$; the actual process connecting initial and final states is irreversible.

This example is *paradigmatic* of processes involving spontaneous increases in entropy. What has changed in the system as a result of the process? Not the energy, nor the number of particles. What has changed is the number of *ways* particles can be located in the increased volume relative to the number of ways they can be located in the original volume. We'll return to this important example in Chapter 7 where we introduce the microscopic interpretation of entropy.

We infer from Eq. (3.10) that U is naturally a function of S and V; $U = U(S,V)$. We're free, however, to take U and S to be functions of any other variables, whatever is convenient for a particular calculation. Let $U = U(T,V)$ and $S = S(T,V)$. With that assumption, the following identities follow from Eq. (3.10):

$$T\left(\frac{\partial S}{\partial T}\right)_V = \left(\frac{\partial U}{\partial T}\right)_V = C_V \qquad (3.11)$$

$$T\left(\frac{\partial S}{\partial V}\right)_T = \left(\frac{\partial U}{\partial V}\right)_T + P, \qquad (3.12)$$

---

[14]Gibbs emphasized the properties of *material systems* in equilibrium rather than the "motive power of heat." In this approach, energy and entropy (state functions) take precedence over quantities that depend on process (work and heat).

[15]Equation (3.10) is sometimes called the "entropy form" of the first law, or the "combined first and second laws" of thermodynamics. I prefer just to call it the first law; it's the first law become exact, that is, written in terms of exact differentials.

where we've used Eq. (1.25) in Eq. (3.11). Likewise, assuming $U = U(T, P)$, $S = S(T, P)$, and $V = V(T, P)$, it's straightforward to show from Eq. (3.10) that

$$T\left(\frac{\partial S}{\partial T}\right)_P = \left(\frac{\partial U}{\partial T}\right)_P + P\left(\frac{\partial V}{\partial T}\right)_P = C_P \qquad (3.13)$$

$$T\left(\frac{\partial S}{\partial P}\right)_T = \left(\frac{\partial U}{\partial P}\right)_T + P\left(\frac{\partial V}{\partial P}\right)_T = -T\alpha V . \qquad (3.14)$$

3.5 CHEMICAL POTENTIAL AND OPEN SYSTEMS

As advertised in Section 1.4, energy is required to change the amount of matter in a system, *chemical work*. *Open systems* are enclosed by permeable boundaries that allow for the flow of matter as well as energy (Table 1.1). Following Gibbs, we take the first law[16] for open systems to be

$$dU = TdS - PdV + \mu dN , \qquad (3.15)$$

where N is the number of particles and $\mu \equiv (\partial U/\partial N)_{S,V}$ is the *chemical potential*, the energy to add another particle at fixed entropy and volume.[17] The extra term in Eq. (3.15) is not simply the energy to change *total* particle numbers, but that to change the number of particles having chemical potential μ, i.e., *μ is specific to a chemical species*. For a system with distinct chemical species (a *multicomponent system*), each has its own chemical potential μ_j and Eq. (3.15) generalizes to

$$dU = TdS - PdV + \sum_j \mu_j dN_j . \qquad (3.16)$$

Chemical potential is often a *negative* quantity. Consider that even if it were possible to add a particle at *zero energy cost*, the energy U of the system would be shared among $N + 1$ particles. Increasing the number of ways to distribute the energy *increases* the entropy. To keep the entropy fixed upon adding another particle (the definition of μ), the energy must be *decreased* slightly. Only if the interactions between particles are sufficiently repulsive does μ become positive. The Fermi energy in semiconductors is an example; the repulsive interaction in this case is the requirement of the Pauli exclusion principle. That entropy plays a primary role in determining μ can be seen from the fact that the ideal gas, which has *no* inter-particle interactions, nevertheless has a nonzero chemical potential (Section 4.6).

3.6 HOMOGENEOUS FUNCTIONS*

A function F of k variables having the scaling property

$$F(\lambda x_1, \cdots, \lambda x_k) = \lambda^p F(x_1, \cdots, x_k) \qquad (3.17)$$

is called a *homogeneous function of order p*. There is a simple theorem on homogeneous functions due to Euler. Differentiate Eq. (3.17) with respect to λ:

$$p\lambda^{p-1} F(x_1, \cdots, x_k) = \frac{dF}{d\lambda} = \sum_{j=1}^k \frac{\partial F}{\partial(\lambda x_j)}\frac{\partial(\lambda x_j)}{\partial\lambda} = \sum_{j=1}^k \frac{\partial F}{\partial(\lambda x_j)} x_j . \qquad (3.18)$$

---

[16]Equation (3.15) is sometimes called *Gibbs's equation*.

[17]Chemical potential is an *energy*, and thus calling it a potential is confusing. We'll see that matter flows from regions of higher μ to regions of lower μ (Section 3.11), justifying the use of the term potential. We show in Section 4.6 that μ is also the energy to add a particle at constant T and P.

Now set $\lambda = 1$ in Eq. (3.18) and we have *Euler's theorem*:

$$pF(x_1, \cdots, x_k) = \sum_{j=1}^{k} (\partial F/\partial x_j)\, x_j \,. \tag{3.19}$$

Take for example $F(x) = x^2$; clearly $2x^2 = (\partial F/\partial x)\, x$.

3.7 EXTENSIVITY OF ENTROPY

Can the entropy of the ideal gas be calculated, the *simplest macroscopic system*? The standard answers are either 1) it is not possible or 2) what we *can* calculate has deficiencies that can only be fixed with quantum mechanics. Let's see how far we can get using the methods at hand.

Adopt the "differentiate-then-integrate" strategy (Section 1.11), where we integrate the differential of S *that we have at our disposal*. Taking T, V, and N as the independent variables, we have for a process in which N does not change,

$$S(T_2, V_2, N) - S(T_1, V_1, N) = \int_1^2 [dS]_N = \int_1^2 \left[\left(\frac{\partial S}{\partial T}\right)_{V,N} dT + \left(\frac{\partial S}{\partial V}\right)_{T,N} dV \right]$$
$$= \int_1^2 \left[C_V \frac{dT}{T} + \left(\frac{\partial P}{\partial T}\right)_{V,N} dV \right],$$

where $[dS]_N$ denotes the variation of S with N held fixed, and where we've used Eq. (3.11) and a Maxwell relation (developed in Section 4.5). For the ideal gas,

$$S(T_2, V_2, N) - S(T_1, V_1, N) = \int_1^2 \left[\frac{3}{2} N k_B \frac{dT}{T} + N k_B \frac{dV}{V} \right]$$
$$= \frac{3}{2} N k_B \ln\left(\frac{T_2}{T_1}\right) + N k_B \ln\left(\frac{V_2}{V_1}\right) = N k_B \ln\left[\frac{V_2}{V_1}\left(\frac{T_2}{T_1}\right)^{3/2} \right]. \tag{3.20}$$

Equation (3.20) is satisfied by an entropy function of the form

$$S(T, V, N) = N k_B \ln\left(T^{3/2} V\right) + k_B \phi(N)\,, \tag{3.21}$$

where $\phi(N)$ is an unknown function. Generally one should beware of equations like Eq. (3.21) in which logarithms of *dimensional* quantities appear. The logarithm, like all transcendental functions, can only be a function of *dimensionless* quantities. Yet, equations like Eq. (3.21), featuring logarithms of dimensional quantities, are common in science and engineering. The only way such equations can be valid is if there are other terms in the equation (perhaps not displayed) that lead to logarithms of dimensionless arguments. What physics can we draw upon to determine $\phi(N)$? The Clausius definition of entropy, Eq. (3.5), involving heat transfers from the environment, applies to *closed systems only*. We show in Chapter 14 that for *open systems* there is a contribution to the entropy from *the flow of matter* not accounted for by the Clausius definition. As noted by Edwin Jaynes: "As a matter of elementary logic, no theory can determine the dependence of entropy on the size N of a system unless it makes some statement about a process where N changes."[17]

Our "statement" (à la Jaynes) is that *entropy is extensive*, as is the internal energy. By scaling extensive variables such as V and N by a factor λ, while keeping intensive variables fixed, the entropy should scale by the same factor of λ. We *require* that S satisfy the scaling law (for $\lambda > 0$)

$$S(T, \lambda V, \lambda N) = \lambda S(T, V, N)\,. \tag{3.22}$$

The extensivity of S is not part of the Clausius definition,[18] which pertains to closed systems. Imposing Eq. (3.22) on Eq. (3.21), we require that $\phi(N)$ be such that

$$\phi(\lambda N) = \lambda \phi(N) - N\lambda \ln \lambda . \tag{3.23}$$

Differentiate Eq. (3.23) with respect to λ and then set $\lambda = 1$:

$$N\phi' = \phi - N . \tag{3.24}$$

The solution of Eq. (3.24) is

$$\phi(N) = \phi(1)N - N \ln N , \tag{3.25}$$

where $\phi(1)$ is an undetermined constant. Combining Eq. (3.25) with Eq. (3.21),

$$S = Nk_B \ln \left(\frac{V}{N} T^{3/2} \right) + Nk_B \phi(1) .$$

This expression is indeed extensive (it satisfies Eq. (3.22)), but we still have the problem of a logarithm of a dimensional quantity. We can fix that by suitably defining the constant $\phi(1)$. Let $\phi(1) = \ln \left(C k_B^{3/2} \right)$ where C is another constant,

$$S = Nk_B \ln \left(\frac{V}{N} (k_B T)^{3/2} C \right) . \tag{3.26}$$

The constant C must be such that $V(k_B T)^{3/2} C$ is dimensionless; C therefore has the dimensions $(L^2 E)^{-3/2}$ (E is energy, L is length). Now, $L^2 E$ has dimensions of (action)^2/(mass); thus C has the dimensions (mass)$^{3/2}$/(action)^3. We can then write Eq. (3.26) as

$$S = Nk_B \ln \left[\frac{V}{N} \left(\frac{mk_B T}{\xi^2} \right)^{3/2} \right] = Nk_B \ln \left(\frac{V}{N} \frac{1}{\Lambda^3} \right) , \tag{3.27}$$

where m is a characteristic mass, ξ is a quantity having dimension of action, and $\Lambda \equiv \xi/\sqrt{mk_B T}$ is an equivalent length.

Equation (3.27) is the best we can do with the tools we have at present. The unknown quantity ξ of course turns out to be Planck's constant, something *outside* the subject of thermodynamics. In trying to calculate the entropy of the ideal gas, we've arrived at a limitation of the theory of thermodynamics. We'll return to this important topic in Chapter 7, where we'll see that with the derivation of the *Sackur-Tetrode* formula, Eq. (7.52), there's more to the story. We have, however, learned some lessons. Extensivity must be imposed as a separate requirement; it's not a consequence of the Clausius definition. The Clausius definition is *incomplete* in specifying the entropy. Herbert Callen in his axiomatic formulation of thermodynamics raised the extensivity of entropy to the status of a *postulate*.[19] The extensivity of S is not *as* obvious as the extensivity of U: Whereas internal

---

[18]It could be said that extensivity of entropy is implicit in $dS = dQ/T$ because the heat added to a system (to raise T by a fixed amount ΔT) scales with the size of the system: The heat capacity is mass-dependent.

[19]Callen[1] presented thermodynamics based on four postulates. Postulate I (corresponding to the first law) asserts the existence of equilibrium states that are characterized by a function U called internal energy. Postulate II (in correspondence with the second law) asserts the existence of a function S called entropy that takes on a maximum value in equilibrium. Postulate III is that entropy is extensive and a monotonically increasing function of energy. (We'll see in Chapter 11 that there are systems which violate the latter proviso.) Postulate IV is that the entropy vanishes at zero temperature, which corresponds loosely with the third law of thermodynamics (we'll see in Chapter 8 that the status of the third law is not as secure as the other laws of thermodynamics, and there are systems for which $S(T \to 0) \neq 0$). While the postulational approach cleanly features the logical structure of the theory, and may be of great value to those already initiated into thermodynamics, it usually proves bewildering to the beginner. The inductive approach we adopt is more instructive as to the options that were faced in developing the theory.

energy is related to a microscopic property of matter, and thus scales with the amount of mass in the system, *entropy is not a microscopic property of matter*; it's a property of the equilibrium state specified by the values of macroscopic state variables.[20]

3.8 GIBBS-DUHEM EQUATION

Thus, we require entropy to be extensive. We infer from Eq. (3.15) that S is naturally a function of the extensive variables (U, V, N). By the scaling relation Eq. (3.22),

$$S(\lambda U, \lambda V, \lambda N) = \lambda S(U, V, N) \,. \tag{3.28}$$

To visualize what's implied by extensivity, consider a system symbolically represented by the rectangle in Fig. 3.5. Imagine the system arbitrarily divided into nine subsystems. Extensive quantities are additive over subsystems. In this example, $V = 9V_{\text{subsystem}}$. We require the same be true for U, S, and N (when the equilibrium values of intensive variables are held fixed). Extensive variables are thus first-order homogeneous functions to which we can apply Euler's theorem. Intensive variables are *zero-order homogeneous functions*, as in $T(\lambda S, \lambda V, \lambda N) = T(S, V, N)$.

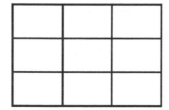

Figure 3.5 The idea behind extensivity: Divide the system into subsystems

Applying Euler's theorem, Eq. (3.19), to $S(U, V, N)$ which satisfies Eq. (3.28),

$$S = \left(\frac{\partial S}{\partial U}\right)_{V,N} U + \left(\frac{\partial S}{\partial V}\right)_{U,N} V + \left(\frac{\partial S}{\partial N}\right)_{U,V} N \,. \tag{3.29}$$

Note the structure of Eq. (3.29): S is expressed as a sum of products of extensive quantities (U, V, N) with intensive quantities (that are obtained from the derivatives of S with respect to these variables). The required derivatives in Eq. (3.29) can be obtained from Eq. (3.15) written in the form $dS = (1/T)\, dU + (P/T)\, dV - (\mu/T)\, dN$. Thus,

$$\left(\frac{\partial S}{\partial U}\right)_{V,N} = \frac{1}{T} \qquad \left(\frac{\partial S}{\partial V}\right)_{U,N} = \frac{P}{T} \qquad \left(\frac{\partial S}{\partial N}\right)_{U,V} = -\frac{\mu}{T} \,. \tag{3.30}$$

Combining Eq. (3.30) with Eq. (3.29), we have as a consequence of Euler's theorem,

$$TS = U + PV - \mu N \,. \tag{3.31}$$

Equation (3.31) is a remarkable result—it relates *seven* thermodynamic quantities, without any constants of integration. We'll refer to Eq. (3.31) as the *Euler relation*.

Now differentiate Eq. (3.31) and combine the result with Eq. (3.15); we obtain

$$N d\mu = -S dT + V dP \,, \tag{3.32}$$

---

[20]Extensivity implies that properties of the system are *additive* over the properties of independent subsystems (Section 1.2). What makes subsystems independent, however? Correlations can arise between parts of a system that would preclude dividing it into independent parts. As long as the length over which correlations exist is microscopic, additivity should apply.

the *Gibbs-Duhem equation*.[21] The first law of thermodynamics is expressed in terms of intensive quantities multiplied by the differentials of extensive variables, (intensive) × d(extensive). The Gibbs-Duhem equation has the converse structure, of extensive quantities multiplied by the differentials of intensive variables, (extensive) × d(intensive). From Eq. (3.32), we infer that $\mu = \mu(T, P)$ and moreover that

$$\left(\frac{\partial \mu}{\partial T}\right)_P = -\frac{S}{N} \qquad \left(\frac{\partial \mu}{\partial P}\right)_T = \frac{V}{N} . \tag{3.33}$$

Note that with the two results in Eq. (3.33) we have derivatives of one intensive variable with respect to another. The right side of the equalities in Eq. (3.33) present another *kind* of intensive variable, what are referred to as *densities* (volume per particle, entropy per particle). We'll reserve the term intensive variable to be the derivative of one extensive variable with respect to another, as in Eq. (3.30). Equation (3.33) is used in the analysis superfluid helium; see Eq. (15.23).

3.9 QUADRATIC FORMS*

A *quadratic form* is a homogeneous quadratic polynomial in any number of variables. In three variables, $ax^2 + by^2 + cz^2 + dxy + exz + fyz$ is a quadratic form for constants $(a, \cdots, f)$. A quadratic form in N variables can be generated by an $N \times N$ symmetric matrix, A, (with matrix elements A_{ij}): $\sum_{ij=1}^N A_{ij} x_i x_j$. A *positive-definite* quadratic form is one that is positive for any nonzero values of its variables. *Sylvester's criterion* for a quadratic form to be positive definite is that all the determinants associated with the upper-left submatrices of A be positive.[18, p52] For example, a positive-definite quadratic form in three variables is such that

$$A_{11} > 0 \qquad \begin{vmatrix} A_{11} & A_{12} \\ A_{21} & A_{22} \end{vmatrix} > 0 \qquad \begin{vmatrix} A_{11} & A_{12} & A_{13} \\ A_{21} & A_{22} & A_{23} \\ A_{31} & A_{32} & A_{33} \end{vmatrix} > 0 . \tag{3.34}$$

3.10 STABILITY OF THE EQUILIBRIUM STATE: FLUCTUATIONS

Entropy has the maximum value it can have consistent with constraints on the system, an *extremum principle* characterizing equilibrium (see Section 4.1). Because of the randomness of molecular motions, *fluctuations* in state variables occur about their equilibrium values.[22] Equilibrium (which persists in time) must be *stable* against fluctuations. In this section we develop the theory of stability of the equilibrium state, the basic ideas of which were introduced by Gibbs.[3, p56]

To do so, we make use of an oft-used theoretical device in thermodynamics, that of the *composite system* (see Fig. 3.6) which consists of subsystems A and B, in contact with each other, surrounded by rigid adiabatic walls. Let A and B be separated by a partition that's moveable, permeable, and diathermic. The subsystems can therefore exchange volume, particles, and energy; they're also assumed to contain the same *type* of particle. Let $V = V_A + V_B$, $N = N_A + N_B$, $U = U_A + U_B$. Let there be small variations in the energy, volume, and particle number of each subsystem, δU_A, δU_B, δV_A, δV_B, δN_A, δN_B. The variations are constrained, however, because these are conserved quantities: $\delta U_A = -\delta U_B$, $\delta V_A = -\delta V_B$, and $\delta N_A = -\delta N_B$. We consider the change in total system entropy ΔS under these variations to first and second order in small quantities, what we denote by $\Delta S = \delta S + \delta^2 S$, where[23] $S = S_A + S_B$.

---

[21]Equation (3.32) is the Gibbs-Duhem equation for a single chemical species in a single thermodynamic phase. In Chapter 6 we generalize the Gibbs-Duhem equation to multicomponent, multiphase systems.

[22]The concept of fluctuation is difficult to formulate in thermodynamics, but it can be done (contrary to what's sometimes said). By invoking the randomness of molecular motions, we're giving a nod to the atomic picture of matter. Below we define thermodynamic fluctuations as *virtual variations* of the system. Fluctuations are more naturally treated in the subject of statistical mechanics. We touch again on this topic in Section 12.3.

[23]Entropy is extensive.

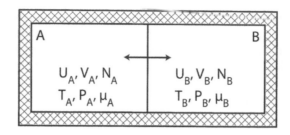

Figure 3.6   The composite system composed of subsystems A and B

Equilibrium between subsystems, $\delta S = 0$

The first-order variation δS is given by

$$
\begin{aligned}
\delta S &= \sum_{\alpha=A,B} \left[\left(\frac{\partial S_\alpha}{\partial U_\alpha} \right)_{V_\alpha, N_\alpha} \delta U_\alpha + \left(\frac{\partial S_\alpha}{\partial V_\alpha} \right)_{U_\alpha, N_\alpha} \delta V_\alpha + \left(\frac{\partial S_\alpha}{\partial N_\alpha} \right)_{U_\alpha, V_\alpha} \delta N_\alpha \right] \\
&= \sum_{\alpha=A,B} \left[\frac{1}{T_\alpha} \delta U_\alpha + \frac{P_\alpha}{T_\alpha} \delta V_\alpha - \frac{\mu_\alpha}{T_\alpha} \delta N_\alpha \right] \\
&= \left(\frac{1}{T_A} - \frac{1}{T_B} \right) \delta U_A + \left(\frac{P_A}{T_A} - \frac{P_B}{T_B} \right) \delta V_A - \left(\frac{\mu_A}{T_A} - \frac{\mu_B}{T_B} \right) \delta N_A ,
\end{aligned}
\tag{3.35}
$$

where we've used Eq. (3.30) in the second line of Eq. (3.35). The total system is isolated, and hence the entropy is a maximum. We require *stationarity* of the entropy, $\delta S = 0$. Because δU_A, δV_A, and δN_A can be varied independently, $\delta S = 0$ in Eq. (3.35) requires *the equality of subsystem temperatures, pressures, and chemical potentials* (the conditions of thermal, mechanical, and compositional equilibrium):

$$
T_A = T_B \qquad P_A = P_B \qquad \mu_A = \mu_B .
\tag{3.36}
$$

A similar analysis could be repeated for a system containing any number of chemical species. As long as the partition can pass a species, $\mu_{j,A} = \mu_{j,B}$ in equilibrium. If, however, the partition cannot pass species i, $\delta N_{i,A} = \delta N_{i,B} = 0$, and such a species cannot come to equilibrium between the subsystems, implying there's no reason for the chemical potentials to be equal,[24] $\mu_{i,A} \neq \mu_{i,B}$.

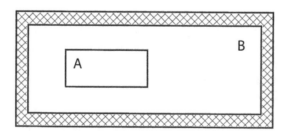

Figure 3.7   Redrawing the composite system so that B surrounds A

We could redraw Fig. 3.6 so that B *surrounds* A—see Fig. 3.7. In that case, B becomes the environment for A. In writing $T dS = dU + P dV - \mu dN$, U, V, and N are *conserved* quantities.

---

[24]We'll see in Chapter 5 (cavity radiation, the *photon gas*), that $\mu = 0$ for photons in equilibrium with matter. Photons are not conserved; there is no external population of photons that the photons in cavity radiation can come to equilibrium with.

Equilibrium, as specified by Eq. (3.36), thus requires *equality between system and environmental intensive variables*, T, P, and μ, those that are *conjugate* to conserved quantities. The equilibrium values of intensive variables are "set" by the environment.

Stability, $\delta^2 S < 0$

Equilibrium in isolated systems is attained when no further increases in entropy can occur by processes consistent with system constraints. As we've shown, $\delta S = 0$ implies the conditions for equilibrium, Eq. (3.36). The requirement $\delta S = 0$, however, is insufficient to determine whether S is a *maximum*. A generic function $S(U, V, N)$ has a maximum at (U_0, V_0, N_0) if $S(U_0, V_0, N_0) > S(U, V, N)$ for all (U, V, N) in a neighborhood of U_0, V_0, N_0. This familiar mathematical idea poses a conceptual problem, however, when applied to entropy. How can we speak of an entropy *surface*[25] associated with *a fixed equilibrium state*, when entropy is defined only in equilibrium, wherein state variables are constrained to have their equilibrium values! That is, we can't venture away from U_0, V_0, N_0 in state space and have the equilibrium entropy function $S(U, V, N)$ continue to be associated with state U_0, V_0, N_0. There *is* a surface defined by the values of S for all equilibrium states specified by general values of (U, V, N), but there cannot be a surface associated with a *given* state (U_0, V_0, N_0) such that $S(U_0, V_0, N_0) > S(U, V, N)$. The system attains equilibrium at U_0, V_0, N_0 through a series of irreversible processes in which $\Delta S > 0$. We need to find a way to characterize that $S(U_0, V_0, N_0)$ is locally a maximum when subjected to *conceivable* variations in state variables that produce $\Delta S < 0$, even though *there are no macroscopic physical processes* that drive $\Delta S < 0$ for isolated systems. We define a *thermodynamic fluctuation* as a *virtual variation* in the state of the system that results in $\Delta S < 0$ by conceptually relaxing the condition of isolation.[26] Fluctuations are virtual processes *inverse* to the real processes that resulted in $\Delta S > 0$ in achieving equilibrium at U_0, V_0, N_0. *Stability* of the equilibrium state requires that fluctuations produce entropy *decreases*, the *stability condition*, because otherwise entropy *increases* imply the evolution of an isolated system to a new equilibrium state. We show that the stability condition places restrictions on the *sign* of response functions. The main results are Eq. (3.46) and Eq. (3.51).

Consider again the composite system of subsystems A and B separated by a moveable porous, diathermic membrane (Fig. 3.6). The change in system entropy up to second order in small quantities is given by the Taylor expansion of a multivariable function:

$$
\Delta S = \sum_{\alpha=A,B} \left[\left(\frac{\partial S_\alpha}{\partial U_\alpha} \right)_{V_\alpha, N_\alpha} \delta U_\alpha + \left(\frac{\partial S_\alpha}{\partial V_\alpha} \right)_{U_\alpha, N_\alpha} \delta V_\alpha + \left(\frac{\partial S_\alpha}{\partial N_\alpha} \right)_{U_\alpha, V_\alpha} \delta N_\alpha \right]
$$

$$
+ \frac{1}{2} \sum_{\alpha=A,B} \left\{ \left(\frac{\partial^2 S_\alpha}{\partial U_\alpha^2} \right) (\delta U_\alpha)^2 + 2 \left(\frac{\partial^2 S_\alpha}{\partial U_\alpha \partial V_\alpha} \right) \delta U_\alpha \delta V_\alpha + 2 \left(\frac{\partial^2 S_\alpha}{\partial U_\alpha \partial N_\alpha} \right) \delta U_\alpha \delta N_\alpha \right. \tag{3.37}
$$

$$
\left. + \left(\frac{\partial^2 S_\alpha}{\partial V_\alpha^2} \right) (\delta V_\alpha)^2 + 2 \left(\frac{\partial^2 S_\alpha}{\partial V_\alpha \partial N_\alpha} \right) \delta V_\alpha \delta N_\alpha + \left(\frac{\partial^2 S_\alpha}{\partial N_\alpha^2} \right) (\delta N_\alpha)^2 \right\}.
$$

The first-order terms in Eq. (3.37) (square brackets) vanish in equilibrium. The form of the second-order terms are the same for A and B (even when we take into account the constraints $\delta U_A = -\delta U_B$, etc), so no need to distinguish the label α. We require that second-order variations

---

[25]Technically a *hypersurface*, defined on page 12.

[26]The idea of fluctuations as virtual processes is similar to the virtual displacements used in mechanics, "mathematical experiments" consistent with existing constraints but conceptually occurring at a fixed time. Gibbs did not use the language of virtual variations, but it's clear from his writings that's what he intended; instead he referred to *all possible* variations.

be *negative* for each subsystem,

$$\delta^2 S = \frac{1}{2}\big[S_{UU}(\delta U)^2 + 2S_{UV}\delta U \delta V + 2S_{UN}\delta U \delta N$$
$$+ S_{VV}(\delta V)^2 + 2S_{VN}\delta V \delta N + S_{NN}(\delta N)^2\big] < 0 , \tag{3.38}$$

where the notation indicates second derivatives; $S_{UV} \equiv \partial^2 S/\partial U \partial V$. The terms comprising $\delta^2 S$ in Eq. (3.38) define a quadratic form in three variables which we want to be *negative definite* for any nonzero variations δU, δV, δN. We should therefore analyze the determinants in Eq. (3.34), except require that they be negative. The derivatives involved in this approach are difficult to evaluate and we take another tack. If we *do* follow this approach, we find that the second derivatives S_{UU}, S_{VV}, S_{NN} are each *negative*, consistent with S having a maximum.

Rewrite the terms in Eq. (3.38):

$$\delta^2 S = \frac{1}{2}\Big\{\delta U\,[S_{UU}\delta U + S_{UV}\delta V + S_{UN}\delta N]$$
$$+\delta V\,[S_{VU}\delta U + S_{VV}\delta V + S_{VN}\delta N]$$
$$+ \delta N\,[S_{NU}\delta U + S_{NV}\delta V + S_{NN}\delta N]\Big\} . \tag{3.39}$$

The key step is to recognize that each term in square brackets is the first-order variation of a *derivative* about its equilibrium value. For example,[27]

$$\delta\left(\frac{\partial S}{\partial U}\right) = S_{UU}\delta U + S_{UV}\delta V + S_{UN}\delta N .$$

Equation (3.39) is therefore equivalent to

$$\delta^2 S = \frac{1}{2}\left[\delta U \delta\left(\frac{\partial S}{\partial U}\right) + \delta V \delta\left(\frac{\partial S}{\partial V}\right) + \delta N \delta\left(\frac{\partial S}{\partial N}\right)\right] . \tag{3.40}$$

Equation (3.40) involves fluctuations of extensive quantities multiplied by fluctuations of intensive quantities that can be obtained from Eq. (3.30). Thus,

$$\delta^2 S = \frac{1}{2}\left[\delta U \delta\left(\frac{1}{T}\right) + \delta V \delta\left(\frac{P}{T}\right) - \delta N \delta\left(\frac{\mu}{T}\right)\right] .$$

By differentiating and recognizing that $T\delta S = \delta U + P\delta V - \mu\delta N$, we find

$$\delta^2 S = -\frac{1}{2T}\left[\delta T\delta S - \delta P\delta V + \delta\mu\delta N\right] . \tag{3.41}$$

The criterion for stability of the equilibrium state against fluctuations is thus

$$\delta T\delta S - \delta P\delta V + \delta\mu\delta N > 0 . \tag{3.42}$$

Equation (3.41) follows from an analysis of how S responds to small variations in the three quantities U, V, and N. Yet there are *six* types of variations in Eq. (3.41), including δS. The variations δU, δV, and δN account for the variation δS *as well as* the variations δT, δP, and $\delta\mu$. The six fluctuations indicated in Eq. (3.41) are therefore not independent. We can work with *any set of three independent variations* as a starting point for a stability analysis.

---

[27]We could define a *variational derivative* $\delta \equiv \delta U\partial/\partial U + \delta V\partial/\partial V + \delta N\partial/\partial N$, which is a kind of directional derivative in thermodynamic state space.

Choose $\delta T, \delta P, \delta N$ as independent variations

With δT, δP, and δN as independent variations, we have for the other variations in Eq. (3.42):

$$\delta S = \left(\frac{\partial S}{\partial T}\right)_{P,N} \delta T + \left(\frac{\partial S}{\partial P}\right)_{T,N} \delta P + \left(\frac{\partial S}{\partial N}\right)_{T,P} \delta N$$

$$\delta V = \left(\frac{\partial V}{\partial T}\right)_{P,N} \delta T + \left(\frac{\partial V}{\partial P}\right)_{T,N} \delta P + \left(\frac{\partial V}{\partial N}\right)_{T,P} \delta N$$

$$\delta \mu = \left(\frac{\partial \mu}{\partial T}\right)_{P,N} \delta T + \left(\frac{\partial \mu}{\partial P}\right)_{T,N} \delta P + \left(\frac{\partial \mu}{\partial N}\right)_{T,P} \delta N . \qquad (3.43)$$

Using Eq. (3.43), we find

$$\delta T \delta S - \delta P \delta V + \delta \mu \delta N = \left(\frac{\partial S}{\partial T}\right)_{P,N} (\delta T)^2 + \delta T \delta P \left[\left(\frac{\partial S}{\partial P}\right)_{T,N} - \left(\frac{\partial V}{\partial T}\right)_{P,N}\right]$$

$$+ \delta N \delta T \left[\left(\frac{\partial S}{\partial N}\right)_{T,P} + \left(\frac{\partial \mu}{\partial T}\right)_{P,N}\right] - \left(\frac{\partial V}{\partial P}\right)_{T,N} (\delta P)^2 \qquad (3.44)$$

$$+ \delta N \delta P \left[\left(\frac{\partial \mu}{\partial P}\right)_{T,N} - \left(\frac{\partial V}{\partial N}\right)_{T,P}\right] + \left(\frac{\partial \mu}{\partial N}\right)_{T,P} (\delta N)^2 .$$

This expression simplifies with the help of some thermodynamic identities. We show in Section 4.5 that[28]

$$\left(\frac{\partial S}{\partial P}\right)_{T,N} = -\left(\frac{\partial V}{\partial T}\right)_{P,N} \qquad \left(\frac{\partial S}{\partial N}\right)_{T,P} = -\left(\frac{\partial \mu}{\partial T}\right)_{P,N} \qquad \left(\frac{\partial \mu}{\partial P}\right)_{T,N} = \left(\frac{\partial V}{\partial N}\right)_{T,P} .$$

These are the three Maxwell relations associated with the Gibbs energy (Table 4.2). With these results, as well as Eq. (3.13) and Eq. (1.20), we have from Eq. (3.44):

$$\delta^2 S = -\frac{1}{2T} \left[\frac{C_P}{T} (\delta T)^2 - 2\alpha V \delta T \delta P + \beta V (\delta P)^2 + \left(\frac{\partial \mu}{\partial N}\right)_{T,P} (\delta N)^2\right] . \qquad (3.45)$$

The terms in square brackets in Eq. (3.45) comprise a quadratic form in δT, δP, and δN that we require to be positive definite. Sylvester's criterion Eq. (3.34) requires that:

$$C_P > 0 \qquad\qquad C_P \beta > \alpha^2 TV > 0 \qquad\qquad \left(\frac{\partial \mu}{\partial N}\right)_{T,P} > 0 . \qquad (3.46)$$

Stability thus requires $C_P > 0$, $\beta > 0$, and $(\partial \mu / \partial N)_{T,P} > 0$.

Choose $\delta T, \delta V, \delta N$ as independent variations

We've shown that $C_P > C_V$, Eq. (1.29), and we've just shown that $C_P > 0$, Eq. (3.46). The two inequalities leave open the possibility that $C_V < 0$, which cannot be right on physical grounds: A negative heat capacity would imply that with positive heat input, the temperature of the object would *decrease*.[29] You could make something arbitrarily cold by heating it! It would be "nice" if

---

[28] Don't you hate it when books do that: Make use of results that haven't been shown yet?

[29] Such a phenomenon occurs with black holes—see Chapter 13. Black holes gobble up mass (energy, $E = mc^2$), increasing their size, decreasing their temperature.

$C_V > 0$ emerged directly from a stability analysis. We can do that if we choose δT, δV, and δN as the independent variations.

We could "start over" and develop expansions for δS, δP and $\delta \mu$, analogous to the expansions in Eq. (3.43). A more direct approach is to first prove the identity

$$\left(\frac{\partial S}{\partial T}\right)_{P,N} = \left(\frac{\partial S}{\partial T}\right)_{V,N} - \left(\frac{\partial P}{\partial V}\right)_{T,N} \left(\frac{\partial V}{\partial T}\right)_{P,N}^2 , \tag{3.47}$$

which is the same as Eq. (1.29) ($C_P = C_V + \alpha^2 TV/\beta$) when combined with Eqs. (3.11), (3.13), and (1.20). Consider that we have a quantity x that's a function of y and z, $x = x(y, z)$. As happens in thermodynamics, assume that x is *also* a function of y and w, $x = x(y, w)$. (Thus, $w = w(y, z)$.) Expand first $x = x(y, z)$ and then $x = x(y, w)$,

$$\mathrm{d}x = \left(\frac{\partial x}{\partial y}\right)_z \mathrm{d}y + \left(\frac{\partial x}{\partial z}\right)_y \mathrm{d}z \qquad \mathrm{d}x = \left(\frac{\partial x}{\partial y}\right)_w \mathrm{d}y + \left(\frac{\partial x}{\partial w}\right)_y \mathrm{d}w .$$

Divide both equations by $\mathrm{d}y$ and equate:

$$\left(\frac{\partial x}{\partial y}\right)_z + \left(\frac{\partial x}{\partial z}\right)_y \frac{\mathrm{d}z}{\mathrm{d}y} = \left(\frac{\partial x}{\partial y}\right)_w + \left(\frac{\partial x}{\partial w}\right)_y \frac{\mathrm{d}w}{\mathrm{d}y} .$$

Now let z be constant. We arrive at the relation between four variables

$$\left(\frac{\partial x}{\partial y}\right)_z = \left(\frac{\partial x}{\partial y}\right)_w + \left(\frac{\partial x}{\partial w}\right)_y \left(\frac{\partial w}{\partial y}\right)_z . \tag{3.48}$$

Let $x \to S$, $y \to T$, $z \to P$, and $w \to V$ in Eq. (3.48):

$$\left(\frac{\partial S}{\partial T}\right)_{P,N} = \left(\frac{\partial S}{\partial T}\right)_{V,N} + \left(\frac{\partial S}{\partial V}\right)_{T,N} \left(\frac{\partial V}{\partial T}\right)_{P,N} .$$

Now reach for a Maxwell relation, $(\partial S/\partial V)_{T,N} = (\partial P/\partial T)_{V,N}$ (Table 4.2),

$$\left(\frac{\partial S}{\partial T}\right)_{P,N} = \left(\frac{\partial S}{\partial T}\right)_{V,N} + \left(\frac{\partial P}{\partial T}\right)_{V,N} \left(\frac{\partial V}{\partial T}\right)_{P,N} .$$

Make use of Eq. (1.21) in this expression, and we have Eq. (3.47).

Combine Eq. (3.47) (or Eq. (1.29), the same thing) with Eq. (3.45). We obtain

$$\delta^2 S = -\frac{1}{2T} \left[\frac{C_V}{T} (\delta T)^2 + \frac{1}{\beta V} [\delta V]_N^2 + \left(\frac{\partial \mu}{\partial N}\right)_{T,P} (\delta N)^2 \right] , \tag{3.49}$$

where

$$[\delta V]_N \equiv \left(\frac{\partial V}{\partial T}\right)_{P,N} \delta T + \left(\frac{\partial V}{\partial P}\right)_{T,N} \delta P \tag{3.50}$$

is the change in volume at constant N. Requiring that the terms in square brackets in Eq. (3.49) be positive definite, we have the stability criteria

$$C_V > 0 \qquad \beta > 0 \qquad \left(\frac{\partial \mu}{\partial N}\right)_{T,P} > 0 . \tag{3.51}$$

The inequalities in Eq. (3.51) are the requirements for *thermal stability*, *mechanical stability*, and *compositional stability*.

3.11 DIRECTION OF FLOW IN THERMODYNAMIC PROCESSES

The stability requirements allow us to establish the *direction* of processes as systems come to equilibrium.[30] Suppose two systems are brought into contact and $\mu_A > \mu_B$. To establish equilibrium, the chemical potentials must adjust so that μ_A decreases and μ_B increases. Because $\delta\mu_A < 0$ and $\delta\mu_B > 0$, to maintain $\partial\mu/\partial N > 0$, we must have $\delta N_A < 0$ and $\delta N_B > 0$, i.e., to achieve equilibrium *matter leaves regions of higher chemical potential and flows to regions of lower chemical potential*. The term chemical *potential* is therefore apt: Matter flows from high to low chemical potential. Likewise, if initially $T_A > T_B$, we must have $\delta T_A < 0$ and $\delta T_B > 0$ to achieve equilibrium. From $\partial S/\partial T > 0$, we must have $\delta S_A < 0$ and $\delta S_B > 0$. Entropy is a quantity that *flows* (see Chapter 14): *Entropy leaves regions of high temperature and flows to regions of low temperature*. Entropy flows from high to low temperature. *Entropy is as real as energy*, both are state variables; if energy can flow, so can entropy. Finally, if initially $P_A > P_B$, in reaching equilibrium $\delta P_A < 0$ and $\delta P_B > 0$. To maintain $\partial P/\partial V < 0$, we require that $\delta V_A > 0$ and $\delta V_B < 0$, which we can think of as *volume being transferred from regions of low pressure to regions of high pressure*.

3.12 JACOBIAN DETERMINANTS*

The maze of derivatives encountered in deriving equations such as Eq. (3.47) is daunting. The process can be systematized using the properties of Jacobian determinants, or simply *Jacobians*. Consider functions $u, v, \cdots, w$, that each depend on the variables $x, y, \cdots, z$. The Jacobian is the determinant of the matrix of partial derivatives. There's a highly practical notation for Jacobians that we'll employ. Let the Jacobian determinant be represented by what resembles a partial derivative:

$$\frac{\partial(u,v,\cdots,w)}{\partial(x,y,\cdots,z)} \equiv \begin{vmatrix} \dfrac{\partial u}{\partial x} & \dfrac{\partial u}{\partial y} & \cdots & \dfrac{\partial u}{\partial z} \\ \dfrac{\partial v}{\partial x} & \dfrac{\partial v}{\partial y} & \cdots & \dfrac{\partial v}{\partial z} \\ \vdots & \vdots & \ddots & \vdots \\ \dfrac{\partial w}{\partial x} & \dfrac{\partial w}{\partial y} & \cdots & \dfrac{\partial w}{\partial z} \end{vmatrix} .$$

Jacobians have the following properties that make them useful in thermodynamic calculations. First, a single partial derivative can formally be expressed as a Jacobian:

$$\left(\frac{\partial u}{\partial x}\right)_{y,\cdots,z} = \frac{\partial(u,y,\cdots,z)}{\partial(x,y,\cdots,z)} = \begin{vmatrix} \dfrac{\partial u}{\partial x} & \dfrac{\partial u}{\partial y} & \cdots & \dfrac{\partial u}{\partial z} \\ 0 & 1 & \cdots & 0 \\ \vdots & & \ddots & \vdots \\ 0 & \cdots & \cdots & 1 \end{vmatrix} \tag{3.52}$$

(after the first row in Eq. (3.52), there are only ones on the diagonal and zeros everywhere else). Thus, symbols common to the numerator and denominator in the notation for the Jacobian "cancel." Second, there's a product rule for Jacobians that resembles the ordinary chain rule from calculus,[31]

$$\frac{\partial(u,v,\cdots,w)}{\partial(x,y,\cdots,z)} = \frac{\partial(u,v,\cdots,w)}{\partial(r,s,\cdots,t)}\frac{\partial(r,s,\cdots,t)}{\partial(x,y,\cdots,z)} . \tag{3.53}$$

---

[30]Systems *out of equilibrium* are not treated in classic thermodynamics. We consider in Chapter 14 the subject of non-equilibrium thermodynamics, an extension of thermodynamics to systems slightly out of equilibrium, such that equilibrium is restored through the flow of matter and energy.

[31]An explicit demonstration of the chain-rule property of Jacobians is shown in Mazenko.[19, p570] Jacobians are also discussed by Landau and Lifshitz.[20, p51]

In fact, Eq. (3.53) *is* the chain rule, combined with the properties of determinants. Consider that one has a set of functions $\{f_i\}$ that each depend on a set of variables $\{x_j\}$, and that the $\{x_j\}$ are functions of another set of variables $\{y_k\}$. Using the chain rule,

$$\frac{\partial f_i}{\partial x_j} = \sum_m \frac{\partial f_i}{\partial y_m} \frac{\partial y_m}{\partial x_j} \, .$$

The chain rule is thus in the form of matrix multiplication of the Jacobian matrices, the determinants of which are embodied in Eq. (3.53). Equation (3.53) is a result of the property of determinants, that if matrices (A, B, C) are connected by matrix multiplication, $A = BC$, the determinant of A equals the product of the determinants of B and C, $|A| = |B| \cdot |C|$. It follows from Eq. (3.53) that

$$\frac{\partial(u, v, \cdots, w)}{\partial(x, y, \cdots, z)} = \left[\frac{\partial(x, y, \cdots, z)}{\partial(u, v, \cdots, w)} \right]^{-1} \, .$$

Let's use Jacobians to derive Eq. (3.47). From Eq. (3.52) and Eq. (3.53),

$$\frac{\partial(S, V, N)}{\partial(T, V, N)} = \frac{\partial(S, V, N)}{\partial(T, P, N)} \frac{\partial(T, P, N)}{\partial(T, V, N)}$$

$$= \begin{vmatrix} \left(\dfrac{\partial S}{\partial T} \right)_{P,N} & \left(\dfrac{\partial S}{\partial P} \right)_{T,N} \\ \left(\dfrac{\partial V}{\partial T} \right)_{P,N} & \left(\dfrac{\partial V}{\partial P} \right)_{T,N} \end{vmatrix} \left(\dfrac{\partial P}{\partial V} \right)_{T,N}$$

$$= \left[\left(\frac{\partial S}{\partial T} \right)_{P,N} \left(\frac{\partial V}{\partial P} \right)_{T,N} - \left(\frac{\partial S}{\partial P} \right)_{T,N} \left(\frac{\partial V}{\partial T} \right)_{P,N} \right] \left(\frac{\partial P}{\partial V} \right)_{T,N} \, .$$

Thus,

$$\left(\frac{\partial S}{\partial T} \right)_{V,N} = \left(\frac{\partial S}{\partial T} \right)_{P,N} - \left(\frac{\partial S}{\partial P} \right)_{T,N} \left(\frac{\partial V}{\partial T} \right)_{P,N} \left(\frac{\partial P}{\partial V} \right)_{T,N}$$

$$= \left(\frac{\partial S}{\partial T} \right)_{P,N} + \left(\frac{\partial V}{\partial T} \right)_{P,N}^2 \left(\frac{\partial P}{\partial V} \right)_{T,N} \, ,$$

where we've used Eq. (1.18) and a Maxwell relation (Table 4.2). Jacobians *almost* make the process mindless.

CHAPTER SUMMARY

This chapter introduced the extensive state variable known as entropy. Most of the theory of thermodynamics concerns entropy in some way. Its most important attribute is that it can only increase in irreversible processes. We touched on its interpretation as a measure of the number of ways a system can be arranged, what we return to in Chapter 7.

- The Clausius inequality specifies that $\oint \mathrm{d}Q/T \leq 0$ for any cycle, with equality holding for reversible cycles. Here T is the absolute temperature at which heat transfer $\mathrm{d}Q$ occurs. The integral is non-positive because the more inefficient the cycle, the greater is the heat expelled at the lower temperature.

- For reversible cycles $\oint (\mathrm{d}Q)_{\mathrm{rev}}/T = 0$, implying the existence of a state function S, with $\mathrm{d}S = (\mathrm{d}Q)_{\mathrm{rev}}/T$. The integrating factor for $(\mathrm{d}Q)_{\mathrm{rev}}$ is T^{-1}. Changes in entropy are given by $\Delta S = \int_A^B \mathrm{d}Q/T$ for any reversible path connecting equilibrium states A and B.

- The Clausius inequality in differential form is $dS \geq đQ/T$, with equality holding for reversible processes. The quantity $đQ/T$ (representing heat transfers with the environment) therefore does not account for all contributions to the entropy change dS, which we can write as $dS = đQ/T + dS_i$, where $dS_i \geq 0$ is the entropy change associated with irreversible processes. Whereas the entropy supplied to a system in the form of heat transfers can be positive, negative, or zero, entropy produced by irreversibility is always positive.

- The entropy of isolated systems never decreases, $dS \geq 0$. The entropy of equilibrium systems is a maximum as a function of the state variables, consistent with constraints. Entropy increases are associated with the removal of constraints. If you're seeking a one-liner for the meaning of entropy: Entropy captures the tendency of physical systems to spread out, to explore all available possibilities. The word *entropy* was coined by Clausius to mean the "transformation" of a system; he deliberately chose the word to be similar to the word *energy*.

- The first law of thermodynamics can be written in terms of exact differentials, $dU = TdS - PdV$, which can be integrated to find ΔU for any change of state, no matter how it's accomplished. We need only find a reversible path connecting the initial and final equilibrium states.

- For open systems the first law is generalized, $dU = TdS - PdV + \mu dN$, what's sometimes called Gibbs's equation, where $\mu \equiv (\partial U/\partial N)_{S,V}$ is the chemical potential. Chemical potential is often a negative quantity. The Gibbs-Duhem equation $Nd\mu = VdP - SdT$ is the first law expressed in terms of differentials of intensive variables.

- Entropy is an extensive quantity, $S(\lambda U, \lambda V, \lambda N) = \lambda S(U, V, N)$. The requirement of extensivity is a postulate imposed on the theory; it's not explicitly contained in the Clausius definition of entropy. Extensivity implies the Euler relation $TS = U + PV - \mu N$.

- Equilibrium requires equality of system and environmental intensive variables conjugate to conserved quantities (μ, T, and P, associated with conservation of N, U, and V).

- For equilibrium to be stable against fluctuations, the stability conditions must be satisfied, with C_V, β (isothermal compressibility), and $\partial \mu/\partial N$ positive. Matter flows from regions of high to low chemical potential; entropy flows from regions of high to low temperature.

- Jacobian determinants systemize the derivation of thermodynamic identities.

EXERCISES

3.1 Derive the Gibbs-Duhem equation, Eq. (3.32).

3.2 Show that $\partial^2 S/\partial U^2 < 0$. Set up the matrix that generates the quadratic form Eq. (3.38). Show that for Eq. (3.38) to be negative definite implies that the second derivative of S with respect to U is negative.

3.3 Fill in the steps leading to Eq. (3.49). Take Eq. (3.47) as given.

3.4 In Eq. (1.20) we defined the isothermal compressibility, $V\beta_T = -(\partial V/\partial P)_T$. We can also define the *isentropic compressibility* as $V\beta_S \equiv -(\partial V/\partial P)_S$. Show that

$$\frac{\beta_T}{\beta_S} = \frac{C_P}{C_V}.$$

Hint: Use the cyclic and reciprocity relations from Section 1.10. You could also derive this relation using the properties of Jacobians.

3.5 The internal energy function, $U = U(S, V, N)$, has a *minimum* value in equilibrium for a fixed value of the entropy (Section 4.1). Repeat the stability analysis for U against fluctuations in ΔS, ΔV, ΔN. Show that the second derivatives U_{SS}, U_{VV}, U_{NN} are each positive (and hence U is a convex function, Section 4.2). Show that stability requires $C_V > 0$, $\beta_S > 0$, and $(\partial \mu / \partial N)_{S,V} > 0$.

3.6 Show that $\dfrac{\partial(u, v)}{\partial(x, y)} = -\dfrac{\partial(v, u)}{\partial(x, y)} = \dfrac{\partial(v, u)}{\partial(y, x)}$. These properties follow by swapping rows and columns of determinants.

3.7 Show that $\dfrac{\partial(P, V)}{\partial(T, S)} = 1$. Thus, by the rules of calculus $dPdV = dTdS$.
Hint: Write

$$\frac{\partial(P, V)}{\partial(T, S)} = \frac{\partial(P, V)}{\partial(T, V)} \frac{\partial(T, V)}{\partial(T, S)}$$

and make use of a Maxwell relation, Table 4.2. This identity is a one liner with the use of Jacobians. Without Jacobians, it's difficult to derive.

3.8 Show that $\dfrac{\partial(U, T)}{\partial(V, P)} = \left(\dfrac{\partial U}{\partial S}\right)_T$. Hint: Make use of Exercise 3.7.

3.9 Use Eq. (3.27) (our tentative formula for the entropy of the ideal gas) to calculate the change in entropy for a free expansion where the volume doubles.

3.10 Show that $\left(\dfrac{\partial U}{\partial V}\right)_S = \left(\dfrac{\partial U}{\partial V}\right)_T - T\left(\dfrac{\partial S}{\partial V}\right)_T$. Hint: $\dfrac{\partial(U, S)}{\partial(V, S)} = \dfrac{\partial(U, S)}{\partial(V, T)} \dfrac{\partial(V, T)}{\partial(V, S)}$.
Then use a Maxwell relation to conclude that

$$\left(\frac{\partial U}{\partial V}\right)_S = \left(\frac{\partial U}{\partial V}\right)_T - T\left(\frac{\partial P}{\partial T}\right)_V .$$

3.11 Give an argument why

$$\frac{\partial}{\partial V}(1/T) = \frac{\partial}{\partial U}(P/T)$$

should hold as a general thermodynamic identity. Hint: dS is exact.

3.12 A metal bar of heat capacity C_M at temperature T is suddenly immersed in a mass of water having heat capacity C_W at temperature T_0, with $T > T_0$. Calculate the change in entropy. Hint: First find the temperature at which the metal bar and the water come to equilibrium.

CHAPTER 4

Thermodynamic potentials

The four ways to say energy

4.1   CRITERIA FOR EQUILIBRIUM

$\mathbf{G}$ IBBS formulated two equivalent criteria for characterizing equilibrium, the *entropy maximum principle* and the *energy minimum principle*, that for systems in equilibrium:[3, p56]

I. Entropy is a maximum for fixed energy, *there are no fluctuations by which* $[\Delta S]_U > 0$;

II. Energy is a minimum for fixed entropy, *there are no fluctuations by which* $[\Delta U]_S < 0$.

Can the extremal nature of S and U be *visualized*? Seemingly not: For systems *not* in equilibrium state variables are not defined, yet for systems *in* equilibrium the entropy cannot increase nor the energy decrease. We can get around this problem using the composite system (Section 3.10), the equilibrium state of which can be represented as a point in a *higher-dimensional* state space. Before showing that, let's introduce some language. For a system with r chemical species the first law can be written $dS = dU/T + PdV/T - T^{-1}\sum_{k=1}^{r} \mu_k dN_k$; see Eq. (3.16). The entropy is thus a function of $r + 2$ extensive variables, $(U, V, \{N_k\}_{k=1}^{r})$, with an equilibrium state represented as a point in a space spanned by these variables. When we work with $S = S(U, V, \{N_k\})$, the formalism is referred to as the *entropy representation*. Equivalently, we can write $dU = TdS - PdV + \sum_{k=1}^{r} \mu_k dN_k$. The internal energy is thus a function of S, V, and $\{N_k\}_{k=1}^{r}$, with an equilibrium state represented as a point in a space spanned by these variables. When we take $U = U(S, V, \{N_k\})$, the formalism is referred to as the *energy representation*. Thermodynamics can be developed in either representation, whatever is convenient for a particular problem. Let subsystems (1) and (2) be in contact through a boundary through which energy and all chemical species may pass, but are otherwise isolated (see Fig. 3.6). In the entropy representation, S for the composite system is a function of $2(2 + r)$ variables $(U^{(1)}, V^{(1)}, \{N_k^{(1)}\}_{k=1}^{r}, U^{(2)}, V^{(2)}, \{N_k^{(2)}\}_{k=1}^{r})$. Equivalently, we can consider S as a function of the $2 + r$ *fixed* quantities $V = V^{(1)} + V^{(2)}$, $U = U^{(1)} + U^{(2)}$, $\{N_k = N_k^{(1)} + N_k^{(2)}\}_{k=1}^{r}$, leaving $2 + r$ extensive variables of one of the subsystems free to vary, which we denote generically as $\{X_j^{(1)}\}_{j=1}^{2+r}$. The quantities $X_j^{(1)}$ are *unconstrained* extensive parameters of subsystem 1. The equilibrium state for fixed energy occurs for the values of $X_j^{(1)}$ for which the entropy is maximized. In the energy representation, the equilibrium state for fixed entropy occurs for the values of $X_j^{(1)}$ for which the energy is minimized.

The equivalence of the two criteria can be shown using the strategy that the falsity of one implies the falsity of the other (see Chapter 2). We're aided in this task by the observation that it's always possible to increase (decrease) the energy and entropy of a system *together* by adding (removing) heat. By adding heat, $\Delta U > 0$ and $\Delta S > 0$, and by removing heat, $\Delta U < 0$ and $\Delta S < 0$. If criterion I is violated, there is a fluctuation[1] by which $\Delta S > 0$ and $\Delta U = 0$. By removing heat from this system *in its varied state*, we can obtain a variation of the original state for which $\Delta S = 0$ and $\Delta U < 0$, in violation of criterion II. Thus, not-I implies not-II, $\overline{I} \Rightarrow \overline{II}$. If criterion II is violated, there is a fluctuation by which $\Delta U < 0$ and $\Delta S = 0$. By adding heat to this system in its varied state, we can obtain a variation of the original state such that $\Delta U = 0$ and $\Delta S > 0$, in violation of criterion I. Not-II implies not-I, $\overline{II} \Rightarrow \overline{I}$. Thus, $\overline{I} \Leftrightarrow \overline{II}$, and hence $I \Leftrightarrow II$.

The two criteria are thus logically equivalent. We now give an argument for the physical reality of criterion II. Assume the energy U of a system does *not* have the smallest possible value consistent with a fixed value of S. If $U > U_{\min}$, reduce the energy by letting the system do adiabatic work. Return the energy in the form of heat, restoring the system to its initial energy state, but *increasing the entropy*. We therefore have a process in which the energy stays fixed, but in which $\Delta S > 0$. As long as the energy is *lowerable* ($U > U_{\min}$), the process can be repeated indefinitely, arbitrarily increasing S for fixed U, which is unphysical. *Minimum energy implies fixed entropy.*

4.2 LEGENDRE TRANSFORMATION*

A function $f(x)$ is *convex* if the chord joining $f(x_1)$ and $f(x_2)$ lies above the function for $x_1 < x < x_2$, and thus $f'' \geq 0$ in this range (see Fig. 4.1). A function $f(x)$ is *strictly convex* if $f'' > 0$

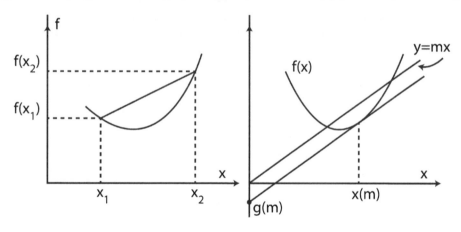

Figure 4.1 Convex function (left) and Legendre transform $g(m)$ (right)

everywhere. A strictly convex function has no more than one minimum. The tangent to a convex function lies below the function. A *concave* function $f(x)$ is such that $-f(x)$ is convex. The entropy $S(U, V, N)$ is a concave function of its arguments; the second derivatives of S with respect to these variables are negative (Section 3.10). The internal energy $U(S, V, N)$ is a convex function of its arguments (the second derivatives of U with respect to these variables are positive, Exercise 3.5).

The *Legendre transformation* of a convex function $f(x)$ is a new function, $g(m)$, of a new variable,[2] m. It's not an *integral transform* (like the Fourier transform), rather it's a geometric construction. Pick a number, call it m. Draw line $y = mx$ (see Fig. 4.1). Define $D(x, m) \equiv mx - f(x)$ as the distance between the line mx and the function $f(x)$. Let $x(m)$ be the point where $D(x, m)$ is maximized, where $f' = m$. Clearly $x(m)$ is the point where the tangent to the curve is parallel

---

[1] We're using the concept of fluctuation as a *virtual variation* in the state of the system (see Section 3.10) in which the condition of thermal isolation is conceptually relaxed. Adding or removing heat to the system in it's varied state is treated as occurring instantaneously.

[2] A nice discussion of the Legendre transformation is given by Arnold.[21, p61] It's shown that the Legendre transformation of the Legendre transformation is the original function.

to the line mx. The Legendre transform is defined as $g(m) \equiv D(x(m), m) = mx(m) - f(x(m))$. Geometrically, $g(m)$ is the number such that $f(x(m)) = -g(m) + mx$. The Legendre transform provides a *duality between points and tangents as equivalent ways of characterizing a function*: Instead of specifying a function pointwise, the same values of f are constructed from the *slope* of the function at that point plus the "offset," the y-intercept, which is the Legendre transform. The Legendre transform of f at point x is the projection onto the y-axis with the slope f' at x. For $f(x) = x^2$, $g(m) = m^2/4$.

4.3   THE FOUR THERMODYNAMIC POTENTIALS

... the whole thermodynamical behavior of a substance is determined by a single characteristic function, a knowledge of which is sufficient once and for all to determine uniquely the conditions of the physical and chemical states of equilibrium which the substance may assume. The form of the characteristic function depends, however, on the choice of independent variables.—Max Planck[11, p272]

Equilibrium is specified by the values of state variables. Whether measured or postulated by the laws of thermodynamics, state variables are not independent of each other. There is considerable *redundancy* in how equilibrium can be described. Depending on the system, there may be variables that are not readily subject to experimental control and yet others that are. Legendre transformations provide a way of obtaining equivalent descriptions of the energy (known as *thermodynamic potentials*) involving variables that may be easier to control. There are many ways to skin a cat.

Three Legendre transformations of $U(S, V)$ can be formed from the products of variables with the dimension of energy: TS and PV (our old friends heat and work).[3] They are:

$$
\begin{aligned}
F &\equiv U - TS && \text{(Helmholtz free energy)} \\
H &\equiv U + PV && \text{(enthalpy)} \\
G &\equiv U - TS + PV = F + PV = H - TS\,. && \text{(Gibbs free energy)}
\end{aligned}
\tag{4.1}
$$

The relationships among these functions are shown in Fig. 4.2 The physical interpretation of the

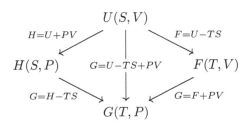

Figure 4.2   Legendre transformations of the internal energy function

potentials (and why they're called potentials) is discussed in the next section. The natural variables for these functions are found by forming their differentials:

$$
\begin{aligned}
dU &= TdS - PdV + \mu dN \\
dH &= TdS + VdP + \mu dN \\
dF &= -SdT - PdV + \mu dN \\
dG &= -SdT + VdP + \mu dN\,.
\end{aligned}
\tag{4.2}
$$

---

[3]The three Legendre transforms of $U(S, V)$ bring to mind the three Legendre transforms in classical mechanics of the generating function of canonical transformations.[22, p373]

Thus, $U = U(S, V, N)$, $H = H(S, P, N)$, $F = F(T, V, N)$, and $G = G(T, P, N)$. The differentials of U, H, F, and G each involve a different "mixture" of differentials of extensive and intensive quantities, multiplied respectively by intensive and extensive quantities. The intensive variables are obtained from the derivatives of U with respect to the extensive variables S and V, $T = \partial U/\partial S$ and $P = -\partial U/\partial V$. The transformation $U(S, V, N) \rightarrow F(T, V, N)$ (for example) shifts the emphasis from S as an independent variable to its associated slope T as the independent variable. *The duality between points and slopes achieved by the Legendre transformation is reflected in the formalism of thermodynamics as a duality between extensive and intensive variables.*

4.4 PHYSICAL INTERPRETATION OF THE POTENTIALS

The Clausius inequality, (3.6), $T dS \geq dQ$ helps to provide a physical interpretation for the thermodynamic potentials. It's important to recognize that U, H, F, and G are not "just" Legendre transforms (possessing certain mathematical properties); they each represent *energies* stored in the system under prescribed conditions. They are, as the name suggests, *potential energies*. Assume in the following that N is fixed.

Combine the Clausius inequality with the first law, Eq. (1.6):

$$T dS \geq dQ = dU - dW .$$ (4.3)

In what follows it's useful to distinguish between the work due to changes in volume ($dW = -P dV$) and *all other forms of work*, denoted dW': $dW \equiv -P dV + dW'$.

Internal energy

Consider a process in which no work is performed on the system. From the inequality (4.3),

$$T dS \geq dU .$$ (4.4)

We can interpret the inequality (4.4) in two ways, in either the entropy or the energy representation:

$$[dS]_{U,V} \geq 0 \quad \text{or} \quad [dU]_{S,V} \leq 0 .$$ (4.5)

If U is fixed and no work is done, then $dQ = 0$ from the first law, and thus by the Clausius inequality (3.6), $[dS]_{U,V} \geq 0$; the entropy of an isolated system never decreases. The other part of (4.5), $[dU]_{S,V} \leq 0$, should be interpreted as $[dQ]_S \leq 0$ because $dS - dQ/T \geq 0$ (Clausius inequality), and thus, because $[dQ]_S \leq 0$, $[dU]_{S,V} \leq 0$. An irreversible process at constant S and V is accompanied by a *decrease* in U. *Heat leaves systems in irreversible processes*[4] such that $dS = 0$. Thus, from the inequalities in (4.5), *the system spontaneously evolves either towards a minimum value of U at constant S or towards the maximum value of S at constant U*—the Gibbs criteria for equilibrium. In reversible processes (the case of equality in (4.3)), $[\Delta U]_S = W$ (first law), which shows the sense in which U is a potential energy, the energy of work performed adiabatically.

Enthalpy

From the inequality (4.3) and $dQ = dH - V dP - dW'$ (first law), we have

$$T dS \geq dQ = dH - dW' - V dP .$$ (4.6)

For a process at constant pressure and no other forms of work ($dW' = 0$),

$$T [dS]_P \geq [dH]_{P,} .$$ (4.7)

---

[4]This statement is consistent with what we developed in Section 3.3. In an irreversible process $dS_i > 0$; to keep $dS = 0$, heat must leave the system.

Like the inequality (4.4), we can interpret (4.7) in the entropy or the energy representation:

$$[dS]_{H,P} \geq 0 \qquad \text{or} \qquad [dH]_{S,P} \leq 0 . \tag{4.8}$$

For constant H and P (and $dW' = 0$), $dH = 0 = dU + PdV$ implies $[dQ]_{H,P} = 0$ (first law); thus, $[dS]_{H,P} \geq 0$ (Clausius inequality) is again the entropy of an isolated system can never decrease.[5] From (4.6), for constant P and $dW' = 0$, $dH = [dQ]_P$; *enthalpy is the heat added at constant pressure without other work performed*. The word enthalpy is based on the Greek *enthalpein* ($\epsilon\nu\theta\alpha\lambda\pi\epsilon\iota\nu$)—"to put heat into." Gibbs referred to enthalpy as the *heat function*.[3, p92] As for the other part of (4.8), $[dH]_{S,P} \leq 0$ is equivalent to $[dQ]_{S,P} \leq 0$ (inequality (4.6)). Again, *heat leaves systems in irreversible processes at constant S*. From (4.8), systems evolve spontaneously until equilibrium is achieved with H having a minimum value. For reversible processes, $[\Delta H]_{S,P} = W'_{ad}$; *enthalpy is the energy of "other" adiabatic work performed on the system.*

Helmholtz and Gibbs energies

By combining $TdS \geq dQ$ with the definitions of F and G, we infer the inequalities

$$dF \leq -SdT - PdV + dW' \qquad\qquad dG \leq -SdT + VdP + dW' .$$

The quantities F and G spontaneously decrease in value until a minimum is achieved for the equilibrium compatible with the constraints on the system ($dW' = 0$):

$$[dF]_{T,V} \leq 0 \qquad\qquad [dG]_{T,P} \leq 0 .$$

For reversible processes $[\Delta F]_{T,V} = W'$ and $[\Delta G]_{T,P} = W'$; F and G are storehouses of the energy of other work performed under the conditions indicated.

The properties of the potentials are summarized[6] in Table 4.1. Additional interpretations of the Gibbs and Helmholtz energies are discussed in Sections 4.6 and 4.7.

Table 4.1 Thermodynamic potentials (N fixed, W' is work other than PdV work)

Potential	Stored energy	Spontaneous change	In equilibrium
Internal energy, $U = U(S,V)$ $dU = TdS - PdV + dW'$	$[\Delta U]_S = (W)_{adiabatic}$	$[\Delta U]_{S,V} \leq 0$	Minimum
Enthalpy, $H = H(S,P)$ $H = U + PV$ $dH = TdS + VdP + dW'$	$[\Delta H]_{S,P} = (W')_{adiabatic}$ $[\Delta H]_P = Q$	$[\Delta H]_{S,P} \leq 0$	Minimum
Helmholtz energy, $F(T,V)$ $F = U - TS$ $dF = -SdT - PdV + dW'$	$[\Delta F]_{T,V} = W'$ $[\Delta F]_T = W$	$[\Delta F]_{T,V} \leq 0$	Minimum
Gibbs energy, $G = G(T,P)$ $G = U - TS + PV$ $dG = -SdT + VdP + dW'$	$[\Delta G]_{T,P} = W'$	$[\Delta G]_{T,P} \leq 0$	Minimum

---

[5]We can treat a system for which H and P are held constant as isolated.

[6]Other books adopt different sign conventions for heat and work; the inequalities in Table 4.1 may be reversed.

4.5 MAXWELL RELATIONS

Maxwell relations were invoked without definition in Chapter 3. These are equalities between certain partial derivatives of state variables that ensue from the *exactness* of differentials of thermodynamic potentials. For $\Phi = \Phi(x, y)$ a thermodynamic potential, $d\Phi = A dx + B dy$ is exact when Eq. (1.1) is satisfied, $\partial A/\partial y = \partial B/\partial x$. The Maxwell relations are none other than the defining condition for a differential to be exact, applied to each thermodynamic potential. There are three Maxwell relations[7] for each of the differential expressions in Eq. (4.2). They are listed in Table 4.2.

Table 4.2 Maxwell relations

$$U = U(S, V, N)$$

$$\left(\frac{\partial T}{\partial V}\right)_{S,N} = -\left(\frac{\partial P}{\partial S}\right)_{V,N} \qquad \left(\frac{\partial T}{\partial N}\right)_{S,V} = \left(\frac{\partial \mu}{\partial S}\right)_{V,N} \qquad \left(\frac{\partial \mu}{\partial V}\right)_{S,N} = -\left(\frac{\partial P}{\partial N}\right)_{S,V}$$

$$H = H(S, P, N)$$

$$\left(\frac{\partial T}{\partial P}\right)_{S,N} = \left(\frac{\partial V}{\partial S}\right)_{P,N} \qquad \left(\frac{\partial T}{\partial N}\right)_{S,P} = \left(\frac{\partial \mu}{\partial S}\right)_{P,N} \qquad \left(\frac{\partial V}{\partial N}\right)_{S,P} = \left(\frac{\partial \mu}{\partial P}\right)_{S,N}$$

$$F = F(T, V, N)$$

$$\left(\frac{\partial S}{\partial V}\right)_{T,N} = \left(\frac{\partial P}{\partial T}\right)_{V,N} \qquad \left(\frac{\partial S}{\partial N}\right)_{T,V} = -\left(\frac{\partial \mu}{\partial T}\right)_{V,N} \qquad \left(\frac{\partial P}{\partial N}\right)_{T,V} = -\left(\frac{\partial \mu}{\partial V}\right)_{T,N}$$

$$G = G(T, P, N)$$

$$\left(\frac{\partial S}{\partial P}\right)_{T,N} = -\left(\frac{\partial V}{\partial T}\right)_{P,N} \qquad \left(\frac{\partial S}{\partial N}\right)_{T,P} = -\left(\frac{\partial \mu}{\partial T}\right)_{P,N} \qquad \left(\frac{\partial V}{\partial N}\right)_{T,P} = \left(\frac{\partial \mu}{\partial P}\right)_{T,N}$$

4.6 GIBBS ENERGY, CHEMICAL POTENTIAL, AND OTHER WORK

Combining the Euler relation Eq. (3.31) with the definition of Gibbs energy Eq. (4.1),

$$G = U - TS + PV = N\mu . \tag{4.9}$$

Thus, we have another interpretation. *The chemical potential is the Gibbs energy per particle*, $\mu = G/N$. From Eq. (4.2), $[dG]_{T,P} = \mu[dN]_{T,P}$ ($[d\mu]_{T,P} = 0$, from Eq. (3.32)). *The chemical potential is the energy to add another particle at constant[8] T and P.* Gibbs energy is also "other work." From Eq. (4.2), $[dG]_{T,N} = [V]_{T,N}dP$, which can be integrated:

$$G(T, P, N) = G^0(T, N) + \int_{P_0}^{P} [V]_{T,N} dP' , \tag{4.10}$$

where $G^0(T, N) = G(T, P_0, N)$ (a function that cannot be determined through the present method of analysis). For the ideal gas,

$$G(T, P, N) = G^0(T, N) + nRT \ln(P/P_0) . \tag{4.11}$$

---

[7]A thermodynamic potential with n independent variables implies $\frac{1}{2}n(n-1)$ Maxwell relations.
[8]We found in Section 3.5 that μ is the energy to add another particle at constant S and V.

The chemical potential of the ideal gas (Gibbs energy per particle) therefore has the form

$$\mu(T, P) = \mu(T, P_0) + RT \ln(P/P_0) . \tag{4.12}$$

What should be seen from Eq. (4.12) is that $\mu \neq 0$, *even though there are no interactions between the atoms of an ideal gas*. Chemical potential is the energy to add another particle at constant entropy, and entropy is not a microscopic property of matter, it's a property of the equilibrium state. Even the ideal gas has to be in thermal equilibrium. An expression for the chemical potential of the ideal gas involving fundamental constants is given in Chapter 7 (see Exercise 7.9).

4.7 FREE ENERGY AND DISSIPATED ENERGY

Free energy and maximum work

By rewriting (4.3), we have the inequality

$$-\dW \leq T\dS - \dU . \tag{4.13}$$

For there to be work done *by* the system (counted as a negative quantity), we must have $\dU < T\dS$, a generalization of $\Delta U = W_{\text{ad}}$: $\Delta U < 0$ if $W_{\text{ad}} < 0$. The *maximum* value of $\dW$ (counted as a negative quantity) is therefore $|\dW|_{\text{max}} = T\dS - \dU$. Undoing the minus sign, $\dW_{\text{max}} = \dU - T\dS = [\dF]_T$. The Helmholtz energy is *the maximum obtainable work at constant T*:

$$W_{\text{max}} = [\Delta F]_T . \tag{4.14}$$

Thus, not *all* of the energy change ΔU is available for work if $\Delta S > 0$, which is the origin of the term *free energy*: *the amount of energy available for work*. For this reason F is called the *work function*. While enthalpy is the heat function, $[\Delta H]_P = Q$, the Helmholtz energy is the work function, $[\Delta F]_T = W$. It's straightforward to show that $[\Delta H]_P + [\Delta F]_T = \Delta U$.

The Gibbs energy also specifies a free energy. With $\dW = -P\dV + \dW'$, the inequality (4.13) implies $-\dW' \leq T\dS - (\dU + P\dV)$. To obtain other work *from* the system, we must have that $T\dS > \dU + P\dV$. The maximum value of $\dW'$ (counted as a negative quantity) is therefore $\dW'_{\text{max}} = \dU + P\dV - T\dS = [\dG]_{T,P}$. The Gibbs energy is the maximum work obtainable from the system in a form other than $P\dV$ work:

$$W'_{\text{max}} = [\Delta G]_{T,P} . \tag{4.15}$$

Energy dissipation

That not all energy is available for work is called the *dissipation* (or *degradation*) of energy. Energy is not *lost* (which would be a violation of the first law), it's diverted into microscopic degrees of freedom. Consider two reservoirs at temperatures T_1 and T_2, with $T_1 > T_2$ (see Fig. 4.3). If we spontaneously (irreversibly) transfer heat Q from 1 to 2, there's a change in entropy:

$$\Delta S = Q \left(\frac{1}{T_2} - \frac{1}{T_1} \right) > 0 .$$

Suppose we use Q (extracted reversibly) to run a Carnot cycle between reservoirs at T_1 and T_0, where $T_1 > T_2 > T_0$. We would obtain work W_1 (using the definition of efficiency):

$$W_1 = Q \left(1 - \frac{T_0}{T_1} \right) .$$

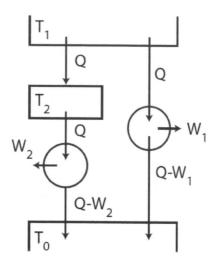

Figure 4.3 Energy dissipation: $W_2 = W_1 - T_0 \Delta S$.

Suppose we use Q (extracted reversibly) to run a Carnot cycle between T_2 and T_0. In that case we would obtain work W_2 :

$$W_2 = Q \left(1 - \frac{T_0}{T_2}\right) < W_1 .$$

Energy has been dissipated ($W_2 < W_1$) because we're able to extract less work from Q when Q has first been transferred irreversibly from 1 to 2. The energy *dissipated* (the energy not available for work) is related to the increase in entropy

$$(\Delta U)_{\mathrm{dis}} \equiv W_1 - W_2 = T_0 \Delta S . \tag{4.16}$$

4.8 HEAT DEATH OF THE UNIVERSE?

The founders of thermodynamics quickly realized the cosmological implications of the second law. Assuming that the universe is an isolated system (OK?), and knowing that isolated systems evolve toward equilibrium characterized by maximum entropy, with all temperatures within the system equalized, the fate of the universe would be a state with all energy dissipated, an ominous state of affairs termed the *heat death of the universe*, a universe in which nothing of human value is likely to survive. Kelvin wrote in 1862[23, p349]

> The second great law of thermodynamics involves a certain principle of *irreversible action in Nature*. It is thus shown that, although mechanical energy is *indestructible*, there is a universal tendency to its dissipation, which produces gradual augmentation and diffusion of heat, cessation of motion, and exhaustion of potential energy through the material universe. The result would inevitably be a state of universal rest and death, if the universe were finite and left to obey existing laws. But it is impossible to conceive a limit to the extent of matter in the universe; and therefore science points rather to an endless progress, through an endless space, of action involving the transformation of potential energy into palpable motion and thence to heat, than to a single finite mechanism, running down like a clock, and stopping for ever.

Kelvin thus postulated an infinite universe to preclude its heat death.[9] The concept of heat death pertains to the long-term future state of the universe. Curiously, however, it also implies something

---

[9]Newton also invoked an infinite universe, to prevent gravitational collapse.

about its past: If the universe were *infinitely old*, heat death would already have occurred! Thus we have an argument from thermodynamics that *the universe must have come into existence within the finite past.*[10] Clausius was not unaware of the cosmological implications of entropy. He wrote, in 1865[15, p364]:

> The second fundamental theorem ...asserts that all transformations occurring in nature may take place in a certain direction .... In fact, if in all the changes of condition occurring in the universe the transformations in one definite direction exceed in magnitude those in the opposite direction, the entire condition of the universe must always continue to change in that first direction, and the universe must consequently approach incessantly a limiting condition. ... If for the entire universe we conceive the same magnitude to be determined, consistently and with due regard to all circumstances, which for a single body I have called *entropy*, and if at the same time we introduce the other and simpler conception of energy, we may express in the following manner the fundamental laws of the universe which correspond to the two fundamental theorems of the mechanical theory of heat.
>
> 1. *The energy of the universe is a constant.*
> 2. *The entropy of the universe tends to a maximum.*

Clausius wrote in 1868 (quoted in Brush[24, p88]):

> The more the universe approaches this limiting condition in which the entropy is a maximum, the more do the occasions of further changes diminish; and supposing this condition to be at last completely obtained, no further change could evermore take place, and the universe would be in a state of unchanging death.

The thought that the universe is inevitably winding down as a result of thermodynamics can lead one to a gloomy place. Bertrand Russell wrote:[25, p107]

> All the labors of the ages, all the devotion, all the inspiration, all the noonday brightness of human genius, are destined to extinction in the vast death of the solar system, and ...the whole temple of man's achievement must inevitably be buried beneath the debris of a universe in ruins—all these things, if not quite beyond dispute, are yet so nearly certain that no philosophy which rejects them can hope to stand.

It's not that bad, however: There are more things in heaven and earth, than are dreamt of in your philosophy. Heat death will occur if thermodynamic equilibrium can be established over cosmological length scales with all inhomogeneities leveled, when matter and energy in all forms are evenly distributed throughout the universe. The temperatures of stellar photospheres range from 3000 to 50,000 K, and the *cosmic microwave background* (CMB, Section 5.4) has a temperature of 2.7 K. A state of *extreme* disequilibrium therefore exists in the universe with regard to electromagnetic radiation. A discussion of the evolution of the universe is not appropriate here, but it may be noted that with the expansion of the universe,[11] the CMB will continue to be redshifted into a state of *even lower temperature*. Disequilibrium will therefore last at least as long as stars shine, and *new stars are being created all the time*. There is more to the universe than stars, however. The lifetime of black holes of mass M is $2 \times 10^{67}(M/M_\odot)^3$ yr, where $M_\odot$ is the solar mass (see Eq. (P13.1)). The question of the time frame for heat death thus concerns the ultimate fate of the universe, the

---

[10]A bit off topic, but *Olbers's paradox* asks why is the night sky dark, why is the sky not uniformly as bright as the sun? If the universe is infinite in extent and is infinitely old, light from an infinite number of stars would have had time to reach Earth. One way "out" is to recognize that stars are not old enough.

[11]Discovered in 1929.

prediction of which is an uncertain enterprise at best. An expanding universe can *never* reach thermodynamic equilibrium: Inhomogeneities become frozen in place when the expansion rate exceeds the rate of processes that maintain equilibrium, and *the universe is accelerating in its expansion*.[12] One can question whether entropy of the universe is well defined. Entropy is a property of systems in equilibrium; $\Delta S \geq 0$ for closed systems refers to *changes* in entropy between equilibrium states. The universe, however, *has never been in equilibrium*; to what extent can we speak of the entropy of such a system? The heat death of the universe, once a bugaboo implication of the second law, shouldn't be taken seriously as a reason that the universe is trying to kill us.[13]

4.9   FREE EXPANSION AND THROTTLING

Joule expansion

In a free expansion (or *Joule expansion*), a gas initially confined to a volume V_1 is allowed to escape into an evacuated chamber of volume V_2 (as in Fig. 3.4). The experiment is thermally insulated so that no heat flows from the environment, $Q = 0$. Because the gas expands freely, no work is performed against an external pressure, $W = 0$. The internal energy of the gas therefore does not change, $\Delta U = 0$. Entropy, however, increases because the expansion takes place irreversibly, $\Delta S > 0$. During the expansion, we cannot apply the methods of thermodynamics because the system is not in equilibrium. We can, however, apply thermodynamics to the initial and final states: ΔU and ΔS are process independent.

Does the temperature change in a free expansion? Because $dU = 0$, we have

$$dU = 0 = \left(\frac{\partial U}{\partial V}\right)_T dV + \left(\frac{\partial U}{\partial T}\right)_V dT .$$

The *Joule coefficient* μ_J is defined as

$$\mu_J \equiv \left(\frac{\partial T}{\partial V}\right)_U = -\frac{(\partial U/\partial V)_T}{(\partial U/\partial T)_V} = -\frac{1}{C_V}\left[T\left(\frac{\partial P}{\partial T}\right)_V - P\right] = -\frac{1}{C_V}\left[T\frac{\alpha}{\beta} - P\right] , \quad (4.17)$$

where we have used Eqs. (1.19), (1.25), and (1.28). For small changes in volume, the change in temperature is $\Delta T = \mu_J \Delta V$.

Using the ideal gas equation of state, we have from Eq. (4.17) that $\mu_J = 0$. For real gases, energy is transferred in the expansion from kinetic energy to the potential energy of the molecules of the gas. For attractive intermolecular forces, there's an *increase* in the potential energy as the molecules move further apart (the potential energy becomes *less negative* with increasing intermolecular separation); this is accompanied by a decrease in the kinetic energy of the molecules. Thus we expect the temperature to decrease as a result of the expansion. The ideal gas ignores inter-particle interactions.

For a van der Waals gas, μ_J can be shown to be given by

$$\mu_J = -\frac{a}{C_V}\left(\frac{n}{V}\right)^2 . \quad (4.18)$$

Note that μ_J is related to the van der Waals parameter a, a measure of inter-particle interactions. For finite changes in volume we can integrate Eq. (4.18),

$$\Delta T = \int_{V_i}^{V_f} \mu_J dV = \frac{an^2}{C_V}\left(\frac{1}{V_f} - \frac{1}{V_i}\right) , \quad (4.19)$$

---

[12]Discovered in 1998, awarded the 2011 Nobel Prize in Physics.

[13]Death by thermodynamics should be the least of our worries. Impacts from asteroids and comets could prove cataclysmic to terrestrial civilization, events far more likely to occur in a human lifespan. Add to that solar flares, supernovae, gamma ray bursts, black holes, and other potential maladies: We have far more to worry about than entropic dissolution because of the second law of thermodynamics.

where we've assumed that C_V is constant. Because $V_f > V_i$, $\Delta T < 0$. Putting in numbers (Exercise 4.13), we conclude that a free expansion is not an effective way to cool a gas.

To calculate the change in entropy, from $dU = TdS - PdV$, because $dU = 0$, $\Delta S = \int_{V_i}^{V_f} (P/T)\, dV$. For an ideal gas,

$$\Delta S = nR\ln\left(V_f/V_i\right), \tag{4.20}$$

in agreement with what we calculated in Section 3.4. Because the system is isolated in a free expansion, the entropy change is due entirely to irreversibility.

Joule-Kelvin effect

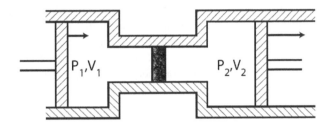

Figure 4.4 Throttling (Joule-Kelvin) process. Enthalpy is conserved.

In the *Joule-Kelvin* (or *throttling*) process, a steady flow of gas is forced through a porous plug under conditions of thermal isolation (see Fig. 4.4). In expanding through the plug the pressure drops from P_1 to P_2. The net work done on the gas is

$$W = -\int_{V_1}^{0} P_1 dV - \int_{0}^{V_2} P_2 dV = P_1 V_1 - P_2 V_2, \tag{4.21}$$

where the pressure is maintained at a constant value in both chambers. Because $Q = 0$, $\Delta U = U_2 - U_1 = W = P_1 V_1 - P_2 V_2$. Enthalpy is therefore conserved in the process, $H_1 = U_1 + P_1 V_1 = U_2 + P_2 V_2 = H_2$. The *Joule-Kelvin coefficient* is defined as

$$\mu_{JK} \equiv \left(\frac{\partial T}{\partial P}\right)_H = \frac{1}{C_P}\left[T\left(\frac{\partial V}{\partial T}\right)_P - V\right] = \frac{V}{C_P}(\alpha T - 1), \tag{4.22}$$

(Exercise 4.8). For small changes in pressure, $\Delta T \approx \mu_{JK}\Delta P$. The Joule-Kelvin coefficient can be of either sign. Because $\Delta P < 0$ in the process, $\Delta T < 0$ for $\mu_{JK} > 0$. For finite pressure changes, $\Delta T = \int_{P_1}^{P_2} \mu_{JK} dP$. For an ideal gas $\mu_{JK} = 0$ because $\alpha = T^{-1}$. For real gases, the Joule-Kelvin process can result in either a heating or a cooling of the gas. As a gas expands it cools, for the reason discussed in the case of a free expansion. In throttling, however, work is done on the gas, Eq. (4.21). From Eq. (4.22), if $\alpha > T^{-1}$, i.e., if the expansivity exceeds that of the ideal gas, cooling occurs even though work is performed on the gas. If $\alpha < T^{-1}$, the gas expands less than the ideal gas and heating occurs. Where μ_{JK} changes sign is called the *inversion temperature*. From Eq. (4.22), this occurs where $(\partial V/\partial T)_P = V/T$. The inversion temperatures (and pressures) of gases are widely tabulated. To calculate the change in entropy, from $dH = TdS + VdP$, because $dH = 0$, we have $\Delta S = -\int_{P_1}^{P_2} (V/T)\, dP$. For an ideal gas,

$$\Delta S = nR\ln\left(P_1/P_2\right). \tag{4.23}$$

For the ideal gas $\Delta T = 0$, but $\Delta S \neq 0$ (irreversibility).

CHAPTER SUMMARY

This chapter introduced the thermodynamic potentials (enthalpy, Helmholtz free energy, and Gibbs free energy), which are the three Legendre transformations of the internal energy function that can be formed from the products of variables TS and PV. We emphasized their physical interpretations as energies stored in the system under controlled conditions.

- Equilibrium can be characterized either as maximum entropy for fixed energy, or as minimum energy for fixed entropy. Internal energy $U(S,V,N)$ is a convex function of S, V, and N.

- Legendre transformations are an equivalent way of characterizing convex functions $f(x)$. Instead of specifying $f(x)$ pointwise, the quantity $g(m)$ gives the value of f at x in terms of its tangent at x and the intercept on the ordinate axis, $f(x) = mx - g(m)$, where $m = f'(x)$.

- Legendre transformations provide representations of the internal energy under conditions that can be experimentally controlled. The Helmholtz function $F(T,V)$ is the Legendre transformation of $U(S,V)$, $F = U - TS$; emphasis is shifted from S to the slope of U as a function of S, $T = (\partial U/\partial S)_V$. The enthalpy $H(S,P)$, $H = U + PV$, shifts emphasis from V to the slope of U, $P = -(\partial U/\partial V)_S$. The Gibbs energy $G(T,P)$ shifts the independent variables from (S,V) to (T,P), $G = H - TS = F + PV = U - TS + PV$.

- Each of the quantities U, H, F, G represent a potential energy under controlled conditions (summarized in Table 4.1): $[\Delta U]_S = (W)_{\text{ad}}$, $[\Delta H]_{S,P} = (W')_{\text{ad}}$, $[\Delta F]_T = W$, and $[\Delta G]_{T,P} = W'$, where W' is the work done on the system in forms other than PdV work.

- Under the conditions indicated, the second law prescribes the direction of spontaneous changes, $[\Delta U]_{S,V} \leq 0$, $[\Delta H]_{S,P} \leq 0$, $[\Delta F]_{T,V} \leq 0$, and $[\Delta G]_{T,P} \leq 0$, where equality holds for reversible processes. The quantities U, H, F, and G are each a minimum in equilibrium.

- Maxwell relations, identities between derivatives of thermodynamic variables, follow from the exactness of the differentials of thermodynamic potentials. Summarized in Table 4.2.

- The chemical potential is the Gibbs energy per particle, $G = N\mu$, the energy to add another particle at constant T and P, $[dG]_{T,P} = \mu[dN]_{T,P}$.

- Not all of the energy change ΔU in a process can be converted into work if $\Delta S > 0$, an effect called the dissipation of energy. Energy is transferred to microscopic degrees of freedom, those associated with an increase in entropy.

- The maximum amount of work that can be obtained from a system, the free energy, is given by $W_{\text{max}} = [\Delta F]_T$ and $W'_{\text{max}} = [\Delta G]_{T,P}$.

- In a free expansion a gas expands into an evacuated, thermally isolated chamber. The internal energy is conserved in such a process, $W = 0$ and $Q = 0$. The temperature change of a non-ideal gas is given by the Joule coefficient, $\mu_J = (\partial T/\partial V)_U$. The temperature decrease in a free expansion is due to the conversion of kinetic energy of the molecules into the work required to separate molecules against attractive intermolecular forces. For an ideal gas, $\mu_J = 0$; the ideal gas ignores intermolecular forces. The entropy change of the ideal gas for a free expansion is $\Delta S = nR\ln(V_f/V_i) > 0$. The increase in entropy is entirely associated with the irreversibility of the process.

- In throttling a steady flow of gas is forced through a porous plug under thermally isolated conditions. Enthalpy is conserved in this process. The temperature change is given by the Joule-Kelvin coefficient, $\mu_{JK} = (\partial T/\partial P)_H$. The temperature of the gas can either increase or decrease in throttling. For the ideal gas the entropy change $\Delta S = nR\ln(P_i/P_f) > 0$ even though $\mu_{JK} = 0$; the entropy change is associated with the irreversibility of the expansion.

EXERCISES

4.1 Consider a film where volume effects are not negligible. The first law for this system is:

$$dU = TdS - PdV + \sigma dA,$$

where σ is the surface tension and A is the surface area. (a) What is the differential of the Gibbs function for this system? (b) Derive the three Maxwell relations associated with the Gibbs function. (c) How can the derivative $(\partial V/\partial A)_{T,P}$ be interpreted? That is, how can the volume of the film depend on the surface area? What is the *system* here?

4.2 Derive the inequalities (4.3), (4.6), (4.9), and (4.10).

4.3 Show that $H = N\mu + TS$ and $F = N\mu - PV$. Hint: Euler relation

4.4 Show that $[\Delta H]_P + [\Delta F]_T = \Delta U$.

4.5 Derive Eq. (4.20).

4.6 Derive the relations:

$$\left(\frac{\partial U}{\partial V}\right)_T = T^2 \frac{\partial}{\partial T}\left(\frac{P}{T}\right)_V = T\left(\frac{\partial P}{\partial T}\right)_V - P.$$

and

$$\left(\frac{\partial U}{\partial P}\right)_T = -T\left(\frac{\partial V}{\partial T}\right)_P - P\left(\frac{\partial V}{\partial P}\right)_T.$$

One could use Maxwell relations to derive these identities, or, what is effectively the same thing, one could invoke the integrability condition for S, U, and V, as functions of T and V or T and P.

4.7 Show that

$$\left(\frac{\partial C_V}{\partial V}\right)_T = T\left(\frac{\partial^2 P}{\partial T^2}\right)_V \qquad \left(\frac{\partial C_P}{\partial P}\right)_T = -T\left(\frac{\partial^2 V}{\partial T^2}\right)_P.$$

Use the results of Exercise 4.6 and for the second part, Exercise 1.7.

4.8 Show that dH can be written

$$dH = C_P dT + \left[-T\left(\frac{\partial V}{\partial T}\right)_P + V\right]dP.$$

Conclude that $C_P = (\partial H/\partial T)_P$ and hence that in any constant-pressure process $[\Delta H]_P = \int_{T_1}^{T_2} C_P dT$. Hint: Let $S = S(T, P)$. Don't forget about Eq. (3.14) and don't forget about the Maxwell relations.

4.9 Show that

$$\left(\frac{\partial T}{\partial V}\right)_S = -\frac{T}{C_V}\left(\frac{\partial P}{\partial T}\right)_V \qquad \left(\frac{\partial T}{\partial P}\right)_S = \frac{T}{C_P}\left(\frac{\partial V}{\partial T}\right)_P.$$

Use the cyclic relation and reach for a Maxwell relation.

4.10 Show that

$$dS = \frac{C_V}{T}dT + \frac{1}{T}\left[P + \left(\frac{\partial U}{\partial V}\right)_T\right]dV,$$

and hence that in any constant-volume process $[\Delta S]_V = \int_{T_1}^{T_2} \frac{C_V}{T}dT$.

4.11 Show that, in analogy with Eq. (4.14),

$$F(V, T_0) = F(V_0, T_0) - \int_{V_0}^{V} [P]_{T_0} \, dV \,.$$

Because $G = G(T, P)$ and $F = F(T, V)$, whatever we can do with G involving P (at constant T), we can find an analogous expression for F involving V (at constant T).

4.12 Derive Eq. (4.22) starting from Eq. (4.21).

4.13 In Eq. (4.19), let $V_f \rightarrow \infty$, so that we can consider the maximum cooling available from a free expansion. What is a typical value of the van der Waals parameter, a? The *Handbook of Chemistry and Physics* is a great source of thermodynamic data. For oxygen, $a = 0.1382 \, \text{Pa m}^6 \, \text{mol}^{-2}$. For nitrogen, $a = 0.1370$ in the same units. Let's take $a = 0.1 \, \text{Pa m}^6 \, \text{mol}^{-2}$. Take $C_V = \frac{3}{2} nR$. Moreover, normalize Eq. (4.19) by the temperature to get a fractional temperature change,

$$\frac{\Delta T}{T} = -\frac{2an}{3RTV_i} \,.$$

Take $n = 1$ mole, $T = 300$ K, and $V_i = 10^{-3}$ m^3. What fractional change in T do you find?

4.14 Instead of using the van der Waals equation of state, suppose we use the virial expansion, Eq. (1.11). Show that to first order in the virial expansion the Joule coefficient is given by

$$\mu_J = -\frac{1}{C_V} \frac{n^2 RT^2}{V^2} B_2'(T) \,.$$

Show using Eq. (1.13) that this result reduces to Eq. (4.22).

4.15 The Joule-Kelvin coefficient for the van der Waals gas is a bit of a chore to derive. Using Eq. (1.12), show that

$$\mu_{JK} = \frac{n}{C_P} \frac{-b + \frac{2a}{RT}\left(1 - \frac{nb}{V}\right)^2}{1 - \frac{2an}{RTV}\left(1 - \frac{nb}{V}\right)^2} \,.$$

In the low-density limit, $n/V \ll b^{-1}$ and $n/V \ll RTa^{-1}$, show that

$$\mu_{JK} \approx \frac{2}{5R}\left(-b + \frac{2a}{RT}\right) \,,$$

where we have used $C_P = \frac{5}{2} nR$. Show that the inversion temperature for the van der Waals gas is given by

$$RT = \frac{2a}{b}\left(1 - \frac{nb}{V}\right)^2 \,.$$

4.16 Show that C_V for the van der Waals gas depends on T but not on V. Use the result of Exercise 4.7. C_V for the ideal gas is a constant.

4.17 Show that an alternate way of stating one of the stability conditions (Section 3.10) is

$$\left(\frac{\partial^2 F}{\partial V^2}\right)_T > 0 \,. \tag{P4.1}$$

Modify the argument used in Section 3.10, that because in equilibrium entropy is a maximum, fluctuations must lead to a momentary decrease in entropy, to the case of the Helmholtz free energy F which is a minimum in equilibrium. Which of the three stability conditions in Eq. 3.51 does Eq. P4.1 correspond to? Hint: $P = -(\partial F/\partial V)_T$.

CHAPTER 5

Thermodynamics of radiation

I T'S often said that the origins of quantum mechanics can be traced to the year 1900 with Planck's theory of black-body radiation. But where did that problem come from, that so ably exposed the limitations of classical physics? The term *black body* as a perfect absorber of radiant energy was introduced in 1860 by Gustav Kirchhoff, who applied thermodynamics to electromagnetic radiation, or heat radiation as it was called then. It's not obvious that thermodynamics pertains to radiation because thermodynamics is concerned with the state of equilibrium, and radiating bodies are not normally in equilibrium with their environment. Kirchhoff examined the singularly important problem of *cavity radiation*, in which matter and radiation *are* in equilibrium. *Cavity radiation is electromagnetic energy contained within a hollow enclosure bounded by thick opaque walls maintained at a uniform temperature.* To observe cavity radiation, a small hole must be made in the walls surrounding the cavity so that some of it can escape.[1] As we'll see, that hole plays a significant role in the theory, and we'll refer to it simply as the "hole." Cavity radiation is closely related to black-body radiation; see Eq. (5.3). In this chapter we consider cavity radiation as a *thermodynamic* system. The purview of thermodynamics is not limited to material systems; it applies to systems *in equilibrium*.

5.1 KIRCHHOFF LAW OF THERMAL RADIATION

All objects emit radiant energy. The *spectral emissive power*, $E(\lambda, T)\mathrm{d}\lambda$, is the energy emitted by a body per time at temperature T for wavelengths between λ and $\lambda + \mathrm{d}\lambda$, *per surface area*.[2] The *spectral absorptivity* $A(\lambda, T)$ is the fraction of incident energy at wavelength λ absorbed by a body at temperature T. *Surfaces* do not emit radiation per se, only atoms do that; it's convenient, however, to think of surfaces as emitting and absorbing radiation— radiation must pass through a surface. Three processes are associated with the interaction of material objects and electromagnetic radiation: reflection, absorption, and emission. Kirchhoff reasoned there must be a connection between absorption and emission for objects in equilibrium with radiation. Darker objects reflect less energy than shiny objects, and hence absorb more energy than shiny objects. As a rule, good absorbers are good emitters, an observation that finds a natural explanation in thermodynamics. The radiant energy absorbed by an object in thermal equilibrium must be re-emitted. That can be demonstrated by assuming the opposite and showing it leads to a contradiction. Consider two objects placed in an enclosed box held at a constant temperature, and suppose one is shiny and the other is dark. Assume the objects are initially at the same temperature. Suppose, contrary to fact, that the shiny object is a good emitter, and the dark object is a poor emitter. The dark object absorbs more energy than the

---

[1] The hole must be small enough so that it has a negligible effect upon the state of radiation in the cavity.
[2] The quantity $E(\lambda, T)$ therefore has units of $\mathrm{W\,m^{-3}}$; do you see why?

shiny object, and hence its temperature must rise if it emits less energy than it absorbs. Conversely, the temperature of the shiny object must fall if it emits more energy than it absorbs. If this were the case, energy would be passed from cold to hot with no other effect—a violation of the second law.[3] To maintain equilibrium in the cavity, the total energy leaving all objects (reflected and emitted) must be the same whether they're shiny or dark. Thus, *there must be a linkage between absorption and emission.* The same applies to the material of the cavity walls. In equilibrium all objects in the cavity attain the temperature of the cavity walls.[4]

Cavity radiation is independent of the *specifics* of the cavity—the size and shape of the cavity or the material of the walls—and depends only on the temperature of the walls. Consider two cavities *A* and *B* (constructed of different materials and having different sizes and shapes) connected by a narrow tube through which radiation may pass. If energy transport from *A* to *B* exceeds that from *B* to *A*, the temperature of *B* would rise, in violation of the second law. The energy emitted per time by both cavities must therefore be the same, *regardless of their specifics*. Other properties of the radiation follow by the same method of reasoning. By repeating the argument with color filters or polarizers placed in the tube, we conclude that cavity radiation is unpolarized, isotropic, and with the radiation in each cavity having the same spectral distribution. Because the energy radiated by each cavity per time into the tube is the same (independent of cavity size), the *density* of electromagnetic energy must be the same in each cavity. As we'll show, there is a simple relation between the energy emitted per time per area into the tube and the energy density in the cavity; see Eq. (5.3).

The processes of emission and absorption are thus linked dynamically: Whatever energy is absorbed is emitted so as to maintain equilibrium. *Emissivity and absorptivity must be proportional,*

$$\frac{E(\lambda, T)}{A(\lambda, T)} = f(\lambda, T) . \tag{5.1}$$

Whatever is the function $f(\lambda, T)$, *it must be universal, the same for all bodies.* While E and A are separately material-specific, their ratio is a universal quantity, *Kirchhoff's law of thermal radiation.* Kirchhoff thus posited the *existence* of a universal function describing the spectral distribution of thermal radiation, but did not give its form, and indeed it's not possible to do so within the framework of thermodynamics. A full 40 years would pass before this function was known.

Kirchhoff defined an object to be *perfectly black* if it totally absorbs, i.e., does not reflect, incident radiation so that $A(\lambda) = 1$. *A black body by definition reflects nothing.* By setting $A(\lambda) = 1$ in Eq. (5.1), the emissive power $E(\lambda, T)$ of a black body is a *universal function.* Once the function $f(\lambda, T)$ is known, we infer from $E(\lambda, T) = A(\lambda, T)f(\lambda, T)$ that *the emissivity of a black body is larger than that for any other body at the same temperature*—good absorbers are good emitters.

The notion of a black body sounds like a typical idealization in physics, yet Kirchhoff gave an operational definition of one. A black body can be very nearly realized by an enclosed cavity whose walls are at a uniform temperature, are sufficiently thick that radiation cannot penetrate, and has a small hole in one of the walls. Given the roughness of cavity surfaces, external radiation incident on the hole has virtually no chance of being reflected back out through the hole. The *hole* therefore is very nearly totally absorbing, with the radiation it emits approximating black-body radiation.[5] A measurement of $E(\lambda, T)$ from the hole is thus a measurement of Kirchhoff's universal function.

An *arbitrary* state of radiation (not in thermal equilibrium) can be maintained in a cavity if it 1) encloses *vacuum* (passes all wavelengths without absorption) and 2) has *perfectly reflecting walls* ("white"). Should a small quantity of matter be introduced into such a cavity (the proverbial speck of dust), the radiation would transform into cavity radiation through the mechanism posited by Kirchhoff's law. In this process the total energy of the radiation (integrated over all wavelengths)

---

[3]Imagine Maxwell's demon (see Chapter 12) sitting between the shiny and dark objects, passing energy from cold to hot.
[4]We're presuming the heat capacity of the material surrounding the cavity far exceeds the heat capacity of the objects in the cavity, i.e., the walls of the cavity act as a heat reservoir.
[5]A cavity connected to a small hole is a "roach motel" for photons: They check in, but they don't (directly) check out.

stays essentially constant, $\Delta U = 0$, because the piece of matter can be made as small as we please; its change in energy is small compared to that of the radiant energy in the cavity.[6] Because the process is irreversible, however, there's an increase in entropy, $\Delta S > 0$.

5.2 THERMODYNAMICS OF BLACK-BODY RADIATION

At the time of Kirchhoff's work it was not known that heat radiation is electromagnetic in origin and travels at the speed of light, c. That fact can be used to derive a simple relation between the emissive power of a black body (the "hole") and the energy density of cavity radiation. Let $u(\lambda, T)\mathrm{d}\lambda$ denote the *energy spectral density* of cavity radiation, the energy per volume contained in the wavelength band $(\lambda, \lambda + \mathrm{d}\lambda)$ in equilibrium at temperature T. (Thus, $u(\lambda, T)$ has units $\mathrm{J\,m^{-4}}$.) Let $\mathrm{d}\sigma$ represent an element of surface area. Because cavity radiation is isotropic, the fraction of radiation emitted from $\mathrm{d}\sigma$ in the direction of a small cone of solid angle $\mathrm{d}\Omega$ is $\mathrm{d}\Omega/(4\pi)$, where $\mathrm{d}\Omega = \sin\theta\mathrm{d}\theta\mathrm{d}\phi$, with θ the angle between the normal to $\mathrm{d}\sigma$ and the direction of radiation. A small *flux tube* oriented parallel to the direction of radiation that intersects $\mathrm{d}\sigma$ has cross-sectional area $\mathrm{d}\sigma\cos\theta$. Let the length of the tube be $c\mathrm{d}t$. The amount of energy at wavelength λ in this tube flowing in the direction of $\mathrm{d}\Omega$ is therefore $\mathrm{d}\Omega\mathrm{d}\sigma\cos\theta c\mathrm{d}tu(\lambda, T)\mathrm{d}\lambda/(4\pi)$. The energy emitted per time $\mathrm{d}t$ per area $\mathrm{d}\sigma$ integrated over all directions is (note the limits of integration)

$$E(\lambda, T)\mathrm{d}\lambda = \frac{c}{4\pi}u(\lambda, T)\mathrm{d}\lambda \int_0^{2\pi}\mathrm{d}\phi \int_0^{\pi/2}\sin\theta\cos\theta\mathrm{d}\theta = \frac{c}{4}u(\lambda, T)\mathrm{d}\lambda. \tag{5.2}$$

Equation (5.2) can be integrated over all wavelengths:

$$E(T) \equiv \int_0^\infty E(\lambda, T)\mathrm{d}\lambda = \frac{c}{4}\int_0^\infty u(\lambda, T)\mathrm{d}\lambda \equiv \frac{c}{4}u(T). \tag{5.3}$$

Equation (5.3) provides a connection between the total emissive power (integrated over all wavelengths) of a black body and the total energy density of radiation in the cavity.

Photons are non-interacting,[7] like the particles of an ideal gas, and thus one might think the photon gas (cavity radiation) is ideal. The energy of the ideal gas is independent of the volume of the container (Joule's law, $(\partial U/\partial V)_T = 0$), and we might expect the same for cavity radiation. For cavity radiation, however, the energy *density* is independent of cavity size and depends only on temperature (argued in Section 5.1). Thus, we have the analog of Joule's law for cavity radiation[8]

$$\left(\frac{\partial u}{\partial V}\right)_T = 0. \tag{5.4}$$

Cavity radiation is not an ideal gas, even though both systems are collections of non-interacting entities. For the ideal gas, $\Delta U = 0$ in an isothermal expansion, Eq. (1.31). For cavity radiation, the *energy density* is independent of volume and thus $\Delta U = u\Delta V$. In the expansion of a gas, the number of particles is *fixed*; for cavity radiation the number of photons is *variable*. Consider a cavity having perfectly reflecting walls, fitted with a piston, that contains a minute speck of material maintained at temperature T through contact with an external reservoir. By the argument of Section 5.1, the radiation in the cavity has the spectral distribution of a black body at the temperature T. If the piston is raised slowly, departures from equilibrium are small. As the piston is raised, the radiation expands into a larger volume. To maintain the same *density* of radiant energy, the speck

---

[6] Adding matter to an evacuated cavity with perfectly reflecting walls is somewhat analogous to a spark in an air-fuel mixture. The energy of the initiating factor (mass or spark) is small in comparison with the energies undergoing transformation.

[7] Photons cannot couple directly to each other because they don't carry charge. At very high intensities, photons can interact by a variety of higher-order processes at the quantum level, e.g., Delbrück scattering.

[8] While the energy density of cavity radiation is independent of V under isothermal conditions, it's not independent of V for isentropic processes; see Eq. (5.32).

emits more energy than it absorbs until the energy density is consistent with that at temperature T. If the piston is lowered, the object absorbs more energy than it emits until equilibrium is re-established. In the isothermal expansion of an ideal gas, heat is absorbed from the reservoir to keep the temperature of the particles fixed. With cavity radiation, heat is absorbed from the reservoir but it goes into creating new photons to keep the energy density fixed.

Electromagnetic radiation carries momentum as well as energy. The momentum density g of the electromagnetic field is, from electromagnetic theory, $g = S/c^2$, where S is the Poynting vector (energy per time per area). Because photons all travel at the same speed, the magnitude of the Poynting vector[9] $S = cu(T)$, and thus the momentum density[10] $g = u/c$. For momentum p incident at a surface at the angle θ relative to the normal, the change in momentum $\Delta p = 2p \cos \theta$. Using the flux-tube argument, the contribution to the pressure P from the momentum of the radiation is

$$dP = \frac{dP}{d\Omega}d\Omega = \left(\frac{d\Omega}{4\pi}\right)\frac{(c\Delta t)(\Delta\sigma\cos\theta)}{\Delta t\Delta\sigma}\frac{u(T)}{c}2\cos\theta = \frac{u(T)}{2\pi}\cos^2\theta d\Omega . \qquad (5.5)$$

Integrating Eq. (5.5) over all directions, we have the total pressure at temperature T

$$P = \frac{u(T)}{2\pi}\int_0^{2\pi}d\phi\int_0^{\pi/2}\sin\theta\cos^2\theta d\theta = \frac{1}{3}u(T) . \qquad (5.6)$$

Contrast Eq. (5.6) with the ideal gas, $P = \frac{2}{3}u(T)$ (use Eq. (1.32) and the ideal gas law).

Equation (5.6) is the equation of state for cavity radiation, *to which the machinery of thermodynamics can be applied*. Combining Eqs. (5.4) and (5.6) with Eq. (1.28), where $U = uV$:

$$u = \frac{T}{3}\left(\frac{\partial u}{\partial T}\right)_V - \frac{u}{3} ,$$

implying that

$$4u = T\frac{du}{dT} , \qquad (5.7)$$

where we can replace the partial derivative with the total derivative. The solution of Eq. (5.7) is:

$$u(T) = aT^4 , \qquad (5.8)$$

where a is the *radiation constant*. The value of this constant ($a = 7.5657 \times 10^{-16}$ J m^{-3} K^{-4}) cannot be obtained using the methods of thermodynamics. It would be a fundamental constant of nature (akin to the gas constant R) were it not for the fact that, using the Planck distribution, it can be evaluated[11] in terms of other fundamental constants, $a = \pi^2 k_B^4/(15\hbar^3 c^3)$. Note that we cannot take the classical limit of the radiation constant by formally letting[12] $h \to 0$. *There is no classical antecedent of cavity radiation*; it's an intrinsically quantum problem from the outset. We have our first instance where thermodynamics anticipates quantum mechanics. Combining Eq. (5.8) with Eq. (5.3), we have the *Stefan-Boltzmann radiation law* for the total emissive power of a black body at temperature T

$$E(T) = \frac{ac}{4}T^4 \equiv \sigma T^4 , \qquad (5.9)$$

---

[9]This result is easily established using a flux-tube argument.

[10]The result $g = u/c$ is $p = E/c$ (from the theory of relativity) turned into a relation between densities.

[11]A standard exercise in statistical physics.

[12]I thank Professor Andrés Larraza for pointing out there is no classical limit of the radiation constant. Note that the radiation constant does not contain a mass or a charge. Photons carry no charge, and have zero rest mass. The radiation constant does, however, contain Boltzmann's constant, which appears in physics through the average kinetic energy per particle of an ideal gas (particles that have mass) at temperature T. As we discussed earlier, measuring absolute temperature in degrees Kelvin artificially introduces a conversion factor between energy and Kelvin. Absolute temperature "should" have the dimension of energy, in which case Boltzmann's constant would be unnecessary.

where $\sigma = ac/4 = 5.6704 \times 10^{-8}$ W m^{-2} K^{-4}, *Stefan's constant*.

Example. The luminosity of the sun $L = 3.85 \times 10^{26}$ W, and the solar radius is $R = 6.96 \times 10^8$ m. Calculate the effective surface temperature of the sun, assuming it to be a black body. From Eq. (5.9), $\sigma T^4 = L/(4\pi R^2)$. We find $T = 5780$ K. The sun is of course *not* a black body, but Stefan's law gives us an effective temperature of its photosphere. The center of the sun is thought to have a temperature $\approx 1.5 \times 10^7$ K, while the corona has a temperature $\approx 5 \times 10^6$ K.

Once we know Eq. (5.8), we know a lot. Combining Eq. (5.8) with Eq. (5.6), the pressure of cavity radiation as a function of absolute temperature T is

$$P = \frac{a}{3} T^4 . \tag{5.10}$$

The radiant energy in a cavity of volume V at temperature T is, from Eq. (5.8) and Eq. (5.4),

$$U = aVT^4 . \tag{5.11}$$

Contrast Eq. (5.11) with Eq. (1.32): For the ideal gas, the extensible quantity is the number of moles (amount of matter at temperature T), whereas for the photon gas the extensible quantity is the volume of the cavity (bounded by walls at temperature T).

By combining Eq. (5.11) with Eq. (1.25), we have an expression for the heat capacity:

$$C_V = \left(\frac{\partial U}{\partial T} \right)_V = 4aVT^3 . \tag{5.12}$$

Note that $C_V \to 0$ as $T \to 0$, as required by the third law of thermodynamics.[13] There's no reason not to expect that Eq. (5.12) holds for all T: Photons do not undergo phase transitions at low temperature. In contrast, the heat capacity of monatomic gases, $C_V = \frac{3}{2} N k_B$, a *constant*, is valid only for temperatures well above liquefaction temperatures. From Eq. (3.11) and Eq. (5.12),

$$\left(\frac{\partial S}{\partial T} \right)_V = \frac{1}{T} C_V = 4aVT^2 ,$$

which can be integrated (at constant V)

$$S(T) = S_0 + \int_0^T \frac{C_V}{T'} dT' = \frac{4}{3} aVT^3 , \tag{5.13}$$

where we have set $S_0 = 0$.

With Eqs. (5.10), (5.11), and (5.13) (expressions for P, U, and S), we have the ingredients to construct the thermodynamic potentials:

$$U = aVT^4 = \frac{3}{4} TS = 3PV \qquad F = U - TS = -\frac{1}{3} aVT^4$$

$$H = U + PV = TS = \frac{4}{3} aVT^4 \qquad G = H - TS = 0 . \tag{5.14}$$

The Gibbs energy is identically *zero*. From Eq. (4.9), $G = N\mu$, and because we have $G = 0$ for any N, we infer that $\mu = 0$, even though (as we argue below) the chemical potential isn't well defined

---

[13]The low-temperature heat capacity of many solids vanishes like T^3 as $T \to 0$, formally the same as the photon gas. Quantized vibrations of a crystal lattice, *phonons*, behave at low temperature like photons; both are bosons and both are non-dispersive with $\omega = ck$ where in the case of phonons, c is the speed of sound. In many respects, a solid at low temperature can be considered a box of phonons, together with electrons (fermions). The low-temperature heat capacity of electrons vanishes linearly with T.

for cavity radiation. Taking $\mu = 0$ implies that it costs no *extra* energy to add photons to cavity radiation but of course it costs energy to *make* photons; photons are a special case in that creating them is the same as adding them to the system.[14] Now that we've brought it up, however, what *is* the number of photons in a cavity? Nowhere has N been specified, and indeed such a quantity is beyond the capabilities of thermodynamics to calculate. Using the Planck distribution, the *average* number of photons in a cavity in thermal equilibrium can be calculated (assuming $\mu = 0$):

$$N = \frac{2\zeta(3)k_B^3}{\pi^2(\hbar c)^3}VT^3 \approx \frac{1}{2.7k_B}aVT^3 \,, \tag{5.15}$$

where $\zeta(3) \approx 1.202$ is the Riemann zeta function of argument 3. Comparing Eqs. (5.13) and (5.15), the entropy of cavity radiation is proportional to the average number of photons,

$$S \approx 3.6k_B N \,. \tag{5.16}$$

It's not possible to change the number of photons keeping entropy fixed, and thus chemical potential is not defined for cavity radiation (see definition of μ in Section 3.5).

However, Eq. (5.15), which is based on the Planck formula, *assumes* $\mu = 0$. Is the assignment $\mu = 0$ for cavity radiation consistent with general thermodynamics? From Table 4.1, $[\Delta F]_{T,V} = W'$, the amount of "other" work. From Eq. (5.14), $[\Delta F]_{T,V} = 0$. The maximum work in *any* form is given by $[\Delta F]_T = W_{\max}$, Eq. (4.14), and from Eq. (5.14), $[\Delta F]_T = -\frac{1}{3}aT^4\Delta V = -P\Delta V$, where we've used Eq. (5.10). Thus, *no forms of work other than PdV work are available to cavity radiation*, which is consistent with $\mu = 0$. As shown in Section 3.10, thermodynamic equilibrium is achieved when the intensive variables *conjugate to conserved quantities* (energy, volume, particle number) equalize between system and surroundings. *Photons are not conserved quantities.* Photons are *created and destroyed* in the exchange of energy between the cavity walls and the radiation in the cavity. There's no population of photons external to the cavity for which those in the cavity can come to equilibrium with. The natural variables to describe the thermodynamics of cavity radiation are T, V, S, P, or U, but not N. Cavity radiation should not be conceived of as an open system (Section 3.5); it's a *closed* system that exchanges energy with its surroundings. The confusion here is that photons are particles of energy, a quintessential quantum concept.

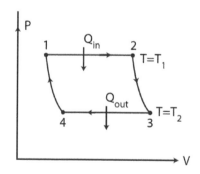

Figure 5.1   Carnot cycle using cavity radiation

Cavity radiation can thus be treated as a macroscopic system in equilibrium.[15] Let's calculate the efficiency of a Carnot cycle with the photon gas as the working substance. (We already know what the answer is, that for any reversible cycle, Eq. (2.13), but let's calculate it explicitly.) From Eq. (5.10), isotherms on a P-V diagram are the locus of constant P (see Fig. 5.1). The change in

---

[14]We can't get into the rest frame of a photon; a photon held in your hand is a destroyed (absorbed) photon.

[15]This point underscores the universality of thermodynamics. Equilibrium is equilibrium.

enthalpy at constant P is the heat absorbed, $[\Delta H]_P = Q$ (Table 4.1), and thus, from Eq. (5.14), $Q = 4P\Delta V = \frac{4}{3}aT^4\Delta V$. Along adiabats, we have from Eq. (5.13) (constant entropy),

$$VT^3 = \text{constant} . \tag{5.17}$$

Combining Eq. (5.17) with Eq. (5.10),

$$PV^{4/3} = \text{constant} . \qquad \text{(adiabatic process)} \tag{5.18}$$

Temperature and pressure of cavity radiation therefore change in adiabatic processes; in particular $\Delta T < 0$ for $\Delta V > 0$. From Eq. (2.1),

$$\eta = 1 - \frac{|Q_{\text{out}}|}{Q_{\text{in}}} = 1 - \frac{T_2^4 (V_3 - V_4)}{T_1^4 (V_2 - V_1)} = 1 - \frac{T_2}{T_1} ,$$

where the last equality follows from applying Eq. (5.17) to each adiabat: $T_2^3 V_3 = T_1^3 V_2$ and $T_2^3 V_4 = T_1^3 V_1$.

At various points in our discussion we've contrasted the thermodynamic properties of the ideal gas with those of the photon gas; these results are summarized in Table 5.1.

Table 5.1   Thermodynamics of the ideal gas and the photon gas

	Ideal gas	Photon gas
Internal energy	$U = \frac{3}{2}nRT$	$U = aVT^4$
Volume dependence of U	$\left(\dfrac{\partial U}{\partial V}\right)_T = 0$	$\left(\dfrac{\partial u}{\partial V}\right)_T = 0$
Equation of state	$P = nRT/V = \frac{2}{3}u$	$P = \frac{1}{3}aT^4 = \frac{1}{3}u$
Heat capacity	$C_V = \frac{3}{2}nR$	$C_V = 4aVT^3$
Adiabatic process	$TV^{\gamma-1} = \text{constant}$	$TV^{1/3} = \text{constant}$
Entropy	$S = Nk_B\left[\dfrac{5}{2} + \ln\left(\dfrac{V}{N\lambda_T^3}\right)\right]$ $\left(\lambda_T \equiv h/\sqrt{2\pi m k_B T}\ \text{(Chapter 7)}\right)$	$S = \frac{4}{3}aVT^3$
Chemical potential	$\mu = -k_B T \ln\left(\dfrac{V}{N\lambda_T^3}\right)$ (Chapter 7)	$\mu = 0$

5.3   WIEN'S DISPLACEMENT LAW

The Stefan-Boltzmann law, Eq. (5.9), specifies the total power emitted per unit area of a black body at absolute temperature T. Useful as that is, it doesn't bring us any closer to the holy grail of Kirchhoff's universal function, the energy spectral density of cavity radiation, $u(\lambda, T)$. An important step in that direction, and the last to be taken using thermodynamics, is Wien's *displacement law*, derived in 1893. Wilhelm Wien showed that $u(\lambda, T)$, presumed a function of two independent variables λ and T, is actually a function of a *single* variable, the *product* λT: $u = u(\lambda T)$. That's a nontrivial accomplishment;[16] how did he do it? We derive Wien's law in this section.

---

[16] Wien received the 1911 Nobel Prize in Physics for this work.

We first give a "thermodynamic" proof of an important result (a result that also follows from Wien's law): *Cavity radiation retains its spectral distribution in reversible adiabatic processes* (even though the temperature changes). The proof consists of assuming the opposite and showing it leads to a contradiction. Let there be thermal radiation in an evacuated chamber with perfectly reflecting walls. Let the volume and entropy of the radiation be V_1 and S_1. Compress quasistatically and adiabatically to volume $V_2 < V_1$ ($\Delta S = 0$ in this process). At this point, either the radiation has the spectral distribution of a black body, or it doesn't. Assume not. By introducing a small amount of matter into the chamber (at volume V_2), the radiation would change to cavity radiation (Section 5.1). This step would be accompanied by an increase in entropy (irreversible), but the energy would be unchanged (Section 5.1). No change in U at constant V implies $\Delta T = 0$: From Eq. (5.11), $[\Delta U]_V = 4aVT^3\Delta T$. From Eq. (5.13), however, $[\Delta S]_V = 4aVT^2\Delta T$. The two are inconsistent: $[\Delta U]_V = 0 \Rightarrow \Delta T = 0$, and $[\Delta S]_V > 0 \Rightarrow \Delta T > 0$. The premise that $\Delta S > 0$ thus leads to a contradiction. The spectral distribution is unaffected by a reversible adiabatic process.

Wien's law follows from a detailed analysis of the means by which $u(\lambda, T)$ would change in a reversible adiabatic process. A simple way to do that is to study the reflection of radiation from a slowly moving mirror. Let a piston comprised of a perfectly reflecting material move at a small velocity v in an evacuated cylinder with reflecting walls. We take v as positive when the piston moves to decrease the volume of the cylinder. Let monochromatic radiation of frequency ν be directed toward the piston at normal incidence. As a result the reflected radiation has frequency ν' given by the Doppler shift formula

$$\nu' = \nu \left(\frac{1 + v/c}{1 - v/c} \right) = \nu \left[1 + 2\frac{v}{c} + O\left(\frac{v}{c}\right)^2 \right].$$

The frequency of reflected radiation is increased for $v > 0$. Now let radiation be incident at angle θ with respect to the direction of the piston's motion. The component of v projected onto the direction of the radiation is $v \cos\theta$, and thus we have a generalization of the Doppler shift formula:

$$\nu' = \nu \left[1 + 2\frac{v}{c} \cos\theta + O\left(\frac{v}{c}\right)^2 \right]. \tag{5.19}$$

Sidebar discussion: Deriving Eq. (5.19) using relativity
The following assumes you're well versed in the ways of relativity. The energy-momentum four-vector of a mirror, in a reference frame where it has velocity $\beta \equiv v/c$ (see Fig. 5.2), is $P^\mu =$

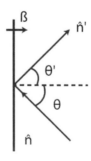

Figure 5.2 Reflection from a moving mirror

$M\gamma c\,(1, \boldsymbol{\beta})$, where M is the mirror mass, $\mu = (0, 1, 2, 3)$, with $\mu = 0$ the time index, and the Lorentz factor $\gamma = (1 - \beta^2)^{-1/2}$. The four-momentum of the incident photon is $Q^\mu = (E/c)\,(1, \hat{\boldsymbol{n}})$, where E is the photon energy (related to the frequency ν through the Planck relation $E = h\nu$), and the unit vector $\hat{\boldsymbol{n}}$ points in the direction of photon propagation. After reflecting from the mirror,

$Q'^\mu = (E'/c)\,(1, \hat{\boldsymbol{n}}')$ and $P'^\mu = M\gamma'c\,(1, \boldsymbol{\beta}')$. Conservation of energy-momentum is expressed as the four-vector equation: $P'^\mu + Q'^\mu = P^\mu + Q^\mu$. Because we're not interested in P'^μ, eliminate it: $P'^\mu = P^\mu + Q^\mu - Q'^\mu$. Now construct the inner product of P'^μ with itself:

$$P'^\mu P'_\mu = P^\mu P_\mu + 2P_\mu Q^\mu - 2P_\mu Q'^\mu - 2Q_\mu Q'^\mu + Q^\mu Q_\mu + Q'^\mu Q'_\mu \,,$$

where the summation convention on repeated indices is employed, the covariant vectors are constructed using the Lorentz metric (which we take to be $(-1, 1, 1, 1)$), and we've used for any four-vectors, $A^\mu B_\mu = A_\mu B^\mu$. The photon four-momentum is a null vector, $Q^\mu Q_\mu = Q'^\mu Q'_\mu = 0$, while that for the mirror is timelike, $P^\mu P_\mu = P'^\mu P'_\mu = -M^2 c^2$. Conservation of energy-momentum is thus equivalent to the expression

$$P_\mu \left(Q'^\mu - Q^\mu \right) = -Q_\mu Q'^\mu \,.$$

With $Q_\mu = (E/c)\,(-1, \hat{\boldsymbol{n}})$ and $P_\mu = M\gamma c\,(-1, \boldsymbol{\beta})$, we find

$$E' = \frac{E(1 - \boldsymbol{\beta} \cdot \hat{\boldsymbol{n}})}{1 - \boldsymbol{\beta} \cdot \hat{\boldsymbol{n}}' + \dfrac{E}{M\gamma c^2}(1 - \hat{\boldsymbol{n}} \cdot \hat{\boldsymbol{n}}')} \approx E\frac{1 - \boldsymbol{\beta} \cdot \hat{\boldsymbol{n}}}{1 - \boldsymbol{\beta} \cdot \hat{\boldsymbol{n}}'} = E\frac{1 + \beta\cos\theta}{1 - \beta\cos\theta'}$$

$$\approx E\left(1 + \beta(\cos\theta + \cos\theta') + O(\beta)^2\right) \,, \tag{5.20}$$

where the photon energy relative to the mass energy of the mirror can be ignored, and we've used $\beta \ll 1$ in the final step. Equation (5.20) differs from Eq. (5.19); they agree at lowest order in β if we take $\theta' = \theta$. The angle θ' *must* differ from θ, however: In changing the energy of the reflected photon, one changes its momentum as well, both in direction and magnitude. By analyzing four-momentum conservation, to the extent that $E/(Mc^2)$ can be ignored, it can be shown that $\sin\theta' = \sin\theta - \beta\sin(\theta + \theta')$. At lowest order in β we can set $\theta' = \theta$ in Eq. (5.20).

The amount of radiant energy[17] of frequency ν, $\delta E(\nu)$, impinging on a small area $\mathrm{d}\sigma$ of the piston in the next $\mathrm{d}t$ seconds from the direction of $\mathrm{d}\Omega$ is (using the flux-tube method)

$$\delta E(\nu) = \frac{\mathrm{d}\Omega}{4\pi} \cdot c\mathrm{d}t \cdot \mathrm{d}\sigma\cos\theta \cdot u(\nu, T)\mathrm{d}\nu \,, \tag{5.21}$$

where $u(\nu, T)$ is the energy spectral density as a function of frequency, defined such that $|u(\nu, T)\mathrm{d}\nu| = |u(\lambda, T)\mathrm{d}\lambda|$. The momentum of the energy in the tube ($\delta p \equiv \delta E/c$) exerts an infinitesimal force $\mathrm{d}F$ on $\mathrm{d}\sigma$: (see Eq. (5.5))

$$\mathrm{d}F = \frac{\Delta p}{\mathrm{d}t} = \frac{2\delta p\cos\theta}{\mathrm{d}t} = \frac{2\cos\theta}{c}\frac{\delta E(\nu)}{\mathrm{d}t}.$$

The infinitesimal work done in time $\mathrm{d}t$ on $\mathrm{d}\sigma$ by the piston in overcoming $\mathrm{d}F$ at speed v is then

$$\mathrm{d}W = \mathrm{d}Fv\mathrm{d}t = 2\frac{v}{c}\cos\theta\delta E(\nu).$$

By the first law, adiabatic work[18] $\mathrm{d}W$ must appear in the energy of the reflected radiation $\delta E(\nu')$, where ν' is the Doppler-shifted frequency, Eq. (5.19). Energy conservation requires

$$\delta E(\nu) + \mathrm{d}W = \left(1 + 2\frac{v}{c}\cos\theta\right)\delta E(\nu) = \delta E(\nu') \,.$$

What's the *net change* in radiant energy at a given frequency? There's a Doppler shift *out of* $(\nu, \nu + \mathrm{d}\nu)$ upon reflection from the moving piston, implying there are frequencies $\nu_1 < \nu$ Doppler

---

[17]The symbol E here denotes energy, not emissivity.
[18]We're assuming the compression occurs under adiabatic conditions.

shifted *into* $(\nu, \nu + d\nu)$ (see Fig. 5.3). The energy *lost* from $(\nu, \nu + d\nu)$ in time dt is found by integrating $\delta E(\nu)$ over solid angle and the area A of the piston. Using Eq. (5.21):

$$dU_{\text{loss}}(\nu) \equiv \frac{c}{4\pi} A dt u(\nu) d\nu \int_0^{2\pi} d\phi \int_0^{\pi/2} \sin\theta \cos\theta d\theta = \frac{c}{4} A u(\nu) d\nu dt \,. \tag{5.22}$$

(Equation (5.22) is consistent with Eq. (5.3) if integrated over all frequencies.) Frequencies $\nu_1(\theta)$

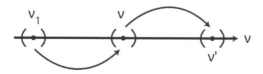

Figure 5.3   Frequencies scattered into and out of $(\nu, \nu + d\nu)$

are Doppler shifted into ν if, from Eq. (5.19),

$$\nu = \nu_1(\theta) \left(1 + 2\frac{v}{c}\cos\theta\right) \,. \tag{5.23}$$

For $v \ll c$, $\nu_1(\theta)$ is restricted to the interval $\nu\,(1 - 2v\cos\theta/c) \le \nu_1(\theta) \le \nu$. The energy *gained* in $(\nu, \nu + d\nu)$ in time dt is given by the integral

$$dU_{\text{gain}}(\nu) = \frac{c}{4\pi} A dt d\nu \int \cos\theta u(\nu_1(\theta)) d\Omega \,. \tag{5.24}$$

For small v, $2\nu\cos\theta v/c$ is small so that a first-order Taylor expansion is accurate,

$$u(\nu_1(\theta)) \approx u(\nu) + (\nu_1(\theta) - \nu)\frac{\partial u(\nu)}{\partial \nu} \approx u(\nu) - 2\frac{v}{c}\nu\cos\theta\frac{\partial u(\nu)}{\partial \nu} \,, \tag{5.25}$$

where we've used Eq. (5.23). Combining Eq. (5.25) with Eq. (5.24),

$$\begin{aligned}
dU_{\text{gain}}(\nu) &= \frac{c}{4\pi} A dt d\nu \int_0^{2\pi} d\phi \int_0^{\pi/2} d\theta \sin\theta \cos\theta \left[u(\nu) - 2\frac{v}{c}\nu\cos\theta\frac{\partial u(\nu)}{\partial \nu}\right] \\
&= \frac{c}{4} A dt d\nu \left[u(\nu) - \frac{4}{3}\frac{v}{c}\nu\frac{\partial u(\nu)}{\partial \nu}\right] \,.
\end{aligned} \tag{5.26}$$

Subtracting Eq. (5.22) from Eq. (5.26), the net change in energy in $(\nu, \nu + d\nu)$ as a result of an adiabatic compression is

$$\Delta U(\nu) \equiv dU_{\text{gain}}(\nu) - dU_{\text{loss}}(\nu) = -\frac{1}{3} A v dt \nu \frac{\partial u(\nu)}{\partial \nu} d\nu = \frac{1}{3} dV \nu \frac{\partial u(\nu)}{\partial \nu} d\nu \,, \tag{5.27}$$

where the change in volume in time dt is $dV = -A v dt$. Because the sign of dV is linked to the sign of v, Eq. (5.27) holds for reversible processes.

The energy change $\Delta U(\nu)$ implies a change in the energy spectral density,

$$d(u(\nu)V) d\nu = \Delta U(\nu) = u(\nu) dV d\nu + V du(\nu) d\nu = \frac{1}{3} dV \nu \frac{\partial u(\nu)}{\partial \nu} d\nu \,,$$

implying that

$$V du(\nu) = dV \left(\frac{\nu}{3}\frac{\partial u(\nu)}{\partial \nu} - u(\nu)\right) \,. \tag{5.28}$$

Equation (5.28) specifies the change in $u(\nu)$ with a change in volume under adiabatic conditions; it's therefore equivalent to the derivative

$$V\left(\frac{\partial u(\nu)}{\partial V}\right)_S = \frac{\nu}{3}\frac{\partial u(\nu)}{\partial \nu} - u(\nu) . \qquad (5.29)$$

Equation (5.29) is a partial differential equation for u as a function of ν and V; *it's the central result in the derivation of Wien's law.*

As can be shown, the solution of Eq. (5.29) can be written in the form[19]

$$u(\nu) = \nu^3 \phi(V\nu^3) , \qquad (5.30)$$

where ϕ is *any function* of a single variable, *Wien's law*. The "trouble" with Eq. (5.29) is that, being based on the Doppler shift and energy conservation, it applies to the spectral distribution of *any* system of radiant energy, not just thermal radiation. Equation (5.30), however, places a constraint on the form that Kirchhoff's function must have.

What do we know about the derivative on the left side of Eq. (5.29)? From Exercise 3.10, we have the general thermodynamic identity

$$\left(\frac{\partial U}{\partial V}\right)_S = \left(\frac{\partial U}{\partial V}\right)_T - T\left(\frac{\partial P}{\partial T}\right)_V . \qquad (5.31)$$

By writing $U = uV$, Eq. (5.31) implies the result *specific to cavity radiation*,

$$\left(\frac{\partial u}{\partial V}\right)_S = -\frac{T}{V}\left(\frac{\partial P}{\partial T}\right)_V = -\frac{4}{3}\frac{u}{V} , \qquad (5.32)$$

where we've used Eqs. (5.4), (5.10), and (5.11). Thus, while the energy density of cavity radiation is independent of volume under *isothermal* conditions, *it's not for adiabatic processes* (temperature changes in adiabatic processes). By integrating Eq. (5.29) over all frequencies ($u = \int_0^\infty u(\nu)\mathrm{d}\nu$),

$$\left(\frac{\partial u}{\partial V}\right)_S = \frac{1}{3V}\int_0^\infty \nu\frac{\partial u(\nu)}{\partial \nu}\mathrm{d}\nu - \frac{u}{V} = \frac{1}{3V}\left[\nu u(\nu)\right]\Big|_0^\infty - \frac{4}{3}\frac{u}{V} , \qquad (5.33)$$

where we've integrated by parts. Consistency between Eqs. (5.32) and (5.33) requires that

$$\lim_{\nu\to\infty} \nu u(\nu) = 0 . \qquad (5.34)$$

Note that Eq. (5.34) is a consequence of *classical* physics.

We can now show directly that cavity radiation retains its spectral distribution under reversible adiabatic processes. In isentropic processes VT^3 = constant (Eq. (5.13)). Equation (5.30) can therefore be written in the form *for cavity radiation*

$$u(\nu, T) = \nu^3 \phi\left(\frac{\nu^3}{T^3}\right) \equiv \nu^3 \psi\left(\frac{\nu}{T}\right) , \qquad (5.35)$$

where ψ is another function of a single variable. If every dimension of the cavity is expanded uniformly, the wavelength of every mode of electromagnetic oscillation would increase in proportion.[20] For a length $L \equiv V^{1/3}$ associated with the cavity, every wavelength λ scales with L, $\lambda \sim L = V^{1/3}$, and because $TV^{1/3}$ is constant, we have that λT = constant in isentropic processes, or, equivalently

---

[19]We stress that Eq. (5.30) applies for isentropic processes.
[20]That is, there would be a "redshift."

$\nu/T = $ constant. From Eq. (5.8), $\int_0^\infty u(\nu, T)d\nu = aT^4$. Assume that in an isentropic expansion the radiation temperature changes from $T_1 \to T_2$. Then we have the equality

$$\frac{1}{T_1^4} \int_0^\infty u(\nu, T_1)d\nu = \frac{1}{T_2^4} \int_0^\infty u(\nu', T_2)d\nu' . \tag{5.36}$$

Under the expansion, for every frequency ν for the system at T_1, there's an associated ("displaced") frequency $\nu' = (T_2/T_1)\nu$ for the system at T_2 (because $\nu/T_1 = \nu'/T_2$). Hence, $d\nu' = (T_2/T_1)d\nu$. The equality of the integrands in Eq. (5.36) implies

$$u(\nu', T_2) = \left(\frac{T_2}{T_1}\right)^3 u(\nu, T_1) . \tag{5.37}$$

Now make use of Eq. (5.35) in Eq. (5.37); we find (using $\nu/T_1 = \nu'/T_2$)

$$\psi\left(\frac{\nu'}{T_2}\right) = \psi\left(\frac{\nu}{T_1}\right) . \tag{5.38}$$

Equation (5.38) shows that the spectral energy density function remains invariant in isentropic processes; one has to "displace" the frequencies $\nu \to \nu' = (T_2/T_1)\nu$.

It's convenient to express the spectral energy density in terms of the wavelength. Using $u(\lambda)d\lambda = u(\nu)d\nu$ along with Eq. (5.35) and $c = \lambda\nu$,

$$u(\lambda, T) = \left|\frac{d\nu}{d\lambda}\right| u(\nu) = \frac{c^4}{\lambda^5}\psi\left(\frac{c}{\lambda T}\right) \equiv \lambda^{-5} f(\lambda T) , \tag{5.39}$$

where f is yet another unknown (but universal!) function of a single variable. The Planck distribution has the *very form predicted by Wien's law* ($\lambda^{-5}f(\lambda T)$):

$$u(\lambda, T) = \frac{8\pi hc}{\lambda^5}\left(\exp\left(\frac{hc}{\lambda k_B T}\right) - 1\right)^{-1} . \tag{5.40}$$

It's clear that $u(\nu) \to 0$ as $\nu \to 0$ (as $\nu \to 0$, $\lambda \to \infty$, and a cavity of finite size cannot support an infinite-wavelength mode of electromagnetic vibration). From Eq. (5.34), $u(\nu) \to 0$ as $\nu \to \infty$. Mathematically, then, the energy density function must have a maximum (Rolle's theorem). Suppose the function $f(\lambda T)$ in Eq. (5.39) has been found. Its maximum with respect to λ is obtained by satisfying the condition $xf'(x) = 5f(x)$, where $x \equiv \lambda T$. The wavelength where maximum energy density occurs (or maximum emissivity) is

$$\lambda_{\max} = \frac{1}{T}\left(\frac{5f}{f'}\right)_{\max} \equiv \frac{b}{T} . \tag{5.41}$$

The quantity b is known as *Wien's constant*; it has the value $b = 2.898 \times 10^{-3}$ m K.

Example. What's the wavelength of maximum emission associated with the black-body temperature of the sun, $T = 5780$ K? Using Eq. (5.41), we find $\lambda_{\max} = 500$ nm, in the green portion of the spectrum, near the peak sensitivity of the human eye.

5.4 COSMIC MICROWAVE BACKGROUND

It's not out of place to discuss the CMB (mentioned in Section 4.8). In 1964, it was found that the universe is permeated by electromagnetic radiation that's not associated with any star or galaxy, that's unpolarized and highly isotropic, what we'd expect of cavity radiation.[21] Is the universe one big "cavity"? The measured spectral energy density of the CMB is found to occur as a Planck distribution associated with absolute temperature $T = 2.7260 \pm 0.0013$ K. The universe thus contains a population of photons in thermal equilibrium at $T = 2.726$ K. Why?

The CMB is taken to be a *thermal relic* of a universe that was at one time considerably hotter than $T = 2.73$ K. By the *Big Bang* theory, with a universe initially in a hot, dense state, photons are scattered by charged particles (Thomson scattering), particularly free electrons, providing an efficient mechanism for establishing thermal equilibrium[22] because the expansion rate is less than the speed of light, $\dot{R} < c$ (R is the scale length of the universe). If hotter conditions prevail at one region of the early universe than in another, photons act to equalize temperatures spatially. In an adiabatic expansion of a photon gas (the universe is a closed system, right?), $VT^3 = $ constant, Eq. (5.13), and thus the temperature of the photon population decreases as the universe expands, $T \propto R^{-1}(t)$. When the temperature cools to approximately 3000 K, neutral atoms can form and stay in that state, stable against disruption by photons. At that point, photons are said to *decouple* from matter: Neutral species are significantly less effective in scattering photons. By photons having been in thermal equilibrium at the time of decoupling, they maintain their black-body spectrum as the universe keeps adiabatically expanding.

CHAPTER SUMMARY

This chapter introduced the application of thermodynamics to electromagnetic radiation that's in equilibrium with matter, what's called cavity radiation or the closely-related black body radiation.

- Cavity radiation is radiant energy in an enclosure bounded by thick opaque walls at a uniform temperature. Its properties are independent of the specifics of the cavity and depend only on the temperature of the walls.

- A black body absorbs incident energy without reflection. A black body can very nearly be realized by a hole connected to a cavity.

- The total emissive power $E(T)$ (integrated over all wavelengths) of a black body is related to the energy density $u(T)$ of cavity radiation by $E(T) = cu(T)/4$. The quantity $u(T) = aT^4$ where a is the radiation constant and $E(T) = \sigma T^4$, where $\sigma = ac/4$ is Stefan's constant. The pressure $P(T) = \frac{1}{3}u(T)$ for cavity radiation.

- The Gibbs energy for cavity radiation is zero, implying that the chemical potential $\mu = 0$. Chemical potential is not really defined for photons, because they are not conserved. The assignment $\mu = 0$ is consistent with the other thermodynamic properties of the photon gas.

- For cavity radiation the energy density is independent of the volume at constant temperature, $(\partial u/\partial V)_T = 0$. The total radiant energy is $U = aVT^4$, where V is the volume of the cavity.

- In reversible adiabatic processes involving cavity radiation, VT^3 is a constant. Cavity radiation retains the spectral distribution of a black body in such processes, even though the temperature changes.

---

[21] The 1978 Nobel Prize in Physics was award to A.A. Penzias and R.W. Wilson for the discovery of the CMB radiation.
[22] With free charges about, the universe has "opaque walls," just what we need for cavity radiation!

- The energy spectral density is a function of a single variable, λT, $u(\lambda, T) = \lambda^{-5} f(\lambda T)$, where f is a universal function (Wien's law). The wavelength where maximum emissivity occurs, $\lambda_{\max} = b/T$ where b is Wien's constant.

EXERCISES

5.1 Derive Eq. (5.18). Derive Eq. (5.32).

5.2 Verify that Eq. (5.30) satisfies Eq. (5.29).

5.3 Show that $u(\nu) = \dfrac{1}{V}\phi(V\nu^3)$ also satisfies Eq. (5.29).

5.4 What is C_P for cavity radiation? Hint: Consider $C_P = (\partial H/\partial T)_P$. Likewise, for cavity radiation what are the expansivity and isothermal compressibility, Eq. (1.20)? In general, what is $(\partial f/\partial x)_x$?

5.5 Assume that the entropy S of cavity radiation is proportional to the average number of photons in the cavity, N, $S = \alpha N$, where α is a constant. (The constant α turns out to be $\approx 3.6 k_B$, but that's not needed here.) Show that

  a. In an isothermal expansion of cavity radiation the fractional change in photon number is equal to the fractional change in volume

  $$\frac{\Delta N}{N} = \frac{\Delta V}{V}.$$

  b. In an isothermal expansion the fractional change in entropy is equal to the fractional change in volume

  $$\frac{\Delta S}{S} = \frac{\Delta V}{V}.$$

5.6 Consider an enclosure (cavity) in which the temperature is uniform throughout. Would an observer in such an enclosure "see" anything at all, such as objects within the enclosure, even if the objects have different reflectances and colors? Would one see shadows? Read the first part of Section 5.1.

5.7 a. Show using the Planck distribution, that Wien's constant $b \sim hc/k_B$ (see Eq. (5.41)).

  b. Show that the energy of the photon associated with the maximum spectral emissivity is given by the expression

  $$E_{\max} = \left(\frac{hc}{b}\right) T \approx 4.97 k_B T.$$

  c. Show that you can estimate the number of photons simply by dividing the total energy aVT^4 by the energy of a photon which is proportional to $k_B T$. Compare with Eq. (5.15).

  d. Define the *average distance* between photons in cavity radiation as $l \equiv (V/N)^{1/3}$. Using Eq. (5.15), show that $l = 1.6 \hbar c/(k_B T)$.

5.8 Show that the entropy of cavity radiation when expressed in terms of its natural variables is $S(U, V) = (4/3)U^{3/4}(aV)^{1/4}$.

5.9 A fun exercise (if you think such things are fun) is to calculate the Joule expansion coefficient for cavity radiation. Show that $\mu_J = -T/(4V)$. This is a formula in search of an experiment, however. There is no vacuum that cavity radiation can expand into that doesn't already have cavity radiation.

CHAPTER 6

Phase and chemical equilibrium

A *phase*, as the term is used in thermodynamics, is a spatially homogeneous material body in a state of equilibrium[1] characterized by certain parameters such as T and P. A body of given chemical composition can exist in a *number* of possible phases. For example, H_2O can exist in the liquid and vapor phases, as well as several forms of ice having different crystal structures.[2] Substances can undergo *phase transitions*, abrupt changes in phase that occur upon variations of state variables. We do not cover phase transitions in this book.[3] Different phases of the same substance can *coexist* in physical contact under certain conditions. Is there a limit to the number of phases that can coexist? An elegant answer is provided by the *Gibbs phase rule*, Eq. (6.20). First, however, time for a math refresher.

6.1   LAGRANGE MULTIPLIERS*

We're often presented with the problem of finding the extremum of a multivariable function when there are relations (*constraints*) among the variables. There is a well known method for handling such problems—*Lagrange's method of undetermined multipliers*—which we review in this section.

A function of N variables $f(x_1, \cdots, x_N)$ has an extremum where $\mathrm{d}f = 0$, i.e., where

$$\mathrm{d}f = \sum_{j=1}^{N} \frac{\partial f}{\partial x_j} \mathrm{d}x_j = 0 \,. \tag{6.1}$$

If the coordinates x_j can be varied independently, Eq. (6.1) is satisfied by finding the solutions of the N equations, $\partial f / \partial x_j = 0$, $j = 1, \cdots, N$. If, however, there are $n < N$ equations of constraint among the coordinates, $\alpha_i(x_1, \cdots, x_N) = 0$, $i = 1, \cdots, n$, the differentials $\mathrm{d}x_j$ are not independent: They're related through the differentials of the equations of constraint

$$\mathrm{d}\alpha_i = \sum_{j=1}^{N} \frac{\partial \alpha_i}{\partial x_j} \mathrm{d}x_j = 0 \,. \qquad i = 1, \cdots, n \tag{6.2}$$

In that case, Eq. (6.1) can't be satisfied by simply requiring $\partial f / \partial x_j = 0$.

---

[1]The existence of phases is implied by the existence of equilibrium. Systems maintain states of equilibrium when kept under fixed conditions (Chapter 1). As phase diagrams show, phases are consistent with a range of thermodynamic parameters. A phase is thus a superset of equilibrium states. States imply phases.

[2]Note that there *is* a phase of ice called Ice IX. Ice-nine appeared as a fictional material in Kurt Vonnegut's novel *Cat's Cradle* that would have the power to destroy all life on Earth.

[3]The topic of phase transitions can be more thoroughly treated in books on statistical mechanics.

You wouldn't mind if we multiplied each of the equations in Eq. (6.2) by a parameter λ_i, the *Lagrange multiplier*, so that

$$\lambda_i \sum_{j=1}^{N} \frac{\partial \alpha_i}{\partial x_j} \mathrm{d}x_j = 0. \qquad i = 1, \cdots, n \qquad (6.3)$$

Form the sum of the n equations implied by Eq. (6.3) (which is still equal to zero), combine with Eq. (6.1), and group together terms with common factors of $\mathrm{d}x_j$:

$$\sum_{j=1}^{N} \mathrm{d}x_j \left[\frac{\partial f}{\partial x_j} + \sum_{i=1}^{n} \lambda_i \frac{\partial \alpha_i}{\partial x_j} \right] = 0. \qquad (6.4)$$

Can we set the terms in square brackets to zero? While that's what we're going to do, we don't "know that" yet. There are N variables and n constraints; thus only $N - n$ variables can considered independent. It's immaterial which are taken as independent; we can choose any $N - n$ of them as independent. We can eliminate the differentials of the n dependent variables in Eq. (6.4) by choosing

$$\frac{\partial f}{\partial x_j} + \sum_{k=1}^{n} \lambda_k \frac{\partial \alpha_k}{\partial x_j} = 0. \qquad j = 1, \cdots, n \qquad (6.5)$$

Equation (6.5) gives us n equations in n unknowns; the Lagrange multipliers $\{\lambda_k\}_{k=1}^{n}$ can therefore be considered known. The remaining $N - n$ differentials in Eq. (6.4) can then be varied independently, implying that the coefficient of each $\mathrm{d}x_j$ is zero,

$$\frac{\partial f}{\partial x_j} + \sum_{k=1}^{n} \lambda_k \frac{\partial \alpha_k}{\partial x_j} = 0. \qquad j = n+1, \cdots, N \qquad (6.6)$$

Equations (6.5) and (6.6) have exactly the same form and can be combined into one equation, that which would be obtained by setting the terms in square brackets in Eq. (6.4) to zero. The specification of which variables are independent is moot; *all* of the coefficients of the $\mathrm{d}x_j$ in Eq. (6.4) can be taken as zero. It's *as if* we've made the replacement $f \to f + \sum_{i=1}^{n} \lambda_i \alpha_i$ in the function we seek the extremum of, with all variables now considered independent.

Example. Find the maximum of $f(x, y) = x + y$ subject to the constraint $x^2 + y^2 = 1$, i.e., x and y must lie on the unit circle. Form the auxiliary function $g(x, y) \equiv x + y + \lambda(x^2 + y^2 - 1)$ where λ is the Lagrange multiplier. Find the maximum of $g(x, y)$ as if x and y are independent variables. From $\partial g / \partial x = 1 + 2\lambda x = 0$ and $\partial g / \partial y = 1 + 2\lambda y = 0$, the extremum occurs at $x = y = -1/(2\lambda)$. What's the value of λ? Substitute into the equation of constraint: $x^2 + y^2 = 1 \Rightarrow \lambda^2 = 1/2$. The maximum of f occurs at $x = y = 1/\sqrt{2}$, with $f_{max} = \sqrt{2}$.

6.2   PHASE COEXISTENCE

Equilibrium conditions

In this section we show that the chemical potential has the same value in each of the phases in which coexistence can occur—see Eq. (6.19). Consider two phases of a substance, I and II. Because matter and energy can be exchanged between phases in contact, phase coexistence is achieved when T and P are the same in both phases, *and when the chemical potentials are equal*, $\mu^I = \mu^{II}$ (Section 3.10). From Eq. (3.32), $\mu = \mu(T, P)$. The chemical potential can therefore be visualized as a surface $\mu = \mu(T, P)$ (see Fig. 6.1). Two phases of the same substance therefore coexist when

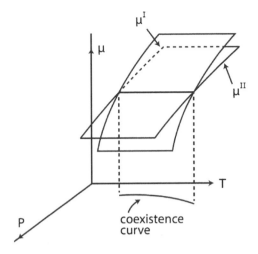

Figure 6.1 Coexistence curve defined by intersection of chemical potential functions

$$\mu^I(T,P) = \mu^{II}(T,P) \,. \tag{6.7}$$

The intersection of the two surfaces defines the locus of points $P = P(T)$ for which Eq. (6.7) is satisfied, the *coexistence curve*. Three phases (I, II, III) coexisting in equilibrium would require the equality of three chemical potential functions,

$$\mu^I(T,P) = \mu^{II}(T,P) = \mu^{III}(T,P) \,. \tag{6.8}$$

In Eq. (6.8) we have two equations in two unknowns; three phases can therefore coexist at a unique combination of T and P known as a *triple point*.[4] By this reasoning, it would not be possible for four phases of a single substance to coexist in equilibrium (what would require three equations in two unknowns). Coexistence of four phases of the same substance is not known to occur in nature.

  Multicomponent phases have more than one chemical species.[5] To analyze multiphase, multicomponent systems, we adopt the notation where μ_j^γ denotes the chemical potential of species j in the γ phase (we use Roman letters to label species and Greek letters to label phases). Let there be k chemical species, $1 \le j \le k$, and π phases, $1 \le \gamma \le \pi$, where π here is an integer. The first law of thermodynamics for multiphase, multicomponent open systems is the generalization of Eq. (3.16):

$$\mathrm{d}U = T\mathrm{d}S - P\mathrm{d}V + \sum_{\gamma=1}^{\pi}\sum_{j=1}^{k}\mu_j^\gamma \mathrm{d}N_j^\gamma \,. \tag{6.9}$$

The extensivity of internal energy implies the scaling property $U(\lambda S, \lambda V, \lambda N_j^\gamma) = \lambda U(S, V, N_j^\gamma)$. By Euler's theorem, therefore, Eq. (3.19),

$$
\begin{aligned}
U &= S\left(\frac{\partial U}{\partial S}\right)_{V,N_j^\gamma} + V\left(\frac{\partial U}{\partial V}\right)_{S,N_j^\gamma} + \sum_{\gamma=1}^{\pi}\sum_{j=1}^{k} N_j^\gamma \left(\frac{\partial U}{\partial N_j^\gamma}\right)_{S,V} \\
&= TS - PV + \sum_{\gamma=1}^{\pi}\sum_{j=1}^{k}\mu_j^\gamma N_j^\gamma \,,
\end{aligned}
\tag{6.10}
$$

---

[4]The triple point of H_2O is used in the definition of the Kelvin temperature scale, Section 2.4.
[5]Mentioned briefly in Section 3.5.

where the derivatives are obtained from Eq. (6.9), with $\mu_j^\gamma \equiv \left(\partial U / \partial N_j^\gamma\right)_{S,V,\overline{N_j^\gamma}}$. The notation $\overline{N_j^\gamma}$ indicates to hold fixed all particle numbers *except* N_j^γ. Equation (6.10) is the generalization of the Euler formula, Eq. (3.31). Applying $G = U - TS + PV$ (Gibbs energy) to Eq. (6.10),

$$G = \sum_{\gamma=1}^{\pi} \sum_{j=1}^{k} \mu_j^\gamma N_j^\gamma \,, \tag{6.11}$$

which generalizes Eq. (4.9). Taking the differential of Eq. (6.10) and using Eq. (6.9), we have the multicomponent, multiphase generalization of the Gibbs-Duhem equation, Eq. (3.32),

$$\sum_{\gamma=1}^{\pi} \sum_{j=1}^{k} N_j^\gamma \mathrm{d}\mu_j^\gamma = -S\mathrm{d}T + V\mathrm{d}P \,. \tag{6.12}$$

By taking the differential of G in Eq. (6.11), and making use of Eq. (6.12),

$$\mathrm{d}G = -S\mathrm{d}T + V\mathrm{d}P + \sum_{\gamma=1}^{\pi} \sum_{j=1}^{k} \mu_j^\gamma \mathrm{d}N_j^\gamma \,, \tag{6.13}$$

and thus the alternate definition of chemical potential,[6]

$$\mu_j^\gamma = \left(\frac{\partial G}{\partial N_j^\gamma}\right)_{T,P,\overline{N_j^\gamma}} \,, \tag{6.14}$$

the energy to add a particle of type j in phase γ holding fixed T, P, and the other particle numbers. The Gibbs energy is a minimum in equilibrium.[7] From Eq. (6.13),

$$[\mathrm{d}G]_{T,P} = \sum_{\gamma=1}^{\pi} \sum_{j=1}^{k} \mu_j^\gamma \left[\mathrm{d}N_j^\gamma\right]_{T,P} = 0 \,. \tag{6.15}$$

If the particle numbers N_j^γ were independent and could be varied freely, one would conclude from Eq. (6.15) that $\mu_j^\gamma = 0$. But the particle numbers are not independent. The number of particles of each species spread among the phases is a constant,[8] $\sum_\gamma N_j^\gamma = \text{constant}$, and thus

$$\sum_{\gamma=1}^{\pi} \mathrm{d}N_j^\gamma = 0 \,. \qquad\qquad j = 1, \cdots, k \tag{6.16}$$

We can incorporate these constraints by introducing Lagrange multipliers λ_j, which when multiplied by Eq. (6.16) and added to Eq. (6.15) leads to

$$\sum_{j=1}^{k} \sum_{\gamma=1}^{\pi} \left(\mu_j^\gamma + \lambda_j\right) \mathrm{d}N_j^\gamma = 0 \,. \tag{6.17}$$

---

[6]See Section 4.6.

[7]A stability analysis of a multicomponent system (requiring entropy to be a maximum in equilibrium), as in Section 3.10, would show that $C_V > 0$ and $\beta > 0$, just as for a single-component phase. The requirement $\partial\mu / \partial N > 0$, Eq. (3.51), is generalized so that the *matrix* of derivatives $\mu_{j,k} \equiv \partial\mu_j / \partial N_k$ is positive definite, with $\sum_{j,k} \mu_{j,k} \delta N_j \delta N_k > 0$. The matrix is symmetric, $\mu_{j,k} = \mu_{k,j}$, which could be considered a Maxwell relation. From the stability condition on S follows that H, F, and G are a minimum in equilibrium.

[8]We're not allowing chemical reactions to take place.

We can now treat the particle numbers as unconstrained, so that Eq. (6.17) implies

$$\mu_j^\gamma = -\lambda_j . \tag{6.18}$$

For coexisting phases in equilibrium ($[dG]_{T,P} = 0$), the chemical potential of each species is *independent of phase*. Equation (6.18) is equivalent to

$$\mu_j^1 = \mu_j^2 = \cdots = \mu_j^\pi . \qquad\qquad j = 1, \cdots, k \tag{6.19}$$

There are $k(\pi - 1)$ equations of equilibrium for k chemical components in π phases.

Gibbs phase rule

How many independent state variables exist in a multicomponent, multiphase system? In each phase there are $N^\gamma \equiv \sum_{j=1}^k N_j^\gamma$ particles, and thus there are $k-1$ independent *concentrations*[9] $c_j^\gamma \equiv N_j^\gamma /$ N^γ, where $\sum_{j=1}^k c_j^\gamma = 1$. Among π phases there are $\pi(k-1)$ independent concentrations. Including P and T there are $2 + \pi(k - 1)$ independent intensive variables.

There are $k(\pi - 1)$ equations of equilibrium, Eq. (6.19). The *variance* of the system, f, is the difference between the number of independent variables and the number of equations of equilibrium,

$$f \equiv 2 + \pi(k - 1) - k(\pi - 1) = 2 + k - \pi . \tag{6.20}$$

Equation (6.20) is the Gibbs phase rule.[3, p96] It specifies the number of intensive variables that can be independently varied without disturbing the number of coexisting phases in equilibrium ($f \geq 0$).

- $k = 1, \pi = 1 \Rightarrow f = 2$: a single substance in one phase. Two intensive variables can be independently varied; T and P in a gas.

- $k = 2, \pi = 1 \Rightarrow f = 3$: two substances in a single phase, as in a mixture of gases. We can independently vary T, P, and one mole fraction.

- $k = 1, \pi = 2 \Rightarrow f = 1$: a single substance in two phases; T or P can be varied along the coexistence curve.

- $k = 1, \pi = 3 \Rightarrow f = 0$: a single substance in three phases; we cannot vary the conditions under which three phases coexist in equilibrium. Unique values of T and P define a triple point.

The reader should appreciate the *generality* of the Gibbs phase rule, which does not depend on the *type* of chemical components, nor on any other physical properties of the system.

Clausius-Clapeyron equation

The *latent heat*, L, is the heat released or absorbed during a phase change. It's specified either as a *molar quantity* (per mole) or as a *specific quantity* (per mass). Two common forms of the latent heat are the *latent heat of vaporization* (boiling) and the *latent heat of fusion* (melting), the energy to transform a given quantity of a substance from one phase to another.[10] The heat of vaporization

---

[9]Ratios of extensive variables are independent of the size of the system; intensive quantities like concentrations are referred to as *densities*; Section 3.8.

[10]There is a latent heat of *sublimation*, where the solid phase converts directly to the gaseous phase, without going through the liquid state.

(or fusion) is also called the *enthalpy of vaporization* (or fusion) because measurements are made at fixed pressure.[11] At a given T,

$$L(T) = h^v - h^l = T\left[s^v(T, P(T)) - s^l(T, P(T))\right] , \qquad (6.21)$$

where v and l refer to vapor and liquid, lower-case quantities such as $s \equiv S/n$ indicate molar values, and $P(T)$ is the pressure along the coexistence curve.[12] The difference in molar entropy between phases is denoted $\Delta s \equiv s^v - s^l$; likewise with Δh. Equation (6.21) is written compactly as $L = \Delta h = T\Delta s$.

Equilibrium between two phases of a single substance requires the equality of chemical potentials, $\mu^I(T, P) = \mu^{II}(T, P)$, Eq. (6.7). As we move along the coexistence curve, T and P vary in such a way as to maintain this equality. Variations in T and P, however, induce changes in the chemical potential functions, $\delta\mu$, and thus along the coexistence curve $\delta\mu^I = \delta\mu^{II}$. From the Gibbs-Duhem equation (3.32), $\mathrm{d}\mu = -s\mathrm{d}T + v\mathrm{d}P$. Along the coexistence curve, therefore, $-s^I\mathrm{d}T + v^I\mathrm{d}P = -s^{II}\mathrm{d}T + v^{II}\mathrm{d}P$, implying that

$$\left(\frac{\mathrm{d}P}{\mathrm{d}T}\right)_{\text{coexist}} = \frac{s^I - s^{II}}{v^I - v^{II}} = \frac{L}{T\Delta v} , \qquad (6.22)$$

where $\Delta v \equiv v^I - v^{II}$ is the change in molar volume and $s^I - s^{II} = L/T$. Equation (6.22) is the *Clausius-Clapeyron equation*; it gives the slope of the coexistence curve. If the changes in molar entropy and volume between the phases are known, Eq. (6.22) can be integrated to give the coexistence curve.

Kirchhoff equation

For the case of the vapor phase of a substance in equilibrium with its liquid phase, we can derive a useful result for the temperature dependence of the latent heat, $L(T)$. Differentiate Eq. (6.21),

$$\frac{\mathrm{d}L}{\mathrm{d}T} = T\frac{\mathrm{d}\Delta s}{\mathrm{d}T} + \Delta s = T\frac{\mathrm{d}\Delta s}{\mathrm{d}T} + \frac{L}{T} . \qquad (6.23)$$

Along the coexistence curve, $s = s(T, P(T))$; the total derivative of Δs is therefore

$$\frac{\mathrm{d}\Delta s}{\mathrm{d}T} = \left(\frac{\partial\Delta s}{\partial T}\right)_P + \left(\frac{\partial\Delta s}{\partial P}\right)_T \frac{\mathrm{d}P(T)}{\mathrm{d}T} = \frac{1}{T}\Delta c_P - \left(\frac{\partial\Delta v}{\partial T}\right)_P \frac{\mathrm{d}P(T)}{\mathrm{d}T} \qquad (6.24)$$

$$= \frac{1}{T}\Delta c_P - \Delta(\alpha v)\frac{\mathrm{d}P(T)}{\mathrm{d}T} = \frac{1}{T}\Delta c_P - \frac{\Delta(\alpha v)}{\Delta v}\frac{L}{T} ,$$

where in the first line we've used Eq. (3.13) (with $\Delta c_P \equiv c_P^v - c_P^l$ the difference in molar heat capacities) and a Maxwell relation, and in the second line, Eq. (1.20) and Eq. (6.22). Now,

$$\frac{\Delta(\alpha v)}{\Delta v} \equiv \frac{\alpha^v v^v - \alpha^l v^l}{v^v - v^l} = \alpha^v \frac{\left(1 - (\alpha^l/\alpha^v)(v^l/v^v)\right)}{1 - (v^l/v^v)} \approx \alpha^v = T^{-1} , \qquad (6.25)$$

where we've used the fact that $v^v \gg v^l$ (approximately a factor of 10^3), $\alpha^l \ll \alpha^v$, and $\alpha^v = T^{-1}$ (Exercise 1.6). Combining Eqs. (6.25) and (6.24),

$$\frac{\mathrm{d}\Delta s}{\mathrm{d}T} = \frac{1}{T}\Delta c_P - \frac{L}{T^2} . \qquad (6.26)$$

Combining Eqs. (6.26) and (6.23), we have *Kirchhoff's equation* for the latent heat:

$$\frac{\mathrm{d}}{\mathrm{d}T}L(T) = \Delta c_P . \qquad (6.27)$$

---

[11]Enthalpy is the heat added at constant pressure; Chapter 4.

[12]Equation (6.21) has been written in terms of the liquid-gas transition, but the form of the equation applies for any transition in which two phases coexist in equilibrium.

6.3 THERMODYNAMICS OF MIXTURES: IDEAL SOLUTIONS

Volume is certainly an extensive variable; for a system consisting of a mixture of k chemical species, $V(\lambda N_1, \cdots, \lambda N_k) = \lambda V(N_1, \cdots, N_k)$. From Euler's theorem,

$$V = \sum_{i=1}^{k} N_i \left(\frac{\partial V}{\partial N_i} \right)_{T,P,\overline{N}_i} \equiv \sum_{i=1}^{k} N_i \overline{V}_i , \qquad (6.28)$$

where $\overline{V}_i$ is the *partial molar volume*. Equation (6.28) would seem to indicate that V is obtained by assembling, for each species, N_i moles of a substance that comes with molar volume $\overline{V}_i$. There's a subtle point here. The quantity $\overline{V}_i$ is the volume per mole of the ith substance *in solution*. The molar volume of the *pure substance* is denoted V_i^0, which is not the same as the molar volume in solution, $\overline{V}_i \neq V_i^0$. The difference

$$\Delta V \equiv V - V^0 = \sum_{i=1}^{k} N_i(\overline{V}_i - V_i^0)$$

is called the *volume of mixing*.

Example. When one mole of water is added to a large volume of water at 25 °C the volume increases by 18 cm^3. The molar volume of pure water V^0 is thus 18 cm^3. When one mole of water is added to a large volume of ethanol, the volume increases by 14 cm^3. The molar volume of water in ethanol $\overline{V}$ is 14 cm^3. The volume occupied by the water molecules *in solution* depends on what the surrounding molecules are. Try adding equal volumes of water and your favorite liquor: $V + V \neq 2V$.

Any extensive property can be expressed in terms of partial molar quantities:

$$H = \sum_i N_i \overline{H}_i \qquad U = \sum_i N_i \overline{U}_i \qquad G = \sum_i N_i \overline{G}_i \qquad F = \sum_i N_i \overline{F}_i \qquad S = \sum_i N_i \overline{S}_i ,$$

where for $X = (H, U, G, F, S)$, $\overline{X}_i \equiv (\partial X/\partial N_i)_{T,P,\overline{N}_i}$. The partial molar quantities for the thermodynamic potentials obey the same combination rules as do the potentials for the pure substances. For example,

$$G - H + TS = 0 \iff \sum_{i=1}^{k} N_i \left(\overline{G}_i - \overline{H}_i + T\overline{S}_i \right) = 0 ,$$

and hence

$$\overline{G}_i = \overline{H}_i - T\overline{S}_i . \qquad (6.29)$$

The entropy, enthalpy, and free energy of mixing are then defined by

$$\Delta S = \sum_i N_i(\overline{S}_i - S_i^0) \qquad \Delta H = \sum_i N_i(\overline{H}_i - H_i^0) \qquad \Delta G = \sum_i N_i(\overline{G}_i - G_i^0) .$$

Because of Eq. (6.29), we have the relation among the quantities of mixing,

$$\Delta G = \Delta H - T\Delta S . \qquad (6.30)$$

The partial molar Gibbs energy deserves special mention because from Eq. (6.14) it's the chemical potential, $\overline{G}_i = \mu_i = (\partial G/\partial N_i)_{T,P,\overline{N}_i}$.

Consider a system of two compartments separated by a rigid, but permeable barrier (see Fig. 6.2). One compartment contains pure hydrogen gas, H_2, and the other contains a mixture of hydrogen and nitrogen gas, $N_2 + H_2$. Palladium is *selectively permeable*: It's permeable to H_2 but not to

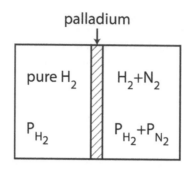

Figure 6.2 Selectively permeable membrane, passes H_2 but not N_2

N_2. Because the barrier is permeable to H_2, the hydrogen pressures on the two sides equalize[13] with the pressure of the pure hydrogen, $P_{H_2\,(pure)} \equiv P_0$, equal to the *partial pressure*[14] of hydrogen in the mixture, $P_{H_2\,(mix)} \equiv xP$, with the total pressure in the right compartment $P = P_{H_2\,(mix)} + P_{N_2}$ and x the hydrogen mole fraction. Thus, $P_0 = xP$. The chemical potentials of hydrogen will equalize, $\mu_{H_2\,(pure)}(T, P_0) = \mu_{H_2\,(mix)}(T, P)$. From Eq. (4.12), the chemical potential of an ideal gas has the form $\mu(T, P) = \mu(T, P_0) + RT \ln(P/P_0)$, where P_0 is a reference pressure. Thus, $\mu_{H_2\,(pure)}(T, P_0) = \mu_{H_2\,(mix)}(T, P) = \mu_{H_2\,(mix)}(T, P_0) + RT \ln(P/P_0)$. With $P_0 = xP$, we have $\mu_{H_2\,(mix)}(T, P) = \mu_{H_2\,(pure)}(T, P) + RT \ln x$, where we have erased the subscript on P_0. This argument could be repeated for a mixture of any number of ideal gases separated from the pure species i by a barrier selectively permeable to substance i. The chemical potential of an ideal gas in a mixture is therefore

$$\mu_i = \mu_i^0(T, P) + RT \ln x_i \,, \tag{6.31}$$

where $\mu_i^0(T, P)$ is the chemical potential of the pure species i. *The chemical potential of a gas in a mixture is less than that of the pure gas for the same total pressure.*[15]

Using Eq. (6.31) we immediately obtain the Gibbs energy of mixing for ideal gases,

$$\Delta G = \sum_{i=1}^{k} N_i \left(\mu_i - \mu_i^0 \right) = RT \sum_{i=1}^{k} N_i \ln x_i = NRT \sum_{i=1}^{k} x_i \ln x_i \,. \tag{6.32}$$

The entropy of mixing follows directly from Eq. (6.32). Using $S = -(\partial G/\partial T)_{P,N}$, it's straightforward to show, working through the definitions, that

$$\Delta S = -\left(\frac{\partial \Delta G}{\partial T} \right)_{P,N} = -NR \sum_{i=1}^{k} x_i \ln x_i \,. \tag{6.33}$$

Note from Eqs. (6.32) and (6.33) that $\Delta G = -T \Delta S$, implying from Eq. (6.30) that the enthalpy of mixing is zero, $\Delta H = \Delta G + T \Delta S = 0$. *There is no heat of mixing associated with the formation of a mixture of ideal gases.* We can also infer from Eq. (6.32) that the volume of mixing vanishes, $\Delta V = (\partial \Delta G/\partial P)_{T,x_i} = 0$. *Mixtures of ideal gases are formed without any volume of mixing.*[16]

Thus, for a mixture of ideal gases there is no volume of mixing, no enthalpy of mixing, and the entropy of mixing is given by Eq. (6.33). An *ideal solution* is defined as any mixture (solution) of liquids, gases, or solids that exhibits these properties.

---

[13] Subsystems come to equilibrium when their intensive variables can equalize, Section 3.10.

[14] Partial pressures are discussed in Section 1.6.

[15] Materials move from high to low values of μ (Section 3.11). Thus Eq. (6.31) makes sense: A system composed entirely of species i would diffuse into a mixture containing i.

[16] Ideal gases don't "know" what's around them!

6.4   LAW OF MASS ACTION

Consider a chemical reaction at constant T and P symbolized by the formula

$$\nu_A A + \nu_B B \rightarrow \nu_C C + \nu_D D \,, \tag{6.34}$$

where $A, \cdots, D$ are the chemical symbols of the substances and $\nu_A, \cdots, \nu_D$ are numbers known as the *stoichiometric coefficients*. If the Gibbs energy G decreases then the reaction proceeds spontaneously in the direction of the arrow, with the reaction proceeding until G reaches a minimum value. Conversely, if G *increases* in the direction shown, the reaction proceeds spontaneously in the opposite direction, again until G reaches a minimum value.

Equation (6.13) implies that, for a single phase,

$$[\mathrm{d}G]_{T,P} = \mu_A \mathrm{d}N_A + \mu_B \mathrm{d}N_B + \mu_C \mathrm{d}N_C + \mu_D \mathrm{d}N_D \,, \tag{6.35}$$

where the changes in mole numbers $\mathrm{d}N_A, \cdots, \mathrm{d}N_D$ are those resulting from the chemical reaction Eq. (6.34). One *unit of reaction* is said to have occurred when ν_A moles of A and ν_B moles of B have been consumed to form ν_C moles of C and ν_D moles of D. Let the reaction proceed by ξ reaction units; at this point the amount of the various substances is given by

$$N_A = N_A^0 - \nu_A \xi \quad N_B = N_B^0 - \nu_B \xi \quad N_C = N_C^0 + \nu_C \xi \quad N_D = N_D^0 + \nu_D \xi \,, \tag{6.36}$$

where N_i^0 are the number of moles of each substance before the reaction advances by ξ units. From Eq. (6.36),

$$\mathrm{d}N_A = -\nu_A \mathrm{d}\xi \quad \mathrm{d}N_B = -\nu_B \mathrm{d}\xi \quad \mathrm{d}N_C = \nu_C \mathrm{d}\xi \quad \mathrm{d}N_D = \nu_D \mathrm{d}\xi \,. \tag{6.37}$$

The variable ξ is called the *extent of reaction* or the *reaction coordinate*. Combining Eq. (6.37) with Eq. (6.35), $[\mathrm{d}G]_{T,P} = (-\nu_A \mu_A - \nu_B \mu_B + \nu_C \mu_C + \nu_D \mu_D)\, \mathrm{d}\xi$, from which

$$\left(\frac{\partial G}{\partial \xi} \right)_{T,P} = \nu_C \mu_C + \nu_D \mu_D - \nu_A \mu_A - \nu_B \mu_B \,. \tag{6.38}$$

If the derivative in Eq. (6.38) is negative, G decreases as the reaction progresses in the direction shown by the arrow; if the derivative is positive, the progress of the reaction would lead to an increase in G, and hence the reaction would spontaneously proceed in the other direction. The reaction is in equilibrium (rates in each direction are balanced) when the derivative vanishes,

$$\left(\frac{\partial G}{\partial \xi} \right)_{T,P,\mathrm{eq}} = 0 \,. \tag{6.39}$$

Equation (6.38) can be written more concisely. Instead of Eq. (6.34), write the chemical reaction as $\sum_i \nu_i M_i = 0$, where M_i is the chemical symbol of the ith species and ν_i is the stoichiometric coefficient, counted as positive if species i is produced in the reaction, and negative if it's consumed.

Example. Carbon monoxide and hydrogen react over a catalyst to produce methanol,

$$CO + 2H_2 \rightleftharpoons CH_3OH \,.$$

The stoichiometric coefficients are $\nu_{CO} = -1$, $\nu_{H_2} = -2$, and $\nu_{CH_3OH} = 1$.

Equation (6.38) can therefore be written as

$$\left(\frac{\partial G}{\partial \xi} \right)_{T,P} = \sum_i \nu_i \mu_i \,.$$

The quantity $\sum_i \nu_i \mu_i$ is called the *chemical affinity*; it plays the role of a driving force for chemical reactions. In equilibrium the affinity vanishes (same as Eq. (6.39)). Thus, there is a relation between the chemical potentials of the constituents in a chemical reaction in equilibrium:

$$\sum_i \nu_i \mu_i = 0 \,. \tag{6.40}$$

For an *ideal solution* we can use Eq. (6.31) to write the affinity

$$\sum_i \nu_i \mu_i = \sum_i \nu_i \left(\mu_i^0 + RT \ln x_i \right) = \sum_i \nu_i \mu_i^0 + RT \sum_i \ln x_i^{\nu_i} \equiv \Delta G^0 + RT \ln(\prod_i x_i^{\nu_i}) \,,$$

so that, in equilibrium, $(\sum_i \nu_i \mu_i = 0)$

$$\prod_i x_i^{\nu_i} = \exp\left(-\Delta G^0 / (RT) \right) \equiv K^*(T, P) \,, \tag{6.41}$$

where K^*, the *equilibrium constant* of the reaction, is a dimensionless function of T and P. The quantity $\sum_i \nu_i \mu_i^0 \equiv \Delta G^0$, termed the *free energy of reaction in the standard state*, is the change in Gibbs energy for one unit of reaction (if it could occur) among the constituents of the reaction *in their pure forms*. In dilute solutions the mole fractions are proportional to the *concentrations*, c_i, and Eq. (6.41) can be written

$$\prod_i c_i^{\nu_i} = K(T, P) \,, \tag{6.42}$$

where $K(T, P)$ is another equilibrium "constant" (having dimensions). Equation (6.42) is the usual form of the *law of mass action*.

Example. For the reaction of carbon monoxide and hydrogen producing methanol (previous example), we have, from Eq. (6.42), that the equilibrium concentrations are related by

$$\frac{c_{CH_3OH}}{c_{CO} \left[c_{H_2} \right]^2} = K(T, P) \,.$$

Example. In semiconductors, free conduction electrons e and holes h (the concentrations of which are denoted n and p) combine to create photons, $e + h \rightarrow \gamma$; likewise, electrons and holes are created in pairs by the absorption of photons, $\gamma \rightarrow e + h$. In equilibrium the rates of the two processes are the same, which can be modeled as a chemical reaction, $e + h \leftrightharpoons \gamma$. With $\nu_e = \nu_h = -1$ and $\nu_\gamma = 0$ (because $\mu = 0$ for photons we can take $\nu_\gamma = 0$), we have from Eq. (6.42) that $np = f(T, P)$, where f is an unknown function of T and P. In the *intrinsic* (undoped) semiconductor, electrons and holes are produced in pairs so that $n_i = p_i$ (subscript i for intrinsic). We can evaluate the function in the intrinsic system with $f(T, P) = n_i^2$. The law of mass action for electrons and holes is thus $np = n_i^2$, a widely-used formula in the theory of semiconductors.

6.5 ELECTROCHEMICAL CELLS

If one knows the temperature dependence of the heat capacity and the latent heat, one can calculate H and S (see Exercise 6.4), and if H and S are known, so are F and G. Is it possible to *measure* ΔG or ΔF, without recourse to heat capacity data? One method makes use of *electrochemical cells*, which we describe in this section. We can give only a cursory account of electrochemical cells here. A knowledge of electrochemical cells (however meager) will help us with the third law of thermodynamics in Chapter 8.

An electrochemical cell has two conductive electrodes, an *anode* and a *cathode* that are immersed in electrolytic solutions containing ions that can move freely. The anode is by definition the electrode that *loses* electrons by chemical reactions (*oxidation*) and the cathode is the electrode that *gains* electrons by chemical reactions (*reduction*). For the configuration shown in Figure 6.3, the

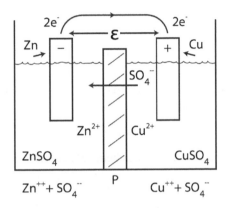

Figure 6.3 Electrochemical cell

anode is a copper bar and the cathode is a zinc bar. In this example the electrolytes consist of an aqueous solution of copper sulphate ($CuSO_4$) and an aqueous solution of zinc sulphate ($ZnSO_4$). The solutions are separated by a semipermeable barrier; for example porous clay allows only SO_4^{--} ions to pass. The transferred charge causes an electromotive force (emf) $\mathcal{E}$ to develop between the electrodes, what's called an *open cell*.[17] If a wire is connected between the electrodes, a current will flow. At the electrodes the following reactions occur:

$$Cu \rightleftharpoons 2e^- + Cu^{++}$$
$$Zn^{++} + 2e^- \rightleftharpoons Zn \ . \tag{6.43}$$

When copper is the anode and zinc the cathode (as shown) we have the *half reactions*

$$Cu \rightarrow 2e^- + Cu^{++} \qquad \text{(oxidation)}$$
$$Zn^{++} + 2e^- \rightarrow Zn \ . \qquad \text{(reduction)}$$

Adding the reactions in Eq. (6.43), the electrons "cancel," leaving the overall reaction involving the electrodes:

$$Cu + Zn^{++} \rightleftharpoons Zn + Cu^{++} \ . \tag{6.44}$$

As the reaction proceeds, Cu and Zn^{++} are consumed to produce Zn and Cu^{++}. Of course, the two electrons are carried internally by the sulphate ion SO_4^{--}

Chemical potential of charged species

How much charge does a *mole of electrons* represent? The *Faraday constant* (conventionally taken as positive), is defined as $\mathcal{F} \equiv eN_A = 96,485$ C mole^{-1}, where $e > 0$ is the magnitude of the electron charge. When dN moles undergo the reaction in Eq. (6.44), there is a charge transfer

---

[17]Electromotive force is a voltage.

$dq = \zeta \mathcal{F} dN$, where ζ denotes the number of valence electrons ($\zeta = 2$ for Zn^{++}). Transporting charge dq against $\mathcal{E}$ requires that work $dW = \mathcal{E} dq$ be performed on the system,

$$dW = \mathcal{E} dq = \zeta \mathcal{F} \mathcal{E} dN \,, \tag{6.45}$$

where, because charge transfer can be done reversibly in electrochemical cells we don't write $\bar{d}W$ in Eq. (6.45).

Equation (6.45) represents an additional energy to be accounted for in the first law of thermo-dynamics,[18]

$$dU = T dS - P dV + \sum_i \mu_i dN_i + \sum_i \zeta_i \mathcal{F} \mathcal{E} dN_i = T dS - P dV + \sum_i (\mu_i + \zeta_i \mathcal{F} \mathcal{E}) \, dN_i \,.$$

Comparing with Eq. (3.16), the chemical potential of a charged species, the *electrochemical potential*, acquires a term related to its charge[19]

$$\bar{\mu}_i \equiv \mu_i + \zeta_i \mathcal{F} \mathcal{E} \,. \tag{6.46}$$

The formal definition of $\bar{\mu}$ is the same as for the chemical potential (Section 3.5): the energy required to add another mole of species i to the system keeping entropy, volume, and the other mole numbers fixed, $\bar{\mu}_i = (\partial U / \partial N_i)_{S,V,\overline{N}_i}$, where the internal energy U contains the electrostatic potential energy of the charges. The Gibbs energy for charged species is therefore described by $dG = -S dT + V dP + \sum_i \bar{\mu}_i dN_i$ and hence $\bar{\mu}_i = (\partial G / \partial N_i)_{T,P,\overline{N}_i}$, formally the same as Eq. (6.14).

Measuring ΔG and ΔF

The emf can be measured using a simple circuit consisting of a battery, a potentiometer, a resistor, and a galvanometer,[20] G. By adjusting the potentiometer the current through the cell can be brought to zero (as indicated by G). Under these conditions the emf of the cell is balanced against the voltage drop across the resistor; in doing so we have measured the emf of an open cell.

Starting with the cell in equilibrium, the potentiometer can be varied causing current to flow *in either direction*, constituting a *reversible* transfer of charge: Small amounts of charge can be made to flow in both directions of the reaction in Eq. (6.44), with the system restorable to its original configuration by reversing the current. Under reversible conditions $[\Delta F]_{T,V} = W'$ and $[\Delta G]_{T,P} = W'$ (Table 4.1), i.e., work other than $P dV$ work is the change in F for constant (T, V) and the change in G for constant (T, P). From Eq. (6.45), *measuring the open-cell emf $\mathcal{E}$ under conditions of constant (T, P) or constant (T, V) is a means of measuring the free energy change, ΔG or ΔF.* The work done in $d\xi$ units of reaction is $dW = \zeta \mathcal{F} \mathcal{E} d\xi$. The change in molar free energy for one unit of reaction is then

$$\Delta G = \zeta \mathcal{F} [\mathcal{E}]_{T,P} \qquad \text{or} \qquad \Delta F = \zeta \mathcal{F} [\mathcal{E}]_{T,V} \,. \tag{6.47}$$

6.6 GIBBS-HELMHOLTZ EQUATIONS

We now derive a set of relations loosely known as the *Gibbs-Helmholtz equations*. There are separate Gibbs-Helmholtz equations for G and F, and we develop them together. Combine $G = H - TS$ and $F = U - TS$ with $S = -(\partial G / \partial T)_P$ and $S = -(\partial F / \partial T)_V$ to obtain:

$$G = H + T \left(\frac{\partial G}{\partial T} \right)_P \qquad \text{and} \qquad F = U + T \left(\frac{\partial F}{\partial T} \right)_V \,. \tag{6.48}$$

---

[18] We're working here with the molar chemical potential; the same ideas apply on a per-particle basis.

[19] In semiconductor devices, the reference energy under "flat-band" conditions is the chemical potential plus whatever is the voltage that's been applied to a given region of the semiconductor—the electrochemical potential.[26, p594]

[20] A galvanometer detects the flow of charge.

The equations in (6.48) are sometimes referred to as the Gibbs-Helmholtz equations; the name is also reserved for Eq. (6.53). The equations in (6.48) are equivalent to:

$$U = F - T\left(\frac{\partial F}{\partial T}\right)_V = -T^2\frac{\partial\,(F/T)_V}{\partial T}$$

$$H = G - T\left(\frac{\partial G}{\partial T}\right)_P = -T^2\frac{\partial\,(G/T)_P}{\partial T}\,. \tag{6.49}$$

Thus, having F implies U, S, P and μ by differentiation; having G implies H, S, V and μ by differentiation.

The equations in (6.48) and (6.49) hold for chemical reactions in terms of the quantities ΔG and ΔF (at the same T),

$$\Delta G = \Delta H + T\left(\frac{\partial\Delta G}{\partial T}\right)_P \qquad\text{and}\qquad \Delta F = \Delta U + T\left(\frac{\partial\Delta F}{\partial T}\right)_V. \tag{6.50}$$

From ΔG and ΔF, we also have the entropy differences

$$[\Delta S]_P = -\left(\frac{\partial\Delta G}{\partial T}\right)_P \qquad\text{and}\qquad [\Delta S]_V = -\left(\frac{\partial\Delta F}{\partial T}\right)_V, \tag{6.51}$$

in terms of which (6.50) can be written

$$T\,[\Delta S]_P = \Delta H - \Delta G \qquad\text{and}\qquad T\,[\Delta S]_V = \Delta U - \Delta F\,. \tag{6.52}$$

The equations in (6.50) are equivalent to

$$\left(\frac{\partial(\Delta G/T)}{\partial T}\right)_P = -\frac{\Delta H}{T^2} \qquad\text{and}\qquad \left(\frac{\partial(\Delta F/T)}{\partial T}\right)_V = -\frac{\Delta U}{T^2}\,. \tag{6.53}$$

The quantities ΔH and ΔU can thus be obtained from the temperature dependence of ΔG and ΔF. We'll use these equations in Chapter 8.

CHAPTER SUMMARY

This chapter introduced applications of thermodynamics to chemistry, notably phase and chemical equilibrium. We covered the Gibbs phase rule, the Clausiu-Clapeyron equation, the thermodynamics of mixing, the law of mass action, and electrochemical cells.

- A thermodynamic phase is a spatially homogencous part of a system, with uniform chemical and physical properties. For phases coexisting in equilibrium the chemical potential of a given substance is independent of the phase.

- The variance f of a system with k chemical species in π phases is $f = 2+k-\pi$ (Gibbs phase rule), the number of intensive variables that can be independently varied without disturbing the number of phases in equilibrium.

- The Clausius-Clapeyron equation gives the slope of the coexistence curve between two phases in equilibrium,

$$\left(\frac{\mathrm{d}P}{\mathrm{d}T}\right)_{\text{coexist}} = \frac{L}{T\Delta v}\,,$$

where L is the latent heat per mole and Δv is the change in molar volume.

- The chemical potential of an ideal gas in a mixture is $\mu = \mu^0 + RT \ln x$ where x is the mole fraction of the gas in the mixture and μ^0 is the chemical potential of the pure gas. The chemical potential of an ideal gas in a mixture is less than that of the pure gas.

- An ideal mixture has zero volume of mixing, zero enthalpy of mixing, and an entropy of mixing

$$\Delta S = -NR \sum_{i=1}^{k} x_i \ln x_i \,,$$

where N is the total number of particles and the x_i are the mole fractions of the species making up the mixture.

- Under equilibrium conditions the chemical potentials μ_i of species M_i undergoing a chemical reaction $\sum_i \nu_i M_i = 0$ (with ν_i the stoichiometric coefficients) are related by $\sum_i \nu_i \mu_i = 0$. The law of mass action is that the equilibrium concentrations of the reactants c_i obey the relation $\prod_i c_i^{\nu_i} = K(P,T)$, where K is the equilibrium constant.

EXERCISES

6.1 What is the maximum number of gaseous chemical species that can coexist in equilibrium?

6.2 Show that $\left(\dfrac{\partial(G/T)}{\partial(1/T)} \right)_P = H$ and $\left(\dfrac{\partial(F/T)}{\partial(1/T)} \right)_V = U$. Hint: Use the Gibbs-Helmholtz equations.

6.3 From Eq. (6.41), $\Delta G^0 = -RT \ln K^*$. Combine this result with one of the equations in Eq. (6.53) to show that

$$\left(\frac{\partial \ln K^*}{\partial T} \right)_P = \frac{\Delta H^0}{RT^2} \,,$$

where $\Delta H^0 \equiv \sum_i \nu_i H_i^0$, with H_i^0 the molar enthalpy of the pure substance. Integrate this formula (at constant pressure) to conclude that

$$\ln K^* = \ln K_0^* + \int_{T_0}^{T} \frac{\Delta H^0}{RT^2} dT \,.$$

The equilibrium constant $K^*(T, P)$ at temperature T can therefore be obtained from knowledge of that at one temperature, $K_0^*(T_0, P)$, and the temperature dependence of ΔH^0, which can be obtained from purely thermal measurements.

6.4 Heat capacity and latent heat data can be used to calculate H and S.

a. Using Eq. (3.13), show that the entropy of the liquid phase of a substance at temperature T can be calculated by heating the solid at a constant pressure P_0 starting at $T = 0$, letting it melt at $T = T_m$, and heating the liquid to a temperature $T > T_m$:

$$S^l(T) = \int_0^{T_m} \frac{C_P^{\text{solid}}(T)}{T} dT + \frac{L(T_m)}{T_m} + \int_{T_m}^{T} \frac{C_P^l(T')}{T'} dT' \,, \tag{P6.1}$$

where $L(T_m)$ is the latent heat at $T = T_m$. The first integral on the right of Eq. (P6.1) is well defined because, from the third law of thermodynamics, $C_P \to 0$ as $T \to 0$.

b. Show that starting from the enthalpy of a reference state $H(T_0, P_0)$, $H(T > T_0, P_0)$ can be obtained from the temperature dependence of $C_P(T)$:

$$H(T, P_0) = H(T_0, P_0) + \int_{T_0}^{T_f} C_P(T, P_0)dT + L_f + \int_{T_f}^{T} C_P(T, P_0)dT$$

where L_f is the latent heat associated with a possible phase transition at $T = T_f$, with $T > T_f > T_0$. The result of Exercise 4.8 may be useful.

c. Show that once we know H, the value of U may be found by subtracting $P_0 V$. Likewise if we know H and S, G may be found from $G = H - TS$, and if U and S are known, $F = U - TS$.

6.5 Some physics students go camping in the desert. They happened to have left out, overnight, an open ice chest containing nothing but a thin layer of water at its bottom. In the morning they notice that the water has frozen, *even though the overnight temperature was always above the freezing point.*

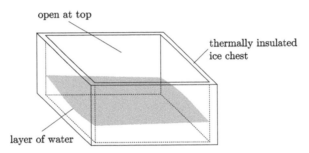

a. Explain this phenomenon. The ice chest is thermally insulated, but it was left open at the top, so it was exposed to the atmosphere. There was no wind that night. The air was still. Also, it was a very clear night; there were no clouds in the sky.

b. Assume that there was one liter of water in the ice chest and that the overnight temperature was 20° C. Ignoring evaporative effects, estimate the time it took for the water to freeze making reasonable assumptions about the experiment.
Facts: $C_p = 4.2$ J g^{-1} K^{-1}, latent heat of fusion for water is $L = 334$ J g^{-1}.

6.6 Winemakers (for reasons best known to winemakers) sometimes want to quickly lower the temperature of grapes that have just been picked, to slow the fermentation process. Cooling equipment is expensive. One could surround the grapes with an ice bath, but this is often not practical. One could add ice cubes to chill the grapes, but then the melted ice adds water to the would-be wine, an undesirable outcome. One possibility is to add *dry ice*, the solid form of carbon dioxide. Dry ice is a substance that *sublimates* at normal atmospheric pressure: It goes directly from the solid phase to the gaseous phase without passing through the liquid phase. Suppose the winemaker has a mass M of grapes, and wants to lower the temperature by $\Delta T = 10 \,°C$. What mass of dry ice, M_D, should the winemaker add? Express your answer as a fraction of the mass of the grapes, M.

The specific heat of grapes $C = 3.43$ J/g $°C$, and the latent heat of sublimation of dry ice is $L = 541$ J/g.

CHAPTER 7

Statistical entropy

From micro to macro

ENTROPY as a state variable was introduced in Chapter 3, with its consequences explored in the ensuing chapters.[1] In this chapter, we turn to the microscopic interpretation of entropy, and in so doing begin to prepare the foundation for statistical mechanics.

7.1 ENTROPY AND PROBABILITY

Several considerations contribute to the view that entropy is related to *orderliness*. Work is an *organized* form of energy, systematic extensions of macroscopic quantities (Section 1.9), and $\Delta S > 0$ *reduces* the effectiveness of energy to perform work (energy dissipation, Section 4.7). Entropy increases upon the removal of constraints (irreversibility), and less constrained systems are less organized (Section 3.3). The entropy of isolated systems is a maximum in equilibrium (Sections 3.3, 3.10, and 4.1). The state of equilibrium should be the *most likely* state, by far: Equilibrium persists unchanged in time; amongst the myriad dynamical microstates of a system, those consistent with equilibrium should occur with far greater likelihood. Disorder, maximum entropy, likelihood: we expect that entropy is related to *probability*.[2] According to Max Planck,[27, p118] "The connection between entropy and probability should be very close."

How to find such a connection? To get started, let W denote the probability of a specified macrostate, where obviously we must define what W represents.[3] We seek $S = f(W)$, where f is a universal function of a single variable. To determine f we are guided by two requirements. First, the entropy of a system of two independent subsystems is the sum of the subsystem entropies, $S = S_1 + S_2$, i.e., S is extensive. Second, by the rules of probability, the probability W for the occurrence of two independent states 1 and 2 is the *product* of the separate probabilities W_1, W_2,

---

[1]It's instructive to review Chapters 4, 5, and 6 with an eye towards the concepts enabled by entropy as a state variable. For example, $\mu = 0$ for photons would not be possible without the concept of free energy.

[2]We're going to invoke concepts from the theory of probability without formally reviewing the subject. Tutorial introductions to probability theory are often given in books on statistical mechanics. In this book, only basic ideas of probability are used, which students seem comfortable with.

[3]We follow tradition and use W for *Wahrscheinlichkeit*, German for probability (W here does not mean work!). Of course W must be defined as a physical concept; the language we use is immaterial. The quantity W is not a traditional probability in the sense that it's a number between zero and one. Planck distinguished *thermodynamic probability* (W) from *mathematical probability*, where the two are proportional but not equal.[27, p120] As we'll see, W is *the number of microstates per macrostate*, an integer. To quote Planck:[28, p226] "Every macroscopic state of a physical system comprises a perfectly definite number of microscopic states of the system, and the number represents the thermodynamic probability or the statistical weight W of the macroscopic state."

i.e., $W = W_1 W_2$. With $S_1 = f(W_1)$ and $S_2 = f(W_2)$, the function f must be such that

$$f(W_1 W_2) = f(W_1) + f(W_2) . \tag{7.1}$$

Equation (7.1) is a *functional equation* (similar to Eqs. (2.8) and (3.23)). Noting that W_1 and W_2 occur in Eq. (7.1) symmetrically, first differentiate Eq. (7.1) with respect to W_1 and then differentiate the result with respect to W_2. We find

$$W_1 W_2 f''(W_1 W_2) + f'(W_1 W_2) = 0 .$$

The solution of the differential equation $z f''(z) + f'(z) = 0$ is $f(z) = k \ln z + f_0$, where k and f_0 are constants. Taking $f_0 = 0$, we have[4]

$$S = k \ln W . \tag{7.2}$$

Equation (7.2) *is one of the great achievements of theoretical physics.*[5] It provides a physical understanding of entropy beyond qualitative notions of disorder; it's the *bridge* between microscopic and macroscopic, it represents the "missing" degrees of freedom not accounted for in macroscopic descriptions. We will distinguish the *statistical entropy*, $S = k \ln W$, from the *thermodynamic entropy*, $\Delta S = \int (\mathrm{d}Q)_{\text{rev}} /T$, Eq. (3.4). Of course, we'd like the two to agree in their predictions.

The scale factor k in Eq. (7.2) (which must be a universal constant) can be established by a simple argument.[6] The *change* in entropy between equilibrium states is, from Eq. (7.2):

$$\Delta S \equiv S_2 - S_1 = k \ln (W_2/W_1) . \tag{7.3}$$

Consider the free expansion of an ideal gas with N particles into an evacuated chamber where the volume doubles in the process (Section 4.9). While we don't yet know how to calculate W_1 and W_2 separately, the *ratio* W_2/W_1 is readily found in this case. After the system comes to equilibrium, it's twice as likely to find a given particle in either volume than it is to find the particle in the original volume. The particles of an ideal gas are non-interacting, and thus the same is true *for each particle*, implying[7] $W_2/W_1 = 2^N$. By Eq. (7.3) therefore, $\Delta S = Nk \ln 2$. In Eq. (4.20) we calculated $\Delta S = nR \ln 2$ for this process. Agreement between the two formulae is obtained by choosing $k = k_B = R/N_A$, the Boltzmann constant (Section 1.6).

Equation (7.2) is encouraging, yet at this point it's on shaky ground: It relates a symbol S to another symbol W. The only properties used in its derivation are that S is additive over subsystems (like entropy), $S = S_1 + S_2$, and W behaves like a probability with $W = W_1 W_2$. Despite the best of intentions, using the symbol S does not make it the entropy. Of course, that begs the question of what *is* entropy. We established the scale factor k by making ΔS from Eq. (7.3) agree with ΔS from thermodynamics for the same process. It's standard practice in theoretical physics to "nail down" unknown parameters by seeking agreement with previously established theories in their domains of validity. Such a practice does not *guarantee* the validity of Eq. (7.2); it merely asserts it's not

---

[4]The disposition of f_0 is actually a pivotal issue. We return to it in Section 7.8 and in Chapter 8.

[5]Equation (7.2) is widely known as the Boltzmann entropy formula (it's inscribed on his tomb in Vienna). Despite the attribution, however, it's not clear that Boltzmann ever wrote down Eq. (7.2). Equation (7.2) appears to have originated in the work of Max Planck, in his 1901 article in which the Planck distribution is derived. In 1872 Boltzmann proved his "H-theorem" (not treated in this book) in which (under reasonable circumstances) the quantity $H(t) \equiv - \iiint f \ln f \, dv_x dv_y dv_z$ never decreases in time, emulating entropy, where $f = f(v_x, v_y, v_z, t)$ is the probability distribution of the velocity components of the atoms of a gas. By the H-theorem, as equilibrium is achieved $f(t)$ evolves into the Maxwell-Boltzmann distribution, Eq. (7.26), which involves the internal energy U and absolute temperature T of the gas. Thus, the *possibility* of identifying the asymptotic value of $H(t)$ with a thermodynamic parameter (entropy) is not surprising. Planck asserted, *based on Boltzmann's work*, that entropy is proportional to the logarithm of a probability, W, what we have in Eq. (7.2).

[6]The scale factor is dependent upon the *base* of the logarithm used in Eq. (7.2). While we've chosen to work with the natural logarithm, $\ln x = \log_e x$, we could work with $\log_b x = \ln x / \ln b$. This distinction shows up in information theory, Chapter 12. The use of the natural logarithm in Eq. (7.2) is standard practice in physics.

[7]We derive this result in Eq. (7.12).

obviously wrong in this one application. If Eq. (7.2) proves to be of general validity, then having determined k in this manner will suffice. Of greater importance, what allows us to call anything (such as S in Eq. (7.2)) entropy? What we know about entropy comes from thermodynamics:

$$\left(\frac{\partial S}{\partial U}\right)_{V,N} = \frac{1}{T} \qquad \left(\frac{\partial S}{\partial V}\right)_{U,N} = \frac{P}{T} \qquad \left(\frac{\partial S}{\partial N}\right)_{U,V} = -\frac{\mu}{T}. \qquad (3.30)$$

If we agree to call "entropy" anything satisfying these identities, then Eq. (3.30) must be imposed as a requirement on Eq. (7.2) before we're entitled to call S in Eq. (7.2), entropy. That implies W must be specified *as a function of macroscopic variables*, $W = W(U,V,N)$, because, from Eq. (7.2), $W = \exp\left(S(U,V,N)/k_B\right)$.

Probability of, what?

> So when gases of different kinds are mixed, if we ask what changes in external bodies are necessary to bring the system to its original state, we do not mean a state in which each particle shall occupy more or less exactly the same position as at some previous epoch, but only a state which shall be undistinguishable from the previous one in its sensible properties. It is to states of systems thus incompletely defined that the problems of thermodynamics relate.—J.W. Gibbs, 1876[3, p166]

Equation (7.2) represents a shift in emphasis, away from thermodynamics, which, to paraphrase Gibbs, can distinguish only the *sensible properties* of macroscopic systems, with ΔS determined by processes, to the view that S can be *calculated* from the microscopic properties of the system.[8] The above quote of Gibbs is apt: *Thermodynamics deals with states incompletely defined.*[9] Equilibrium states are specified by a handful of state variables, leaving unspecified an enormous number of microscopic degrees of freedom. Planck defined a useful term, an *elemental chaos* as "any process containing numerous elements not in themselves measurable."[27, p117] There is what can be measured, and there is the rest. Thermodynamics deals with the measurable properties of macroscopic systems. The quantity W in Eq. (7.2) is a bridge between micro- and macro-descriptions, *the number of microstates per macrostate*, in essence the *size* of Planck's elemental chaos. The word *probability* is perhaps unfortunate; think W for *ways*, the number of ways a macrostate can be realized from the microstates of a system that are consistent with constraints (such as V or N). Still, entropy being a maximum in equilibrium implies that equilibrium is the state having the maximum number of microscopic possibilities for its realization, i.e., *the most probable state*. To use Eq. (7.2), *we must learn to calculate W*, what we do in the upcoming sections.

The derivation of Eq. (7.2) relies on the idea of *mutually independent* microstates. In order to write $W = W_1 W_2$, it must be the case that every microstate underlying macrostate 1 occurs independently of every microstate underlying macrostate 2. *This idea must be taken as a foundational assumption*, the principle of *equal a priori probabilities*, that all microstates associated with a system are equally likely to be realized by the system in the course of its dynamical time evolution. To *prove* this is the province of *ergodic theory*, a branch of mathematics concerning the long-time properties of dynamical systems. Most physicists accept Eq. (7.2) by the success of its predictions. To quote G.S. Rushbrooke:[29, p14]

> Now W, by definition, is the number of *a priori* equally probable complexions of the assembly corresponding to given values of U, V, and N. In the absence of any indication to the contrary we can only assume that *all conceivable different micromolecular*

---

[8]To extend the metaphor of Chapter 3, Boltzmann and Planck opened the black box even more.

[9]Classical mechanics deals with *precisely specified* quantities, such as initial conditions, or parameters. Suppose the spring constant of a harmonic oscillator is only known approximately. How would you handle that? Thermodynamics incorporates what *is* known about a system to make predictions in the face of uncertainty.

states of the assembly (complexions) corresponding to the same values of U, V, and N are equally probable. This hypothesis underlies the whole of statistical mechanics and, like [Eq. (7.2)], must ... be regarded as justified *a posteriori*, by the success of the theory based on it.

Example. A deck of playing cards usually has 52 cards. The probability of drawing any particular card is the same as that for any other card, namely $\frac{1}{52}$. Because there is nothing we know of that favors one card over another, the probabilities are a priori equal.

7.2 COMBINATORICS: LEARNING TO COUNT*

Combinatorics is a branch of mathematics devoted to *counting* the number of elements in various sets. In this section we introduce some basic combinatorial ideas. Suppose there are N numbered balls in a bag (and hence which are distinguishable), and you pull them out one at a time. How many different ways are there of performing that experiment? There are N ways to choose the first ball, $N - 1$ ways to choose the second, and so on: There are $N!$ *permutations* of N items.

How many ways are there to bring out *two* balls, where we don't care about the order in which they are displayed? That is, pulling out ball 3 and then ball 7 is the same (by assumption) of pulling out 7 and then 3. There are N ways to pull out the first ball and $N - 1$ ways to pull out the second. Once the two balls are out, there are 2! ways of arranging them, both of which we deem to be equivalent. Thus there are

$$\frac{N(N-1)}{2!} = \frac{N!}{2!(N-2)!}$$

distinct ways of choosing two balls out of N when order is not important.

Example. Four choose two. There are four items, numbered $(1, 2, 3, 4)$. There are $N(N - 1) = 12$ ways of pulling two items out of the collection of four: $(1, 2), (1, 3), (1, 4), (2, 1), (2, 3), (2, 4),$ $(3, 1), (3, 2), (3, 4), (4, 1), (4, 2),$ and $(4, 3)$. If we don't care about the order in which the pairs are displayed, so that, for example, $(4, 2)$ is considered the same as $(2, 4)$, there are six distinct ways of choosing two items out of the four: $(1, 2), (1, 3), (1, 4), (2, 3), (2, 4),$ and $(3, 4)$.

Generalizing to the number of ways of choosing k balls out of N when order is not important, we define the *binomial coefficient* ("N choose k")

$$\binom{N}{k} \equiv \frac{N!}{k!(N-k)!} . \tag{7.4}$$

The number of *combinations* of N items taken k at a time is given by Eq. (7.4); it's called the binomial coefficient because the same term shows up in the binomial formula,

$$(x + y)^N = \sum_{k=0}^{N} \binom{N}{k} x^{N-k} y^k . \tag{7.5}$$

We note the special result that follows from Eq. (7.5) with $x = 1$ and $y = 1$:

$$2^N = \sum_{k=0}^{N} \binom{N}{k} . \tag{7.6}$$

We now ask a different question. How many ways are there to choose N_1 objects from N (where order is immaterial), followed by choosing N_2 objects from the remainder, where these objects are kept separate from the first collection of N_1 objects? Clearly that number is:

$$\frac{N!}{N_1!(N-N_1)!} \times \frac{(N-N_1)!}{N_2!(N-N_1-N_2)!} = \frac{N!}{N_1!N_2!(N-N_1-N_2)!} .$$

Generalize to the number of ways of choosing from N objects, N_1 kept separate, N_2 kept separate, $\cdots$, N_r kept separate, where $\sum_{k=1}^{r} N_k = N$ exhausts the N objects. The number of ways of distributing N objects among r containers, each containing N_k objects with $\sum_{k=1}^{r} N_k = N$, is given by the *multinomial coefficient*

$$\binom{N}{N_1, N_2, \ldots, N_r} \equiv \frac{N!}{N_1!N_2!\ldots N_r!} . \tag{7.7}$$

Stirling approximation

We will encounter equations that call for factorials of large integers, and it behooves us to have an approximate *formula* for the value of a factorial. In its simplest form, *Stirling's approximation* is

$$\ln N! \approx N \ln N - N . \tag{7.8}$$

By exponentiating Eq. (7.8) we have the approximation

$$N! \approx (N/e)^N . \tag{7.9}$$

Equation (7.8) is surprisingly easy to derive:

$$\ln N! = \sum_{k=1}^{N} \ln k \approx \int_1^N \ln x \, dx = (x \ln x - x) \big|_1^N = N \ln N - N + 1 \approx N \ln N - N .$$

Stirling's approximation is one of those results that should only work for $N \to \infty$, but is reasonably accurate even for small values of N. *Asymptotic analysis* is a branch of mathematics in which improvements to approximate formulas such as Eq. (7.9) can be established. It's found that:[10]

$$N! \overset{N \to \infty}{\sim} \sqrt{2\pi N} \left(\frac{N}{e}\right)^N \left[1 + \frac{1}{12N} + \frac{1}{288N^2} + \cdots\right] . \tag{7.10}$$

The terms in square brackets are rarely required in physics applications.[11]

Distributing particles between compartments

Consider two compartments A and B, each the same size, each containing identical, yet distinguishable particles that can pass between them. At an instant of time there are N_A (N_B) particles in A (B), where $N_A + N_B = N$ is a fixed quantity.[12] The number of ways that compartment A can have N_A particles is, from either Eqs. (7.4) or (7.7),

$$\binom{N}{N_A} = \frac{N!}{N_A!(N-N_A)!} = \frac{N!}{N_A!N_B!} . \tag{7.11}$$

---

[10]*Asymptotic equivalence $f \sim g$ means that for functions $f(x)$ and $g(x)$, the ratio $f(x)/g(x) \to 1$ as x approaches some value, usually zero or infinity. Example: $\sinh x \sim \frac{1}{2}e^x$ as $x \to \infty$.

[11]The factor of $\sqrt{2\pi N}$ is a logarithmically small correction, as can be seen by taking the logarithm of $N!$ in Eq. (7.10) and comparing with Eq. (7.8). The terms in square brackets are an even smaller correction for large N.

[12]In this subsection, N_A does not mean Avogadro's number.

The *total number of ways* W that particles can be situated in the two compartments is found by summing Eq. (7.11) over all possible values of N_A. Using Eq. (7.6),

$$W = \sum_{N_A=0}^{N} \binom{N}{N_A} = 2^N . \qquad (7.12)$$

From Eq. (7.2), the "entropy" of this system is:[13]

$$\ln W = N \ln 2 . \qquad (7.13)$$

Which arrangement of particles occurs in the *maximum number of ways*? Consider, starting from an initial distribution of N_A^0 particles in A and N_B^0 in B ($N_A^0 + N_B^0 = N$), the effect of transferring ξ particles between compartments, with $N_A = N_A^0 - \xi$ and $N_B = N_B^0 + \xi$, $-N_B^0 \le \xi \le N_A^0$. The number of ways to transfer ξ particles is

$$w(\xi) = \frac{N!}{(N_A^0 - \xi)!(N_B^0 + \xi)!} .$$

The logarithm of this number is (using the Stirling approximation)

$$\ln w(\xi) = \ln N! - \ln(N_A^0 - \xi)! - \ln(N_B^0 + \xi)!$$
$$\approx N \ln N - (N_A^0 - \xi) \ln(N_A^0 - \xi) - (N_B^0 + \xi) \ln(N_B^0 + \xi) . \qquad (7.14)$$

As can be shown, the derivative of $\ln w(\xi)$ with respect to ξ is[14]

$$\frac{\partial}{\partial \xi} \ln w(\xi) = \ln \left(\frac{N_A^0 - \xi}{N_B^0 + \xi} \right) . \qquad (7.15)$$

The maximum occurs for $\xi = \xi^* \equiv \frac{1}{2}\left(N_A^0 - N_B^0\right)$. The configuration having maximum entropy is thus for A and B to each have $N/2$ particles. The maximum entropy is found by substituting $\xi = \xi^*$ into Eq. (7.14):

$$\ln w(\xi^*) = N \ln 2 . \qquad (7.16)$$

Equation (7.16), the entropy of the configuration having maximum entropy is, within the accuracy of the Stirling approximation, the *same* as Eq. (7.13), the entropy of the system obtained by including *all possible arrangements of particles*.[15] Is this an accident? Can the value of the sum in Eq. (7.12) (for W) be approximated by just *one term* in the series, the number of configurations associated with maximum entropy (what we have called $w(\xi^*)$)? In short, the answer is *yes*. We cannot give a general proof, but for N sufficiently large,[16] there is almost always a configuration of the system that occurs in such a predominantly large number of ways, that the sum over system configurations[17] can be replaced with the largest term in the series. This behavior is exemplified in the formula for $\ln w(\xi)$, Eq. (7.14). From Eq. (7.15), the second derivative evaluated at the configuration ξ^* has the value $\partial^2 \ln w(\xi)/\partial \xi^2|_{\xi^*} = -4/N$. Thus, for system configurations ξ in the vicinity of ξ^*, we have the Taylor expansion

$$\ln w(\xi) \approx N \ln 2 - \frac{2}{N} (\xi - \xi^*)^2 + \cdots , \qquad (7.17)$$

---

[13]Equation (7.13) strictly does not qualify as entropy because it cannot satisfy the requirements given in Eq. (3.30).

[14]In evaluating the derivative in Eq. (7.15), we're treating ξ as if it's a continuous variable, whereas in actuality it's an integer. This can be justified when N is sufficiently large.

[15]The "state" of the system of two compartments with N particles distributed between them has not been specified with any further refinement; we must include in W all possible configurations consistent with the specification of the system.

[16]Such as we have for macroscopic systems.

[17]Configurations compatible with the macroscopic specification of the system.

or, equivalently,

$$w(\xi) \approx 2^N \exp\left(-\frac{2}{N}(\xi - \xi^*)^2\right) . \tag{7.18}$$

For large N, configurations relatively close to ξ^*, those with $\xi = \xi^* \pm \sqrt{N}$, make a negligible contribution to the entropy in comparison to the entropy of the most frequently occurring configuration.

7.3 COARSE-GRAINED DESCRIPTIONS OF A CLASSICAL GAS

To calculate W, the microstates of the system must have a property that's *countable*, something *discrete* in how we describe states. Ultimately that discreteness will be given to us in the form of Planck's constant,[18] h. In this section we consider two ways of *coarse graining* the description of a classical gas of N atoms so that its microstates are specified in such a way as to be countable.

Space-only description of a gas

We start with an instructive, yet artificial treatment of a gas of N particles. Subdivide the volume occupied by the gas into N_c identical cells. At an instant of time, the kth cell contains N_k atoms, with $\sum_{k=1}^{N_c} N_k = N$. We don't ask for the coordinates of each atom or for their velocities, all we ask is how many atoms are in each cell. The "states" of this system are a specification of the *occupation numbers* $\{N_k\}_{k=1}^{N_c}$, i.e., a list of numbers $(N_1, \cdots, N_{N_c})$. For example, consider 10 atoms in 7 cells. One particular state has $N_1 = 1, N_2 = 2, N_3 = 0, N_4 = 0, N_5 = 1, N_6 = 4, N_7 = 2$. We could "go Dirac" and write $|1200142\rangle$ to denote this state. The number of ways of distributing N atoms with N_k in each cell is, using Eq. (7.7),

$$W[N_1, N_2, \cdots] = \frac{N!}{\prod_k N_k!} . \tag{7.19}$$

For the state $|1200142\rangle$ of 10 balls in 7 cells,

$$W_{|1200142\rangle} = \frac{10!}{1!2!0!0!1!4!2!} = 37,800 .$$

There are $37,800$ ways to distribute 10 balls among 7 cells with the cell numbers $N_1 = 1, N_2 = 2, N_3 = 0, N_4 = 0, N_5 = 1, N_6 = 4, N_7 = 2$.

Assume that N is large enough so that all cells have an appreciable number of atoms. In that case we can apply Stirling's approximation to each of the factorials in Eq. (7.19). We find:

$$W[\{N_k\}] = \prod_k \left(\frac{N}{N_k}\right)^{N_k} \equiv \prod_k \left(\frac{1}{p_k}\right)^{N_k} , \tag{7.20}$$

where the probability p_k that an atom selected at random is in the kth cell is $p_k = N_k/N$, with $\sum_k p_k = 1$. The formula for the entropy is then, combining Eq. (7.20) with Eq. (7.2),

$$S[\{p_k\}] = -Nk_B \sum_k p_k \ln p_k . \tag{7.21}$$

Equation (7.21) is a central result in the theory of statistical entropy.[19] With it, we have the entropy of *any* system for which the probabilities $\{p_k\}$ are known.

---

[18]Another instance where the requirements of thermodynamics, in this case the ability to count states, are such as to *anticipate the development of quantum mechanics*. It's actually easier to calculate the entropy of quantum systems than it is for classical systems, because quantum states naturally have a countable feature, the discreteness of energy eigenvalues.

[19]Equation (7.21) is the starting point for *information theory*; see Section 12.4.

What probability distribution *maximizes* the entropy (apropos of thermodynamic equilibrium)? Because of the constraint $\sum_k p_k = 1$, add zero to Eq. (7.21):

$$S\left[\{p_k\}\right] \equiv -Nk_B \sum_k p_k \ln p_k + \alpha k_B \left(\sum_k p_k - 1\right), \qquad (7.22)$$

where α is a Lagrange multiplier, made dimensionless by including the factor of k_B. The quantities p_k can now be varied as if they're independent: $\delta S = k_B \sum_k \delta p_k \left[\alpha - N\left(1 + \ln p_k\right)\right]$. An extremum ($\delta S = 0$) is achieved for $p_k = e^{(\alpha/N-1)}$. Thus, the distribution of cell probabilities that maximizes S is for p_k to be *independent* of the cell label k. *Maximum entropy is associated with a uniform distribution of atoms*, with each cell occupied with equal likelihood. The quantity α can be determined from the equation of constraint, $1 = \sum_k p_k = e^{(\alpha/N-1)} N_c$, implying that

$$p_k = N_c^{-1}. \qquad (7.23)$$

Combining Eqs. (7.23) and (7.21), the *maximum* value of S for a system specified by particles in N_c cells is (the generalization of Eq. (7.13))

$$S = Nk_B \ln N_c. \qquad (7.24)$$

The entropy per particle, $S/N = k_B \ln N_c$, is thus the logarithm of the number of equivalent "choices" available to a particle, the number of cells N_c. The same result occurs in the theory of information; see Eq. (12.6). The level of description that leads us to Eq. (7.24) is one where we have discarded most of the physical features that make a gas a gas; it's not rich enough to permit the connection with thermodynamics required by Eq. (3.30). The only constraint we have imposed is the number of particles, $\sum_k N_k = N$.

Phase-space description: First pass

We now seek to add more realism by including the *energy* of an ideal gas, and the challenge is to do so in such a way that the microstates are described in a countable fashion. *Such a calculation is considerably more difficult* than that based on the space-only description.[20]

Phase space is a mathematical space in which the states of particles can be represented, with each possible state corresponding to a single point in the space.[30] Phase space for a single classical point particle is a six-dimensional space of all possible positions and momentum values. Let's try to calculate the entropy of a gas by emulating the approach of the previous subsection and introduce cells in phase space.[21] Define a phase-space cell volume g:

$$g = \int_{cell} dxdydzdp_xdp_ydp_z \equiv \int_{cell} d^3rd^3p \equiv \int_{cell} d\sigma,$$

where $d\sigma$ denotes an infinitesimal volume element of phase space. We'll find that in contrast to N_c in the space-only description, which we're free to choose, *the size of phase-space cells cannot be chosen arbitrarily*, at least if we want the calculated entropy to agree with experimental results. A technical issue regarding phase space is that whereas the spatial volume V accessible to the molecules of a gas is bounded, *momentum space*[22] is *unbounded* because the momentum of particles

---

[20]We'll see that the approach adopted in this subsection can only take us so far. Figuring out *why* it ultimately doesn't work is quite instructive. The right way to do this calculation is presented in Section 7.6.

[21]The phase space of a single particle is referred to as μ-space, whereas the phase space associated with N particles is referred to as Γ-space. The deeper reason why the approach attempted in this subsection ultimately doesn't work is because it conceives of the gas in terms of N single-particle states in six-dimensional μ-space, rather than (as experience teaches) a single state in $6N$-dimensional Γ-space (see Section 7.6). An important but subtle point!

[22]Phase space for a point particle has the mathematical structure of $\mathbb{R}^6 = \mathbb{R}^3 \times \mathbb{R}^3$, the direct product of *configuration space* and momentum space. Each point of "real space" has associated with it a three-dimensional momentum space.

can have indefinitely large values.[23] We can't suppose, therefore, that the entropy of the gas will consist of all cells equally occupied (as in the space-only description) because that would imply an infinite amount of energy. We have to anticipate that the occupation of cells in momentum space is nonuniform.

In the space-only description (previous subsection), we could introduce *discrete* probabilities $p_k = N_k/N$ that cells are occupied because the occupation numbers N_k were taken *as specified*.[24] We can't do that in phase space; we can't specify N_k from a "snapshot" of a six-dimensional system—the momentum of particles is not a spatial quantity. Instead, we presume an underlying *phase-space probability density function*, $\rho(r, p)$, such that $\rho(r, p)\mathrm{d}\sigma$ is the probability of finding a particle within $\mathrm{d}\sigma$ about the point in phase space having spatial location r and momentum p; it's normalized such that $\int_\Omega \rho(r, p)\mathrm{d}\sigma = 1$, where $\int_\Omega$ indicates an integration over all of phase space.[25]

With $\rho(r, p)$ presumed known, we can define the probability p_k of cell k being occupied through[26] $p_k \equiv \int_k \rho(r, p)\mathrm{d}\sigma$, where $\int_k$ denotes an integral over cell k. The quantities $\{p_k\}$ are referred to as a *coarse-grained* probability distribution, normalized such that $\sum_k p_k = 1$. For an N-particle gas, there are $N_k = Np_k$ atoms in cell k. The internal energy U of an ideal gas (of particles of mass m) is the sum of the kinetic energies of the particles,

$$U = \frac{1}{2m}\sum_k N_k P_k^2 = \frac{N}{2m}\sum_k p_k P_k^2 , \tag{7.25}$$

where[27] $P_k^2 \equiv \left(P_x^2 + P_y^2 + P_z^2\right)_k$ is the magnitude squared of the momentum vector of an atom in cell k; it's presumed that cells can be chosen small enough that all atoms in a cell ostensibly have the same momentum.

With these assumptions, add another term to Eq. (7.22) reflecting the constraint of the energy,

$$S\left[\{p_k\}\right] \equiv -Nk_B \sum_k p_k \ln p_k + Nk_B\alpha \left(\sum_k p_k - 1\right) - \beta k_B \left(\frac{N}{2m}\sum_k p_k P_k^2 - U\right) ,$$

where the Lagrange multipliers α and β have been parameterized for convenience. Seek the condition for the extremum: $\delta S = Nk_B \sum_k \delta p_k \left[-(1 + \ln p_k) + \alpha - \beta P_k^2/(2m)\right] = 0$. The probability distribution that maximizes the entropy is thus

$$p_k = \mathrm{e}^{\alpha-1}\mathrm{e}^{-\beta P_k^2/(2m)} . \tag{7.26}$$

The equilibrium distribution is *independent of spatial location* and is *isotropic in momentum space*. Equation (7.26) is, up to normalization factors, the *Maxwell-Boltzmann distribution function*.[28]

The two Lagrange multipliers can be found from the two equations of constraint. From conservation of probability, using Eq. (7.26),

$$\sum_k p_k = 1 \implies \mathrm{e}^{\alpha-1}\sum_k \mathrm{e}^{-\beta P_k^2/(2m)} = 1 \implies \mathrm{e}^{\alpha-1} = \left(\sum_k \mathrm{e}^{-\beta P_k^2/(2m)}\right)^{-1} \equiv Z^{-1} ,$$

---

[23]In Newtonian mechanics the momentum of particles becomes infinite as speeds $v \to \infty$; the momentum of relativistic particles becomes infinite as $v \to c$.

[24]A *discrete* probability distribution specifies the probabilities associated with discrete variables, such as the two outcomes of tossing a coin—heads or tails. For continuous quantities, such as the speeds of the molecules of a gas, we must work with *probability densities*, the probability that a continuous variable occurs *within a specified range of values*. In quantum mechanics $|\psi(x)|^2$ is often said to be a probability, whereas actually $|\psi(x)|^2$ is a probability density function, with $|\psi(x)|^2\,\mathrm{d}x$ the probability of finding a particle between x and $x + \mathrm{d}x$.

[25]One would think that $\rho(r, p)$ for a point particle is a six-dimensional Dirac delta function. Such would be the case if one knew precisely the *initial conditions* for the trajectory of the particle. As it is, however, $\rho(r, p)$ is a *statistical* quantity, governed by the Boltzmann transport equation (not treated in this book). The phase-space density function is a fundamental object of study in non-equilibrium statistical mechanics. For our purposes here, consider $\rho(r, p)$ as in-principle known.

[26]The quantities $\{p_k\}$ are a discrete probability distribution, not a probability density.

[27]We use the symbol P_k^2 for momentum squared to distinguish it from p_k for probability.

[28]There are other ways to derive the Maxwell-Boltzmann distribution from the approach adopted here.

where

$$Z(\beta) = \sum_k e^{-\beta P_k^2/(2m)} \tag{7.27}$$

is known as the *partition function*. We've related one Lagrange multiplier (α) to the other, β. Combining Eqs. (7.27) and (7.26), the most probable distribution function has the form

$$p_k = \frac{1}{Z} e^{-\beta P_k^2/(2m)} \,. \tag{7.28}$$

The quantity β can be found from the energy constraint, Eq. (7.25). To do so we rely on an oft-used trick. Differentiate Eq. (7.27):

$$\frac{\partial Z}{\partial \beta} = -\frac{1}{2m} \sum_k P_k^2 e^{-\beta P_k^2/(2m)} \,. \tag{7.29}$$

Equation (7.29) is quite similar to the sum we want in Eq. (7.25), and in fact

$$U = -\frac{N}{Z} \frac{\partial Z}{\partial \beta} \,. \tag{7.30}$$

If we knew $Z = Z(\beta)$, we'd know the internal energy.

To evaluate Z using Eq. (7.27), convert the sum to a phase-space integral:

$$Z = \sum_k e^{-\beta P_k^2/(2m)} \longrightarrow \frac{1}{g} \int_\Omega e^{-\beta P^2/(2m)} d\sigma = \frac{V}{g} \int_0^\infty 4\pi P^2 e^{-\beta P^2/(2m)} dP$$
$$= \frac{V}{g} \left(\frac{2m\pi}{\beta} \right)^{3/2} \,, \tag{7.31}$$

where we've used isotropy in momentum space to adopt spherical coordinates. It's straightforward to show, combining Eq. (7.31) with Eq. (7.30), that β is related to the energy:

$$U = 3N/(2\beta) \,. \tag{7.32}$$

To calculate the entropy, convert Eq. (7.21) into an integral, as in Eq. (7.31). It can be shown that (see Exercise 7.6)

$$S = Nk_B \left(\frac{3}{2} + \ln \left[\frac{V}{g} \left(\frac{2m\pi}{\beta} \right)^{3/2} \right] \right) = Nk_B \left(\frac{3}{2} + \ln \left[\frac{V}{g} \left(\frac{4m\pi U}{3N} \right)^{3/2} \right] \right) \,, \tag{7.33}$$

where we've used Eq. (7.32) in the second equality.

With Eq. (7.33) we have an expression for S having the desired property that it involves the macroscopic variables U, V, and N. But is it right? Ultimately that's a matter of comparison with experiment to decide. For the statistical entropy to agree with the thermodynamic entropy, the relations in Eq. (3.30) must be satisfied. Using Eq. (7.33),

$$\frac{1}{T} = \left(\frac{\partial S}{\partial U} \right)_{V,N} = \frac{3Nk_B}{2U} = \beta k_B \,,$$

where we have used Eq. (7.32). Thus, we have another result for the Lagrange multiplier, $\beta = (k_B T)^{-1}$, implying from Eq. (7.32),

$$U = \frac{3}{2} Nk_B T \,. \tag{7.34}$$

The measured value of C_V for noble gases is $1.50nR$ (Section 1.11); Eq. (7.34) is consistent with that result. Combining Eq. (7.33) with the second equation in Eq. (3.30),

$$\frac{P}{T} = \left(\frac{\partial S}{\partial V}\right)_{U,N} = \frac{Nk_B}{V}, \tag{7.35}$$

the equation of state of the ideal gas. These two results, Eqs. (7.34) and (7.35) (which are independent of g), confirm that Eq. (7.33) *is a valid formula as far as its dependence on U and V is concerned.*

What about the third equation in Eq. (3.30),

$$-\frac{\mu}{T} = \left(\frac{\partial S}{\partial N}\right)_{U,V} ?$$

We cannot test this identity (yet) because Eq. (7.33) does not explicitly display all of its dependence on the variable N. It turns out that the cell size g must be N-dependent, with $g \propto N$. To allow for the possibility that all particles *could* be concentrated into one cell, it would be "nice" if $g \propto N$, but there's no reason from classical physics that the size of a phase-space cell *must* depend on the number of particles.[29] If g *does* scale with N, Eq. (7.33) becomes an extensive expression for S, what's taken as axiomatic in the postulational approach to thermodynamics (see Section 3.7), and perhaps that's reason enough at the macroscopic level. We show in the next section that agreement with experiment requires $g = Nh^3$, *a result that cannot be obtained from thermodynamics.* While coarse-graining phase space seemed like an idea worth trying, it's led us to a place where thermodynamics has no more cards to play, what we may call *Physics Finisterre*, the edge of the world of classical physics. To make progress, we require input from experiment and guidance from the new world of quantum mechanics.

Let's accept that $g \propto N$, as it suggests that *each particle is a cell in phase space*; each particle is associated with an intrinsic phase-space volume. The volume of phase space for point particles has dimensions of (action)^3. Let's write $g = N\xi^3$, where ξ is a quantity having the dimension of action; in this way Eq. (7.33) becomes (with $\beta = (k_B T)^{-1}$)

$$S = Nk_B \left(\frac{3}{2} + \ln\left[\frac{V}{N}\left(\frac{2m\pi k_B T}{\xi^2}\right)^{3/2}\right]\right). \tag{7.36}$$

The quantity $(\xi^2/(2m\pi kT))^{3/2}$ has the dimension of volume (it must for the argument of the logarithm to be dimensionless), and thus $\xi/\sqrt{2\pi mkT}$ is a *length*. Define the *thermal wavelength*

$$\lambda_T \equiv \frac{\xi}{\sqrt{2\pi mkT}}, \tag{7.37}$$

in terms of which Eq. (7.36) can be written:

$$S = Nk\left[\frac{3}{2} + \ln\left(\frac{V/N}{\lambda_T^3}\right)\right]. \tag{7.38}$$

Equation (7.38) makes contact with Eq. (7.24), sort of. Atoms are constrained to "cages" of size V/N (volume per particle); if particles bring to the party an intrinsic volume λ_T^3, the number of "choices" it has is $(V/N)/\lambda_T^3$. Equation (7.38) is in the form of Eq. (3.27), the expression for S derived from thermodynamics.[30]

---

[29]In classical mechanics *Liouville's theorem* states that the volume of a region of phase space remains constant under its dynamical evolution as governed by Hamilton's equations, what can be seen as a classical antecedent of the Heisenberg uncertainty principle, and what would lend credence to the idea that $g \propto N$ once a fundamental volume of phase space per particle is established in line with the uncertainty principle.

[30]The factor of $\frac{3}{2}$ in Eq. (7.38) can be put inside the logarithm: $\frac{3}{2} = \ln e^{3/2}$.

The denominator in Eq. (7.37) is closely related to the mean momentum of a particle (Exercise 7.7), and action divided by momentum is a length. The question becomes: Is there an *intrinsic* quantity of action associated with material particles? If so, λ_T^3 *is a natural spatial cell size associated with a gas at temperature* T. The cell size would then be specified by the physics of the problem (and not chosen artificially as in Eq. (7.24)). Is there a way to find ξ^3, this "natural" phase-space volume per particle? The question is *extra-thermodynamic*, beyond the scope of thermodynamics. We need a reality check from experiment.[31]

7.4 SACKUR-TETRODE EQUATION

Can entropy be *measured*? While one cannot purchase an entropy meter, the entropy of the gas phase can be measured *relative* to that of the liquid phase through the latent heat, $L = T(s^v - s^l)$, Eq. (6.21). If the entropy of the liquid is known—from heat capacity data, Exercise 6.4—and if the latent heat is known, the entropy of the vapor is $s^v = s^l + L/T$. If thermodynamic measurements can be made sufficiently accurately (they can), Eq. (7.38) could be *tested* against Eq. (6.21) in its predictions for the entropy of the vapor. Such a test was made in 1912, independently by H. Tetrode and O. Sackur. We outline the Sackur-Tetrode analysis in this section, because, as we'll see, a fundamental discovery was made in the process. In comparing the results of Eqs. (6.21) and (7.38), we're demanding *consistency between the thermodynamic and statistical entropies*. A substance for which comprehensive thermodynamic data exists is mercury; we'll test Eq. (7.38) using the data for mercury.

Entropy from thermodynamics

Integrate Kirchhoff's equation, $\mathrm{d}L(T)/\mathrm{d}T = \Delta c_P$, Eq. (6.27), using $c_P = \frac{5}{2}R$ for the vapor (for mercury vapor $c_P = \frac{5}{2}R$ to three decimal places),

$$L(T) = L(T_0) + \tfrac{5}{2}R(T - T_0) - \int_{T_0}^{T} c_P^l(T')\mathrm{d}T' , \qquad (7.39)$$

where T_0 is a reference temperature. Between 273 K and 373 K the heat capacity of liquid mercury varies by less than 2%.[31, p6-179] In a first approximation we can treat c_P^l as constant. Thus, from Eq. (7.39), for $T > T_0$

$$L(T) = L(T_0) + \left[\tfrac{5}{2}R - c_P^l\right](T - T_0) . \qquad (7.40)$$

For mercury at $T_0 = 298.15$ K, $L(T_0) = 61.38 \pm 0.04$ kJ mol^{-1}[32] and the vapor pressure in equilibrium with liquid mercury is $P(T_0) = 0.2613$ Pa.[31, p6-181]

We need the entropy of the liquid phase. From Eq. (P6.1),

$$s^l(T) = \int_{0}^{T_m} \frac{c_P^{\text{solid}}(T)}{T}\mathrm{d}T + \frac{L(T_m)}{T_m} + \int_{T_m}^{T} \frac{c_P^l(T')}{T'}\mathrm{d}T' , \qquad (\text{P6.1})$$

where T_m is the temperature at which solid mercury melts. These integrals have mostly been done for us; thermodynamic properties of key chemical substances are tabulated, where, for example, the molar entropy in the liquid phase is given at a reference temperature T_0. Let's denote the tabulated value as $s_0 \equiv s^l(T_0)$. In that case we can write Eq. (P6.1) as, for $T > T_0 > T_m$

$$s^l(T) = s_0(T_0) + \int_{T_0}^{T} \frac{c_P^l(T')}{T'}\mathrm{d}T' . \qquad (7.41)$$

---

[31]One should observe from this section that in trying to provide a coarse-grained description of a *classical* gas, we've been led to the necessity of Planck's constant. It seems one can't scratch the surface of thermodynamics very deeply without finding quantum mechanics lurking beneath.

For mercury, $s_0 = 76.90 \pm 0.12$ J K^{-1} mol^{-1} at $T_0 = 298.15$ K.[32] Making the approximation that c_P^l is constant, we have from Eq. (7.41)

$$s^l(T) = s_0(T_0) + c_P^l \ln \left(\frac{T}{T_0} \right) . \tag{7.42}$$

By combining Eqs. (7.42) and (7.40) with Eq. (6.21), we have *the entropy of the vapor expressed entirely in terms of measured quantities*:

$$s^v(T) = \frac{1}{T} \left[L(T_0) + (\tfrac{5}{2}R - c_P^l)(T - T_0) \right] + s_0(T_0) + c_P^l \ln (T/T_0) . \tag{7.43}$$

Entropy from the statistical theory

Let's now calculate the entropy of the vapor using Eq. (7.38). First, however, let's parameterize that expression. Denote the factor of $\frac{3}{2}$ in Eq. (7.38) as α, and let $\xi = zh$ where z is a number. The molar entropy from Eq. (7.38) can then be written

$$s^v(T) = R \left[\alpha + \ln \left(\frac{kT}{P} \left(\frac{2\pi mkT}{h^2} \right)^{3/2} \frac{1}{z^3} \right) \right] ,$$

where we've used the ideal gas equation of state. Next, scale the temperature and pressure in terms of their values at the reference temperature T_0 by writing $T = tT_0$ and $P = pP(T_0)$. In this way,

$$s^v(T) = R \left[\alpha - 3\ln z + \ln \left(\frac{kT_0}{P(T_0)} \left(\frac{2\pi mkT_0}{h^2} \right)^{3/2} \right) + \ln \left(\frac{t^{5/2}}{p} \right) \right] . \tag{7.44}$$

Using $T_0 = 298.15$ K and $P(T_0) = 0.2613$ Pa, we have the constant

$$A \equiv \ln \left(\frac{kT_0}{P(T_0)} \left(\frac{2\pi mkT_0}{h^2} \right)^{3/2} \right) = 31.3992 , \tag{7.45}$$

where we've used the atomic weight of mercury 200.59.

Concordance?

Equate Eqs. (7.44) and (7.43), the two expressions for the entropy of the vapor. We have:

$$\alpha - 3\ln z = \frac{1}{RT} \left[L(T_0) + (\tfrac{5}{2}R - c_P^l)(T - T_0) \right] + \frac{1}{R} \left(s_0(T_0) + c_P^l \ln t \right) - A - \ln \left(t^{5/2}/p \right) . \tag{7.46}$$

Consistency requires that the right side of Eq. (7.46) be independent of temperature.
    First evaluate Eq. (7.46) for $t = 1$, $p = 1$. We'll need

$$\frac{L(T_0)}{RT_0} = 24.762 \qquad\qquad \frac{s_0(T_0)}{R} = 9.129 . \tag{7.47}$$

Combining Eqs. (7.47), (7.46), and (7.45), we find

$$\alpha - 3\ln z = 2.492 . \qquad\qquad (T = 298.15 \text{ K}) \tag{7.48}$$

    Now evaluate Eq. (7.46) at another temperature; choose $T_1 \equiv 373.15$ K. The value of c_P^l is almost constant for $T_0 < T < T_1$; pick the value of c_P^l as close to the middle of the temperature range as possible. At $T = 333.15$ K, $c_P^l = 28.688$ J K^{-1} for mercury.[31, p6-179] Using this value of c_P^l, from Eq. (7.44) $L(T_1) = 60.86$ kJ mol^{-1}. Note that the heat of vaporization goes *down* with an increase in temperature. From Eq. (7.42),

$$s^l(T_1) = 76.90 + 28.688 \ln \left(\frac{373}{298} \right) = 82.11 \text{ J K}^{-1} , \qquad (7.49)$$

i.e., the entropy of the liquid goes *up* with increased temperature. We then have

$$\frac{L(T_1)}{RT_1} = 19.618 \qquad \frac{s^l(T_1)}{R} = 9.876 .$$

To complete the calculation, we need the vapor pressure $P(T_1) = 38.21$ Pa.[31, p6-181] The remaining term in Eq. (7.46) has the value -4.398. Assembling the numbers, we find

$$\alpha - 3 \ln z = 2.493 . \qquad (T = 373.15 \text{ K}) \qquad (7.50)$$

The close numerical agreement between Eqs. (7.48) and (7.50) gives us considerable confidence in the statistical theory of entropy. If we take $z = 1$ (a physically appealing choice), then we must accept the value of $\alpha \approx 2.49$ as being prescribed by experiment, and *not* $\frac{3}{2}$ as derived in Eq. (7.38). In Section 7.6 we'll see how $\alpha = \frac{5}{2}$ emerges naturally. The difference between $\frac{5}{2}$ and $\frac{3}{2}$ would seem to be small, but actually it's *huge*: It implies an increase of W in Eq. (7.2) by a factor of e^N. By accepting $z = 1$ as consistent with experimental measurements, it would appear to be the first time in the logical development of physics of associating Planck's constant with material particles. Before this point in the history of physics, Planck's constant was associated only with photons. The implication of $z = 1$ is that, from Eq. (7.37), the thermal wavelength is

$$\lambda_T = \frac{h}{\sqrt{2\pi m k T}} \sim \frac{h}{\bar{p}} , \qquad (7.51)$$

where $\bar{p}$ is the mean momentum, Exercise 7.7. Equation (7.51) *presages the de Broglie wavelength*. It also indicates that h^3 is a natural cell size in phase space for a single particle *as prescribed by fundamental physics*. In the attempt to calculate the entropy of the ideal gas, we've been led to the picture provided by the Heisenberg uncertainty principle, that we can't "peer" into a region of phase space $\Delta x \Delta p$ smaller than Planck's constant. Just as with black body radiation, thermodynamics has led us to the frontier separating the classical and quantum descriptions of the physical world.

Example. The thermal wavelength associated with a hydrogen atom at $T = 300$ K is, from Eq. (7.51), $\lambda_T = 0.1$ nm.

Sackur-Tetrode equation

If we take $z = 1$, then $\alpha \approx 2.49$, well within 1% of $\frac{5}{2}$, and if we take $\alpha = \frac{5}{2}$, then $z \approx 1.003$, well within 1% of $z = 1$. Taking $\alpha = \frac{5}{2}$ and $z = 1$ as consistent with experimental measurements, we have the *Sackur-Tetrode formula* for the entropy of the ideal gas,

$$S = Nk \left[\frac{5}{2} + \ln \left(\frac{V}{N} \left(\frac{4\pi m U}{3Nh^2} \right)^{3/2} \right) \right] . \qquad (7.52)$$

All parts of this expression have been tested against experiment: heat capacity consistent with Eq. (7.34), equation of state, Eq. (7.35), and entropy of the vapor phase, Eq. (7.43). We show in Section 7.6 how to derive Eq. (7.52) starting from first principles.

There are now two independent lengths in the problem (Fig. 7.1): the average distance between particles, $(V/N)^{1/3}$, and the thermal wavelength, λ_T. When $(V/N)^{1/3} \gg \lambda_T$, the particles behave independently (that is, classically); when $(V/N)^{1/3} \lesssim \lambda_T$, the particles are strongly interacting and their quantum nature (Fermi or Bose statistics) becomes apparent.

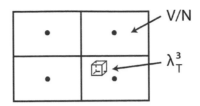

Figure 7.1 Two lengths: Average distance between particles and the thermal wavelength

7.5   VOLUME OF A HYPERSPHERE*

We show that the volume $V_n(R)$ of a *hypersphere* of radius R in n-dimensional Euclidean space $(\sum_{k=1}^{n} x_k^2 = R^2)$ is given by

$$V_n(R) = \frac{\pi^{n/2}}{\Gamma(\frac{n}{2}+1)} R^n \,, \tag{7.53}$$

where $\Gamma(z)$, the gamma function, is by definition $\Gamma(z) \equiv \int_0^{\infty} e^{-t} t^{z-1} dt$. The gamma function satisfies the recursion relation $\Gamma(z+1) = z\Gamma(z)$ (integrate by parts). For integers $\Gamma(n) = (n-1)!$. By direct evaluation, $\Gamma(\frac{1}{2}) = \sqrt{\pi}$. From Eq. (7.53), $V_3(R) = (\pi^{3/2}/\Gamma(1+\frac{3}{2}))R^3$. Using the recursion relation, $\Gamma(1+\frac{3}{2}) = \frac{3}{2}\Gamma(\frac{3}{2}) = \frac{3}{2}\Gamma(1+\frac{1}{2}) = \frac{3}{4}\Gamma(\frac{1}{2}) = 3\sqrt{\pi}/4$. Thus, $V_3(R) = 4\pi R^3/3$. For $n=2$ and $n=1$, Eq. (7.53) gives $V_2(R) = \pi R^2$ and $V_1(R) = 2R$.

To derive Eq. (7.53), we first define a function of n variables, $f(x_1,\cdots,x_n) \equiv \exp\left(-\frac{1}{2}\sum_{i=1}^{n} x_i^2\right)$. In Cartesian coordinates, the integral of f over all of $\mathbb{R}^n$ is

$$\int_{\mathbb{R}^n} f(x_1,\cdots,x_n)dV = \prod_{i=1}^{n}\left(\int_{-\infty}^{\infty} dx_i \exp(-\tfrac{1}{2}x_i^2)\right) = (2\pi)^{n/2} \,, \tag{7.54}$$

where dV is the n-dimensional volume element, and we've used $\int_{-\infty}^{\infty} e^{-ax^2} dx = \sqrt{\pi/a}$. Now recognize that $\sum_{i=1}^{n} x_i^2 \equiv r^2$ defines the radial coordinate in n dimensions. Let $dV = A_n(r)dr$, where $A_n(r)$ is the surface area of an n-dimensional sphere. The surface area of an n-sphere can be written $A_n(r) = A_n(1)r^{n-1}$. The same integral in Eq. (7.54) can then be expressed

$$\int_{\mathbb{R}^n} f dV = \int_0^{\infty} A_n(r) \exp(-\tfrac{1}{2}r^2)dr = A_n(1)\int_0^{\infty} r^{n-1}\exp(-\tfrac{1}{2}r^2)dr \tag{7.55}$$

$$= 2^{(n/2)-1} A_n(1) \int_0^{\infty} t^{(n/2)-1} e^{-t} dt = 2^{(n/2)-1} A_n(1)\Gamma(\tfrac{n}{2}) \,,$$

where in the second line we have changed variables $t = \frac{1}{2}r^2$. Comparing Eqs. (7.55) and (7.54), we conclude that $A_n(1) = 2\pi^{n/2}/\Gamma(\frac{n}{2})$. This gives us the surface area of an n-sphere of radius R:

$$A_n(R) = \frac{2\pi^{n/2}}{\Gamma(\frac{n}{2})} R^{n-1} \,. \tag{7.56}$$

From Eq. (7.56), $A_3(R) = 4\pi R^2$, $A_2(R) = 2\pi R$, and $A_1(R) = 2$. By integrating $dV = A_n(r)dr$ from 0 to R, we obtain Eq. (7.53). Note that $A_n(R) = (n/R)V_n(R)$.

7.6   LEARNING TO COUNT WITH PHYSICS

If it disagrees with experiment, it's wrong. In that simple statement is the key to science. It doesn't make any difference how beautiful your guess is, it doesn't matter how smart you are, who made the guess, or what his name is. If it disagrees with experiment, it's wrong. That's all there is to it.—Richard Feynman[33, p150]

Accepting that we've been led through agreement with experiment to h^3 as the intrinsic volume of phase space per particle ($z = 1$ in Section 7.4), can the Sackur-Tetrode formula be *directly* derived? We have to learn to *correctly* count configurations, in a way that leads to predictions in agreement with experiment. Our method of coarse graining in Section 7.3 led to the factor of $\frac{3}{2}$ in Eq. (7.38), which we now know isn't right. As with any idea in physics, its veracity is ascertained by its consequences, with experiment being the arbiter of truth. We therefore *abandon* the approach adopted in Section 7.3. Instead, we count the number of configurations of N identical particles of a gas as follows:

$$W = \frac{1}{h^{3N} N!} \times \text{(volume of accessible } N\text{-particle phase space)} . \qquad (7.57)$$

We're now talking about a phase space of $6N$ dimensions, that for N particles *taken as a whole*. The factor of h^{3N} "quantizes" phase space for N particles, providing the discreteness we require to be able to count. The road to Eq. (7.57) was tortuous (Sections 7.3 and 7.4), but once we learn that physics is telling us that each particle has attached to it an intrinsic volume of phase space h^3, then Eq. (7.57) seems natural. The factor of $N!$ in Eq. (7.57) recognizes that *permutations of identical particles are not distinct states*. Apparently Gibbs was the first to include the factor of $N!$:

> If two phases differ only in that certain entirely similar particles have changed places with one another, are they to be regarded as identical or different phases? If the particles are regarded as indistinguishable, it seems in accordance with the spirit of the statistical method to regard the phases as identical.[34, p187]

By *phase* here, Gibbs means a point in phase space. By introducing the idea that identical particles should be treated as *indistinguishable*, Gibbs made a break with classical physics wherein identical particles are *distinguishable*. Entropy is a property of the *entire* system, with the interchange of identical particles having no consequence. Without knowing it, Gibbs was straddling classical and quantum physics; another instance where the requirements of thermodynamics seem to demand the existence of quantum mechanics. By the volume of phase space *accessible* to the system in Eq. (7.57), we mean the volume of phase space subject to constraints on the system, for example, the constraints of the spatial volume V that the system occupies and its total energy U.

The Sackur-Tetrode formula can quickly be derived starting from Eq. (7.57). Let

$$W = \frac{1}{h^{3N} N!} \int d^3 x_1 \int d^3 p_1 \cdots \int d^3 x_N \int d^3 p_N = \frac{V^N}{h^{3N} N!} \int d^3 p_1 \cdots \int d^3 p_N ,$$

where the integration over momentum space must be consistent with the total energy, $2mU = \sum_{i=1}^{3N} p_i^2$. The energy constraint specifies a $3N$-dimensional hypersphere of radius $\sqrt{2mU}$, the area of which is given by Eq. (7.56). Thus,

$$W = \frac{V^N}{h^{3N} N!} \frac{2\pi^{3N/2}}{\Gamma(\frac{3N}{2})} (2mU)^{(3N-1)/2} . \qquad (7.58)$$

The Sackur-Tetrode formula follows by combining Eq. (7.58) with Eq. (7.2) and approximating $\ln\left(N! \Gamma(3N/2)\right)$ with the Stirling approximation. Use Eq. (7.8) for $\ln(N!)$ and approximate $\ln\left(\Gamma(z)\right) \sim z \ln(z) - z$ for large z. In the limit of large N we obtain Eq. (7.52) (ignore the difference between $3N - 1$ and $3N$). Equation (7.58) is a gateway into statistical mechanics.

7.7 GIBBS'S PARADOX, NOT

> If we should bring into contact two masses of the same kind of gas, they would mix, but there would be no increase in entropy.—J.W. Gibbs, 1876[3, p166]

> It has always been believed that Gibbs's paradox embodied profound thought.
> —Erwin Schrödinger, 1952[35, p61]

> A paradox is a conflict between reality and your feeling of what reality ought to be.
> —Richard Feynman[36, p18-9]

The term *Gibbs's paradox* refers to several related issues pertaining to entropy, extensivity, and the *sameness* of particles. Extensive quantities are additive over independent subsystems (Sections 1.2 and 3.7). Internal energy is extensive, but it might not be so if there are long-range interactions or if the system is so small that surface effects dominate. Is entropy extensive? While entropy isn't tied to a microscopic property of atoms, it scales with the number of particles. Implicit in the derivation of $S = k \ln W$ is the very condition for extensivity (Section 7.1). Formulas such as Eq. (7.36) therefore exhibit extensivity. For thermodynamic entropy, however, extensivity must be imposed as an additional requirement (Section 3.7). Therein lies part of the confusion behind what's referred to as Gibbs's paradox.

A simple test of extensivity is provided by the following thought experiment. Consider the act of removing a partition that separates two samples of the same ideal gas having the same temperature and pressure (Fig. 7.2). The system is thermally isolated; no heat can flow from the environment.

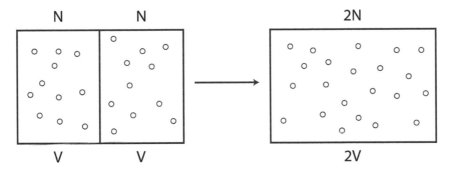

Figure 7.2 Remove the partition: Does the entropy change?

Does entropy change when the partition is removed? Using Eq. (3.27) (thermodynamics) or Eq. (7.52) (Sackur-Tetrode formula), the answer is *No*:

$$\Delta S = 0 . \qquad \text{(extensive entropy)} \qquad (7.59)$$

No change in entropy ensues upon combining two identical parts of the same system if entropy is extensive. What if, however, we were to use Eq. (7.33) as an expression for the entropy, which has the form

$$S = Nk \left[\ln V + \frac{3}{2} \ln \left(\frac{U}{N} \right) + X \right] , \qquad (7.60)$$

where X is a constant? Equation (7.60) (non-extensive) predicts

$$\Delta S = 2Nk \ln 2 . \qquad \text{(non-extensive entropy)} \qquad (7.61)$$

Entropy of isolated systems increases only through irreversible processes (Section 3.3). Sliding back in the partition restores the system to its original *macroscopic state* and is thus a reversible process. Equation (7.61) cannot be correct. *The entropy of an ideal gas must be extensive.*[32]

Where's the paradox? There isn't one. Gibbs's paradox is often "ginned up" by starting with a non-extensive *thermodynamic entropy* and arguing that it leads to a violation of the second law, as in Eq. (7.61). The "resolution" is then presented that one must use the (extensive) Sackur-Tetrode equation, of which, it's touted, one needs quantum mechanics to understand. Invoking quantum mechanics is a red herring: *Quantum mechanics does not resolve a paradox because there is no paradox.* It's true that the standard calculation in thermodynamics leads to a non-extensive entropy in the form of Eq. (7.60), e.g., Eq. (3.20). Extensivity is not a consequence of the Clausius definition; it must be imposed as a separate requirement.[33] The "constant" X in Eq. (7.60) is actually a function of N which can be determined by imposing extensivity, Eqs. (3.26) or (3.27). There is no paradox; the Clausius definition is incomplete (applies to closed systems only). As we've shown, the thermodynamic entropy can be made extensive. All you have to do is ask for it. Paradox lost.

A stronger version of Gibbs's paradox is as follows: Remove the partition separating *dissimilar* ideal gases having the same temperature and pressure[34]—see Fig. 7.3. Using either Eq. (3.26) or

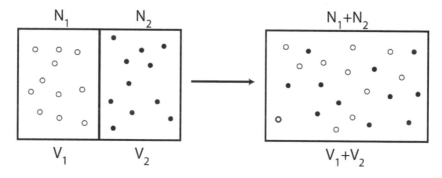

Figure 7.3 Remove the partition: Does the entropy change?

Eq. (7.52), we find for the process depicted in Fig. 7.3,

$$\Delta S = kN_1 \ln \left(\frac{V}{V_1} \right) + N_2 k \ln \left(\frac{V}{V_2} \right) = -Nk \left[x_1 \ln x_1 + x_2 \ln x_2 \right] , \qquad (7.62)$$

where $V \equiv V_1 + V_2$, $N = N_1 + N_2$, and $x_{1,2} \equiv (N_{1,2}/N) = V_{1,2}/V$ are the respective mole fractions. Equation (7.62) expresses the *entropy of mixing* (Section 6.3), although the term is misleading because it's not the mixing *per se* that leads to an increase in entropy, it's the change in volume for each gas. When $V_1 = V_2$, Eq. (7.62) becomes

$$\Delta S = (N_1 + N_2)k \ln 2 . \qquad (7.63)$$

Consider the formal step in Eq. (7.63) of allowing dissimilar atoms to become identical: Let $N_1 \equiv N_2 = N$. *We do not recover* Eq. (7.59). This point was noted by Gibbs in 1876, and ever since then, resolving the "paradox" has been a topic of debate. It's remarkable that *Gibbs did not see a problem with what's known as Gibbs's paradox.* Let's let Gibbs speak for himself:

---

[32]That doesn't imply, however, that entropy is extensive for *all* systems. The possibility and implications of non-extensive entropies for certain systems has become an active field of research.[37]

[33]We noted in Section 3.7 that in axiomatic formulations of thermodynamics, extensivity of entropy is taken as a postulate.

[34]All gases become ideal at sufficiently low pressure, Section 1.6. Thus, there is not simply *one* idealized gas, we can meaningfully speak of dissimilar ideal gases.

It is noticeable that the value of this expression [Eq. (7.63)] does not depend on the kinds of gas ... except that the gases which are mixed must be of different kinds. If we should bring into contact two masses of the same kind of gas, they would also mix, but there would be no increase of entropy. ... When we say that when two different gases mix by diffusion, ..., the energy of the whole remains constant, and the entropy receives a certain increase, we mean that the gases could be separated and brought to the same volume and temperature which they had at first by means of certain changes in external bodies .... But when we say that when two gas-masses of the same kind are mixed ... there is no change of energy or entropy, we do not mean that the gases which have been mixed can be separated without change to external bodies. On the contrary, the separation of the gas is entirely impossible. We call the energy and entropy of the gas-masses when mixed the same as when they were unmixed, because we do not recognize any difference in the substance of the two masses.[3, p166]

Parsing Gibbs, he's saying that mixing dissimilar gases is irreversible, while mixing identical gases is not. A mixture of dissimilar gases can be unmixed by means of "certain changes in external bodies"—through the use of selectively permeable membranes. The gases can be unmixed but the environment will not have been restored to its original configuration (Section 1.7). On the other hand, mixing similar gases produces no change in the *macroscopic* state of the gas. Far from being a paradox, our confidence in the theory should be bolstered that the same extensive expressions give no entropy change for mixing similar gases, but a finite value for dissimilar gases.[35]

7.8 SUBTLETIES OF ENTROPY

Is entropy absolute?

Is there a unique value of S for a given equilibrium state? Consider the analogous question: Is there a unique value of the electrostatic potential at a given point in space? Charged particles respond to *gradients* of the potential (electric field), not its value. Similarly, thermodynamic processes are associated with entropy *differences*. A constant S_0 added to the entropy of every equilibrium state has no physical consequence. Intensive quantities such as P, T, and μ occur as derivatives of S. There is no sense to the notion of absolute entropy.

Yet, the issue is confounded by the two types of entropy: thermodynamic, defined as a *differential*, $dS = đQ/T$, and statistical, specified by a *formula*, $S = k \ln W$. The question becomes *is there a unique value of W for a given equilibrium state?* Extending the analogy with electrostatics, the potential is specified relative to zero, "ground," which we are free to place wherever we want. What *state of matter* should be associated with zero entropy? In Section 7.1 we found a second-order differential equation for S, the solution of which is $S = k \ln W + S_0$. We established the scale, $k = k_B$, but we took $S_0 = 0$ without fanfare, leaving us with Eq. (7.2). From Eq. (7.3), a nonzero value of S_0 would drop out of an entropy difference, $\Delta S = k \ln(W_1/W_2)$. We could have, however, taken $S_0 = -k \ln W_0$, so that instead of Eq. (7.2) we would have its generalization

$$S = k \ln(W/W_0) . \tag{7.64}$$

Entropy in the form of Eq. (7.64) also leads to Eq. (7.3).

Thus, we have two ways of writing the statistical entropy [Eqs. (7.2) and (7.64)], both of which lead to Eq. (7.3). Each indicates that $S = 0$ is possible: $W = 1$ in Eq. (7.2) and $W = W_0$ in Eq. (7.64). It would be natural to associate the special value $S = 0$ with the state of matter at $T = 0$, and it's commonly held (incorrectly it turns out) that entropy vanishes as $T \rightarrow 0$. The third law

---

[35]One can almost hear Yogi Berra saying: Dissimilar gases are dissimilar, and as long as they are dissimilar, they are dissimilar. Don't expect Eq. (7.63) to reduce to Eq. (7.59) as one cannot smoothly let dissimilar objects become the same; nature doesn't let us do that experiment.

of thermodynamics is concerned with the low-temperature properties of matter, and we discuss in Chapter 8 systems[36] for which $S(T = 0) \neq 0$. The issue for us here, continuing the question of whether W is unique: Is *phase space* unique? Chemical elements exist in the form of different isotopes, each of which has its own nuclear spin state. Should we, in counting microstates, enlarge phase space to include nuclear spin or other subatomic variables? *Degrees of freedom not subject to change in experiments, do not contribute to changes in entropy.*[37] We were careful to note on page 10 that entropy is a quantitative *measure* of the number of microstates associated with a given macrostate, not the *absolute number* of such states. With each substratum of matter discovered by physics, there are progressively more microscopic degrees of freedom that could be counted as contributing to W. Absolute entropy is not well founded: A state that we assign zero entropy can have a large number W_0 of associated configurations of microscopic variables. We revisit this topic in Section 8.3.

The issue of the degrees of freedom that contribute to the entropy, or not, is in some ways a tempest in a teapot. Categorize the number of degrees of freedom that are in no way affected by experimental processes of interest as *irrelevant*, W_{irr}, and the number that could possibly become accessible in experiments as *relevant*, W_{rel}. For every irrelevant degree of freedom, there is the entire range of relevant degrees of freedom.[38] The total number of degrees of freedom is thus $W = W_{rel}W_{irr}$. With changes in entropy described by Eq. (7.3), irrelevant degrees of freedom drop out. From $S = k_B \ln W$, changes in entropy are, from Eq. (7.3), with $W_1 = W$ and $W_2 = W_1 + \Delta W$:

$$\Delta S = k_B \ln(1 + \Delta W/W) , \tag{7.65}$$

i.e., changes in entropy are controlled by the relative number of additional microstates that become accessible in a given process. Entropy represents the *potentialities* inherent in a system, or as Clausius put it, the *transformational content* of a system.

Is entropy real?

> But what mechanical or electrodynamical quantity represents the entropy?
> —Max Planck[27, p117]

Does entropy have objective existence?[39] We can measure heat flow and we can measure temperature; thus there's nothing *subjective* about entropy as defined by $dS = (\dj Q/T)_{rev}$, which specifies the *change* in entropy between states connected by reversible paths—if entropy exists for one state, it exists for all states connected to the first by a reversible path.[40] From $TdS = dU + PdV - \mu dN$ we have that the quantities determining dS are differentials of U, V and N, and no one doubts their reality.[41] For extensive entropy, we have the Euler relation $TS = U + PV - N\mu$, Eq. (3.31), *an*

---

[36]The entropy of ice at low temperature, for example, is nonzero which reflects the number of ways H atoms can be arranged in a crystalline network of H_2O molecules.

[37]This point leads some to conclude that entropy is subjective, because it would appear to be a property not of physical systems, but of the experiments we choose to perform on them. See Chapter 12 for Maxwell's take on subjectivity associated with the second law. One can't help notice similarities between entropy and quantum mechanics. Does the quantum state vector, $|\psi\rangle$, objectively exist in nature? The question should be turned around: What objectively-existing quantity in nature does $|\psi\rangle$ describe? A school of thought is that $|\psi\rangle$ is a property not of individual systems, but rather of an *ensemble* of systems. Entropy is a property not of individual particles, but of many.

[38]For this idea to work, there must be a clear separation between the two types of degrees of freedom, such as we have for molecular degrees of freedom (translational, configurational, rotational, vibrational, and electronic) and sub-atomic, i.e., nuclear and sub-nuclear, degrees of freedom.

[39]Without venturing unduly into philosophy, we use the term *objective* as it's used in science, to refer to objects that exist, or processes that occur, independently of the presence of human beings.

[40]Can every point in state space be reached by a reversible path? In the Carathéodory formulation of the second law (Chapter 10), the question is posed differently: Starting from a given point in state space, are there points *not reachable by adiabatic processes*, those with $\dj Q = 0$? It's shown that entropy exists if there are states *inaccessible* by adiabatic processes.

[41]Let's stipulate the objectivity of U, V, and N. Does the number of particles exist if you're not there to count them?

even more direct connection between S and other quantities whose reality is not in question. Thermodynamic entropy (with extensivity imposed) and statistical entropy are *two different theoretical approaches* to the same thing, which agree once we've learned to count states in a way consistent with quantum mechanics (Section 7.6), underscoring the reality of entropy. We must accord S the status of physical reality.

Yet, entropy is *unlike* other physical quantities (such as energy) because it's not related to a property of the individual components of the system. Rather, *it's a property of the equilibrium state*, which is characterized by a set of macrovariables $\{X_i\}_{i=1}^{n}$. Should we discover a new macrovariable X_{n+1}, our specification of the entropy will change. For that reason, it's said by some that entropy is subjective because the observer gets to choose the relevant macrovariables, and thus entropy has no objective meaning, it being a construct of the mind of the experimenter.[42] Consider that in the special theory of relativity, measurable quantities stand in relation to the observer, with length and time measurements not having independent existence because they depend on the choice of inertial reference frame. The same is true of the electric and magnetic fields—not independently existing, but reference-frame dependent. Would one say that space and time and the E and B fields are subjective? Astronomers measure the distance to stars *based on the knowledge they have*; no one would argue that stars don't *have* a distance from Earth. When new astronomical facts are discovered, distances are revised, and no one says that stellar distances are subjective. Incomplete information is not the same as subjectivity.[43]

Questions of subjectivity inevitably arise from the tension between *phenomenological* theories, categorizations of our experience, what *we* observe of the world, and *microscopic* theories that attempt to explain what we observe as a consequence of "impersonal" (read objectively existing) interactions between the microscopic constituents of matter. Science is *always* based on limited, incomplete information, subject to change; scientific theories represent our best effort at understanding the physical world, theories that can and do change. Science is a human activity, and is in that sense, subjective.

CHAPTER SUMMARY

This chapter introduced the microscopic interpretation of entropy, $S = k_B \ln W$, what quantifies the "missing" degrees of freedom not accounted for in macroscopic descriptions. The quantity W represents the number of microstates consistent with a specified equilibrium state. That number would be prodigiously large if every presently-known subatomic variable were included. "Frozen" degrees of freedom, those that don't change in the course of experiments, do not contribute to changes in entropy.

- Statistical entropy is motivated by entropy being a maximum in equilibrium and from the consideration that of the myriad microstates of a system, those compatible with equilibrium occur most frequently, with high probability W. The statistical entropy follows by requiring that entropy be additive over independent subsystems (extensive), $S = S_1 + S_2$, for which probability is multiplicative, $W = W_1 W_2$. These requirements determine $S = k_B \ln W$.

- W is the number of microstates compatible with a given state of equilibrium; the greater the number of microstates, the larger the entropy. We require that S obtained from $S = k_B \ln W$ must satisfy thermodynamic identities such as $T^{-1} = (\partial S/\partial U)_{V,N}$, and thus W must be found as a function of the state variables characterizing equilibrium, $W = W(U, V, N)$.

---

[42]Edwin Jaynes wrote:[38] "The entropy of a thermodynamic system is a measure of the degree of ignorance of a person *whose sole knowledge about its microstate consists of the values of the macroscopic quantities X_i which define its thermodynamic state*. This is a completely *objective* quantity, in the sense that it is a function only of the X_i, and does not depend on anyone's personality. There is then no reason why it cannot be measured in the laboratory."

[43]This point is argued in the book by Denbigh and Denbigh.[39]

- For N identical items distributed over N_c cells, the statistical entropy is (for N sufficiently large): $S = -Nk_B \sum_k p_k \ln p_k$, with p_k the probability that an item selected at random is in the kth cell. The probability distribution that maximizes S is for each cell equally likely, $p_k = N_c^{-1}$. The value of the entropy corresponding to the uniform distribution is $S = Nk_B \ln N_c$. This coarse-grained description is not rich enough to make contact with thermodynamics through $T^{-1} = \partial S / \partial U$; for that purpose we have to include the energy of the system.

- Phase space is a mathematical space in which all states of a system can be represented, with each possible state corresponding to a single point in the space. Phase space for a point mass is a six-dimensional space of all possible positions r and momenta p. That for N point masses is $6N$-dimensional.

- The phase-space probability density function that emerges upon maximizing the entropy subject to the constraints that it be normalized and that the energy U be the sum of the kinetic energies, is the Maxwell-Boltzmann distribution function

$$\rho(r, p) = \frac{1}{V} \left(\frac{\beta}{2m\pi} \right)^{3/2} \exp(-\beta p^2/(2m)),$$

where V is the volume of the gas and $\beta = (k_B T)^{-1}$. The value of the entropy one obtains from the Maxwell-Boltzmann probability distribution is

$$S = Nk_B \left(\frac{3}{2} + \ln \left[\frac{V}{g} \left(\frac{4m\pi U}{3N} \right)^{3/2} \right] \right),$$

where g is the unknown cell size in phase space. With this expression, the thermodynamic identities $T^{-1} = (\partial S / \partial U)_{V,N}$ and $P/T = (\partial S / \partial V)_{U,N}$ reproduce the results for the ideal gas

$$U = \frac{3}{2} Nk_B T \qquad PV = Nk_B T.$$

We cannot apply the thermodynamic identity $\mu/T = -(\partial S / \partial N)_{U,V}$ because we do not know the cell size, g. We argued that g scales with N, $g = N\xi^3$, where ξ has dimension of action. From thermodynamic measurements it is found that $\xi = h$, Planck's constant. This approach leads to the (incorrect) expression for the entropy of the ideal gas, Eq. (7.33)

$$S = Nk_B \left(\frac{3}{2} + \ln \left[\frac{V}{N} \left(\frac{4m\pi U}{3Nh^2} \right)^{3/2} \right] \right). \qquad \text{(incorrect)}$$

- It's found, by comparison with thermodynamic measurements, that Eq. (7.33) is not correct, but that the Sackur-Tetrode formula, Eq. (7.52), correctly accounts for the entropy of the vapor phase,

$$S = Nk_B \left(\frac{5}{2} + \ln \left[\frac{V}{N} \left(\frac{4m\pi U}{3Nh^2} \right)^{3/2} \right] \right). \qquad \text{(correct)}$$

- Writing $\lambda_T = h/\sqrt{2\pi m k_B T}$ as a length (thermal wavelength), the Sackur-Tetrode equation is:

$$S = Nk_B \left[\frac{5}{2} + \ln \left(\frac{V}{N\lambda_T^3} \right) \right].$$

When $V/N \gg \lambda_T^3$, the atoms act independently and a classical description is apt; the particles are strongly interacting when $V/N \ll \lambda_T^3$ and their quantum nature must be taken into account. The Sackur-Tetrode formula provides experimental confirmation (from thermodynamic measurements) of the picture that emerges from quantum mechanics that each particle carries with it a fundamental volume of phase space h^3.

- To derive the Sackur-Tetrode formula, we count microstates as in Eq. (7.57):

$$W = \frac{1}{h^{3N} N!} \times (\text{volume of } N\text{-particle phase space subject to constraints}),$$

where h^{3N} quantizes phase space for N particles and the factor of $N!$ recognizes that permutations of identical particles are not counted as distinct microstates.

EXERCISES

7.1 Show that for any system, the quantity W in Eq. (7.2) must satisfy the differential equation

$$\ln W = U \left(\frac{\partial \ln W}{\partial U} \right)_{V,N} + V \left(\frac{\partial \ln W}{\partial V} \right)_{N,U} + N \left(\frac{\partial \ln W}{\partial N} \right)_{U,V}.$$

Hint: Euler relation.

7.2 a. Plot $(N \ln N - N)/(\ln N!)$ versus N. For what value of N are you comfortable replacing $\ln N!$ with Stirling's approximation in the form of Eq. (7.8)?

   b. Plot (from Eq. (7.10)) $\left(\sqrt{2\pi N}(N/e)^N \right)/(N!)$ versus N. To what extent does the extra factor improve the approximation?

7.3 There is a lottery in the Province of Ontario, Canada known as 649: Pick six numbers out of 49. How many ways are there to do this, when the order in which the numbers are chosen is immaterial? What are odds of winning?

7.4 Use the Maxwell-Boltzmann distribution function, Eq. (7.28), combined with Eq. (7.31) to derive the expression for the energy of the ideal gas, Eq. (7.32), starting from Eq. (7.25).

7.5 Fill in the steps leading to Eq. (7.30).

7.6 Use Eq. (7.28) combined with Eq. (7.31) to derive the expression for the entropy, Eq. (7.33), starting from Eq. (7.21).

7.7 Show that the mean value of p, the magnitude of a particle's momentum, can be derived from the momentum distribution function $f(p)$ in Eq. (7.31), with the result $\bar{p} = \sqrt{8mkT/\pi}$.

7.8 Derive the Sackur-Tetrode equation Eq. (7.52) by combining Eq. (7.58) with Eq. (7.2) and invoking the Stirling approximation.

7.9 Derive an expression for the chemical potential of the ideal gas. Use Eq. (7.52) with

$$-\frac{\mu}{T} = \left(\frac{\partial S}{\partial N} \right)_{U,V}. \qquad A: \mu = -kT \ln \left[\frac{V}{N} \left(\frac{2\pi mkT}{h^2} \right)^{3/2} \right] = -kT \ln \left[\frac{V}{N\lambda_T^3} \right]$$

Show that this expression for μ can be put into the form of Eq. (4.12). Equation (4.12) is the Gibbs energy per mole; ignore the difference between R and k.

7.10 Derive an expression for the chemical potential when the entropy is given by

$$S = Nk \left[\frac{3}{2} + \ln \left(\frac{V}{N} \left(\frac{4\pi m U}{3Nh^2} \right)^{3/2} \right) \right],$$

which is what would be obtained by ignoring the indistinguishability of identical particles (omit the factor of $N!$ in Eq. (7.57)). Does your answer agree with that from the previous problem? Would such an expression for the entropy lead to a correct prediction of the vapor pressure? (Section 7.4)

7.11 Show how an extensive entropy leads to $\Delta S = 0$ in Eq. (7.59).

7.12 Derive Eq. (7.62). Treat each gas as if it is freely expanding into the newly increased volume.

7.13 Show that the chemical potential of the ideal gas found in Exercise 7.9 is what you would obtain from Eq. (3.31).

7.14 Derive an expression for the Helmholtz free energy of the ideal gas.

$$A: \qquad F(T, V, N) = -NkT \left(1 + \ln \left(V/(N\lambda_T^3) \right) \right).$$

Obtain an expression for the chemical potential from $\mu = (\partial F/\partial N)_{T,V}$. Is it the same as that derived in Exercise 7.9? The change in F for a reversible isothermal process is the work performed, $[\Delta F]_T = W$, Table 4.1. Show that this interpretation holds for the ideal gas using the result of this problem.

7.15 Show, starting from Eq. (7.31), and using the results obtained in this chapter, that we can write

$$Z = \frac{V}{N\lambda_T^3}.$$

This result is sometimes written

$$Z = \frac{n_Q}{n},$$

where $n \equiv N/V$ and $n_Q \equiv \lambda_T^{-3}$ is the *quantum concentration*, one particle in a volume of λ_T^3.

7.16 a. The Sackur-Tetrode formula for the entropy of an ideal gas, Eq. (7.52), has the form $S = Nk_B \left[\frac{5}{2} + \ln (\cdot) \right]$. Estimate the size of the logarithm term for a typical gas. For example, consider one mole of hydrogen gas in a volume of one liter at $T = 300$ K. In this case, $\ln \left(V/(N\lambda_T^3) \right) \approx 7.4$. The entropy of this gas would then be $S \approx 10Nk_B$.

b. A crude *estimate* of the entropy of a system can be had by assuming $S \sim Nk_B$. Using this rule of thumb, estimate the entropy in a glass of wine. Make reasonable guesses for whatever it is you need to know. Taking $S \sim N$, useful for purposes of guesstimates, cannot be taken literally. If S varies precisely linearly with N, the chemical potential is identically zero.

c. Using the $S \sim Nk_B$ rule, estimate the entropy of the sun. Clearly we need to know the total number of particles in the sun. The sun (of mass 2×10^{30} kg) is composed of approximately 75% H and 25% ^{4}He. Don't forget electrons in the particle number count! Because the sun is a plasma (ionized gas), for every proton, there's an electron (charge neutrality). Show that $N \approx 10^{57}$.

CHAPTER 8

The third law

You can't get to $T = 0$

THE third law of thermodynamics is concerned with the properties of physical systems in the limit $T \to 0$. Before taking up the third law proper, we consider how low temperatures are produced, because the two are related.

8.1 ADIABATIC DEMAGNETIZATION

The simplest means for cooling a system is to place it in contact with (immerse it in, surround it by) a lower-temperature reservoir. We can see this formally by taking $T = T(S, V)$, in which case temperature variations can be expressed

$$\mathrm{d}T = \left(\frac{\partial T}{\partial S}\right)_V \mathrm{d}S + \left(\frac{\partial T}{\partial V}\right)_S \mathrm{d}V = \frac{T}{C_V}\mathrm{d}S - \frac{\alpha T}{\beta C_V}\mathrm{d}V \,, \tag{8.1}$$

where we've used Eq. (3.11) and the result of Exercise 4.9. The coefficient of $\mathrm{d}S$ in Eq. (8.1) is always positive, whereas the coefficient of $\mathrm{d}V$ is negative if $\alpha > 0$. With $\mathrm{d}V = 0$ (constant volume), entropy is reduced by an outflow of heat ($\mathrm{d}S < 0$), lowering the temperature.

If placing the system in contact with colder objects is not an option, a means of cooling must be devised that utilizes the thermodynamic properties of the system. Equation (8.1) indicates that cooling would result from a reversible adiabatic expansion ($\mathrm{d}S = 0$, $\mathrm{d}V > 0$). A cooling process (not a cycle) could be devised that combines a series of adiabatic expansions with isothermal *compressions* wherein we "suck out the entropy" because, from Eq. (8.1),

$$\left(\frac{\partial S}{\partial V}\right)_T = \frac{\alpha}{\beta} = \left(\frac{\partial P}{\partial T}\right)_V \,, \tag{8.2}$$

where we've used Eq. (1.21). At constant T, $\mathrm{d}S < 0$ for $\mathrm{d}V < 0$. An isothermal compression lowers the entropy by reducing the available volume; a subsequent adiabatic expansion lowers the temperature by converting kinetic energy into potential energy. Both processes (expansion and compression) rely on the cxpansivity of the material, α. At low temperatures, however, thermal expansivities become vanishingly small.[40] Indeed, as we'll show, *a consequence of the third law is that the expansivity vanishes as $T \to 0$*. Cooling by compression-expansion is therefore not a practical means of producing low temperatures, but it suggests that alternatives can be found based on the same principle of reversible adiabatic processes combined with isothermal reductions in entropy.

An effective method of cooling at low temperatures involves the *adiabatic demagnetization* of paramagnetic materials,[1] a discussion of which requires knowledge of the thermodynamics of magnetic substances.[2] As always, start with the first law. The work required to change the magnetization M of a sample in a field H is, from Eq. (1.14), $dW = \mu_0 H dM$. In what follows, we leave off the factor of μ_0 for simplicity. (It can be restored by letting $H \to \mu_0 H$.) With PdV work not playing a significant role at low temperatures, we take as the first law

$$dU = TdS + HdM. \tag{8.3}$$

Define a magnetic Gibbs energy, $G = U - TS - HM$, for which

$$dG = -SdT - MdH. \tag{8.4}$$

Equation (8.4) implies a Maxwell relation

$$\left(\frac{\partial M}{\partial T}\right)_H = \left(\frac{\partial S}{\partial H}\right)_T. \tag{8.5}$$

For relatively small values of M, the equation of state for paramagnets is given by *Curie's law*,

$$M = C\frac{H}{T}, \tag{8.6}$$

where $C > 0$ is the *Curie constant*, the value of which is material specific.[3] Curie's law does not take into account interactions between the dipole moments of paramagnetic atoms (usually ions), and thus, in analogy with the ideal gas, can be said to describe the *ideal paramagnet*.

Combining Eq. (8.6) with Eq. (8.5), we have for the ideal paramagnet

$$\left(\frac{\partial M}{\partial T}\right)_H = -C\frac{H}{T^2} = \left(\frac{\partial S}{\partial H}\right)_T. \tag{8.7}$$

Now consider the formal device of

$$\frac{\partial S}{\partial (H^2)} = \frac{\partial S}{\partial H}\frac{\partial H}{\partial (H^2)} = \frac{\partial S}{\partial H}\left(\frac{\partial (H^2)}{\partial H}\right)^{-1} = \frac{1}{2H}\frac{\partial S}{\partial H},$$

implying from Eq. (8.7),

$$\left(\frac{\partial S}{\partial (H^2)}\right)_T = -\frac{C}{2T^2}. \tag{8.8}$$

An increase in magnetic field decreases the entropy at constant T; there's less disorder among the dipoles of paramagnetic atoms with increasing field strength.

Take the temperature to be a function of S and H, $T = T(S, H)$, and thus

$$dT = \left(\frac{\partial T}{\partial S}\right)_H dS + \left(\frac{\partial T}{\partial H}\right)_S dH. \tag{8.9}$$

---

[1] The 1949 Nobel Prize in Chemistry was awarded to William Giauque for his work on the behavior of substances at low temperatures, including the development of adiabatic demagnetization.

[2] The exercises at the end of this chapter provide plenty of practice with the thermodynamics of magnetism.

[3] The Curie constant does not have a universal value like the gas constant, but is material specific. In thermodynamics M is an extensive quantity, the net dipole moment of the sample (Section 1.9), yet the right side of Eq. (8.6) involves intensive quantities. The Curie constant must have the dimension of volume. In electromagnetic theory, magnetization, call it M_{em}, is the net dipole moment *per volume*. Thus, $M = M_{em}V$. In the SI system, M_{em} has the same dimension as H (amperes per meter), and C is a multiple of the quantities $\mu_0\mu_B^2/k_B$, where $\mu_B \equiv e\hbar/(2m) = 9.27 \times 10^{-24}$ J · T^{-1} is the *Bohr magneton* (T denotes Tesla). When the smoke clears, C has the base units of K · m^3.

To evaluate the derivatives required in Eq. (8.9), it's straightforward to show, using Eq. (8.3), that an expression for the heat capacity at constant magnetic field, C_H, is (see Exercise 8.5)

$$C_H = \left(\frac{\partial U}{\partial T}\right)_H - H\left(\frac{\partial M}{\partial T}\right)_H = T\left(\frac{\partial S}{\partial T}\right)_H. \tag{8.10}$$

The other derivative in Eq. (8.9) can be found using the cyclic relation:

$$\left(\frac{\partial T}{\partial H}\right)_S = -\left(\frac{\partial S}{\partial H}\right)_T \left(\frac{\partial S}{\partial T}\right)_H^{-1} = -\frac{T}{C_H}\left(\frac{\partial M}{\partial T}\right)_H = \frac{C}{C_H}\frac{H}{T}, \tag{8.11}$$

where we've used Eq. (8.7) and Curie's law. Equation (8.9) can therefore be written

$$dT = \frac{T}{C_H}dS + \frac{C}{2C_H T}d\left(H^2\right). \tag{8.12}$$

Equation (8.12) is the analog of Eq. (8.1), and Eq. (8.8) is the analog of Eq. (8.2).

Integrating Eq. (8.12) at constant S (assuming that C_H is approximately constant),

$$\Delta(T^2) = \frac{C}{C_H}\Delta(H^2), \tag{8.13}$$

while integrating Eq. (8.12) at constant T implies

$$\Delta S = -\frac{C}{2T^2}\Delta(H^2). \tag{8.14}$$

Equations (8.13) and (8.14) provide the basis for a cooling process. Turning on the field at constant temperature (magnetizing the sample, $\Delta(H^2) > 0$) decreases the entropy, per Eq. (8.14). At this point, the sample is shielded and the field is turned off ($\Delta(H^2) < 0$) under adiabatic conditions, demagnetizing the sample and decreasing the temperature per Eq. (8.13). The process is shown in Fig. 8.1 as a function of T, S, and H. The right portion of the figure shows the process projected onto the S-T plane, assuming that the entropy curves are as shown (see Section 8.5).[4]

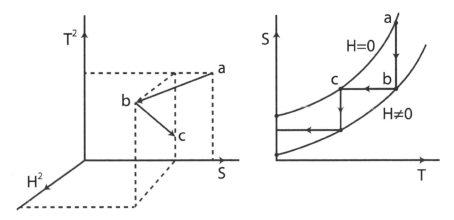

Figure 8.1 Cooling by adiabatic demagnetization. Field turned on isothermally $a \rightarrow b$, lowering entropy; field turned off adiabatically, $b \rightarrow c$, lowering temperature

---

[4]Cooling by adiabatic demagnetization of paramagnetic salts can achieve milli-Kelvin temperatures. Cooling by nuclear demagnetization, removing the entropy of the magnetic dipoles of nuclei, can achieve micro-Kelvin temperatures.

8.2 NERNST HEAT THEOREM

The equilibrium constant K^* of a chemical reaction (Section 6.4) can be calculated from thermodynamic data, either from $\Delta G^0 = \sum_i \nu_i \mu_i^0$, the net change in Gibbs energy of the reactants in their pure forms, Eq. (6.41), or from the heats of reaction $\Delta H^0 = \sum_i \nu_i H_i^0$, Exercise 6.3, where H_i^0 is the enthalpy of the ith pure substance. At a given temperature T we have, by definition,

$$\Delta G^0 = \Delta H^0 - T \Delta S^0 , \tag{8.15}$$

where $\Delta S^0 \equiv \sum_i \nu_i S_i^0$ is the net entropy difference of the reactants. Let the entropy of a substance at temperature T be denoted S_T^0, and write

$$S_T^0 = S_0^0 + S_{0 \to T}^0 , \tag{8.16}$$

where S_0^0 is the entropy at $T = 0$ and $S_{0 \to T}^0$ is the entropy of the substance that has been taken from $T = 0$ to $T > 0$, as in Eq. (P6.1). Using Eq. (8.16) we can write for the chemical reaction

$$\Delta S^0 = \Delta S_0^0 + \Delta S_{0 \to T}^0 , \tag{8.17}$$

where, in an abuse of notation, $\Delta S_0^0 = \sum_i \nu_i S_{0,i}^0$, with $S_{0,i}^0$ the entropy of the ith substance in pure form at $T = 0$. Combining Eq. (8.17) with Eq. (8.15) and using Eq. (6.41),

$$R \ln K^* = \Delta S_0^0 + \Delta S_{0 \to T}^0 - \frac{\Delta H^0}{T} . \tag{8.18}$$

In Section 7.8 we argued that no meaning can be ascribed to absolute entropy. Yet, in Eq. (8.18) we have a measurable quantity, K^*, that seems to involve the zero-temperature entropies of the reactants, constants that are supposedly arbitrary.

To eliminate the possibility that ΔS_0^0 could contribute to a measurable quantity, Walther Nernst postulated in 1906 that $\Delta S \to 0$ as $T \to 0$, what's known as the *Nernst heat theorem*:

At $T = 0$, the entropy change accompanying any thermodynamic process vanishes.

The Nernst heat theorem is a *version* of the third law of thermodynamics, perhaps the best.[5] As we'll see, *the third law has been stated in several forms*.[6] The Nernst theorem is not based on theoretical reasoning, however, but on a detailed study of chemical reactions using electrochemical cells. The open-cell emf under conditions of constant (T, P) or constant (T, V) (see Section 6.4) provides a measurement of the reversible work of the reaction, ΔG or ΔF, Eq. (6.47). From this information, ΔH or ΔU can be calculated using the Gibbs-Helmholtz equations, Eq. (6.53). Nernst found that at low temperatures, ΔG approaches ΔH, $\Delta G \to \Delta H$, and ΔF approaches ΔU, $\Delta F \to \Delta U$.

Using Eq. (6.52), we conclude from this finding that the entropy differences

$$T [\Delta S]_P = \Delta H - \Delta G \qquad \text{and} \qquad T [\Delta S]_V = \Delta U - \Delta F ,$$

tend to zero as $T \to 0$. Given the factor of T in $T [\Delta S]_P$ and $T [\Delta S]_V$, we should not be surprised that these difference vanish as $T \to 0$. Nernst found from his data, however, that the *slopes* vanish as $T \to 0$,

$$\lim_{T \to 0} \left(\frac{\partial \Delta G}{\partial T} \right)_P = \lim_{T \to 0} \left(\frac{\partial \Delta H}{\partial T} \right)_P = \lim_{T \to 0} \left(\frac{\partial \Delta F}{\partial T} \right)_V = \lim_{T \to 0} \left(\frac{\partial \Delta U}{\partial T} \right)_V = 0 .$$

---

[5] Nernst was awarded the 1920 Nobel Prize in Chemistry for his work on the third law.

[6] In contrast to the second law, where the Clausius and Kelvin statements are logically equivalent, no two statements of the third law are the same. For that reason, it's maintained by some that the third law of thermodynamics is not a law of nature but rather a collection of useful ideas.

Using Eq. (6.51), therefore, the Gibbs-Helmholtz equations, we conclude that

$$\lim_{T\to 0}[\Delta S]_{T,P} = \lim_{T\to 0}[\Delta S]_{T,V} = 0. \qquad (8.19)$$

This result provides the experimental foundation for the Nernst heat theorem, and hence the third law: $\lim_{T\to 0}\Delta S = 0$. One does not have to produce heroically low temperatures to arrive at this conclusion; Nernst found that the trends are evident beginning at $T \approx 40$ K. We can now state the Nernst law more precisely. Equation (8.19) refers to entropy differences that occur for processes connecting equilibrium states *at the same temperature*. A refinement of the Nernst heat theorem is:

> The change in entropy associated with isothermal processes between states of equilibrium vanishes as $T \to 0$.

This statement can be taken as a secure version of the third law. Its justification (like the second law) rests on the fact that of all consequences derived from it, none have been found in contradiction with experimental results.[7]

8.3   OTHER VERSIONS OF THE THIRD LAW

Planck version

The Nernst heat theorem says that $\Delta S \to 0$ as $T \to 0$ for isothermal reactions, yet it tells us nothing about the entropies of the individual constituents of the reaction. In 1911 Planck offered a seemingly stronger version of the third law:[11, p274]

> As the temperature diminishes indefinitely, the entropy of a chemically homogeneous body of finite density approaches indefinitely near to the value zero.

In other words, $\lim_{T\to 0} S = 0$. Basically, Planck would erase the factors of Δ in the Nernst heat theorem, that instead of entropy *differences* vanishing at low temperature, entropy *itself* vanishes as $T \to 0$. If correct, it would naturally account for the Nernst heat theorem. The Planck hypothesis, however, cannot be a fundamental law of physics: 1) It's not true—there are materials with nonzero entropies at low temperature, what's termed the *residual entropy*; and 2) It *can't* be true. Planck was motivated by the statistical entropy, $S = k\ln W$. If all systems attain *a single microstate* as $T \to 0$, $W = 1$, and entropy would indeed vanish. Here we have another instance of thermodynamics anticipating quantum mechanics. Planck's version of the third law is the assertion that the *ground state of every system is nondegenerate*. While plausible, it begs the question: *Do* all systems reach a unique state at $T = 0$? No. Consider the solid state of carbon monoxide (CO) at low temperature. In a perfect crystal all molecules would have the same orientation, of, for example, C atoms "up" with O atoms "down." In actuality, which atom is up occurs almost at random. It's as if solid CO is a mixture of two kinds of molecules, CO and OC. If the mixing of the two "species" is random, we would have from Eq. (7.62) an entropy of mixing $\Delta S = R\ln 2 \approx 0.69R$. The measured residual entropy of solid CO is $\approx 0.55R$.[41] Ice is another substance with a nonzero entropy at low temperature (Section 8.6).

The state to which we ascribe zero entropy has no fundamental significance; it's a *convention*, an *agreement* (see Section 7.8). Even for systems where molecules achieve an ordered configuration at low temperature, there are other sources of entropy independent of the state of *crystalline* order. The flaw in Planck's theory is that it assumes there is an absolute sense by which we can say that $W = 1$ for *the* microstate. There are always microscopic degrees of freedom that we *decide* whether to take into account. By convention, entropy is taken to be zero when translational, configurational,

---

[7]It's useful to keep in mind that science does not establish true propositions, but disproves false ones. Consider Einstein's statement that no amount of experimentation could ever prove him right, but a single experiment could prove him wrong.

rotational, vibrational, and electronic contributions to the entropy are all zero. Contributions from nuclear spin and isotopic mixing are ignored. By raising the specter of absolute entropy, Planck's version becomes a sideshow of what shall conventionally be considered zero entropy. Nature doesn't care about our definitions, however; the third law is about physical *processes* at low temperatures.

Nevertheless, it's *convenient* to conventionally assign zero entropy to substances in well defined low-temperature states. By an unfortunate choice of terminology, tabulated values of the (conventional) entropy of substances at standard temperatures (such as 25 C) are referred to as *absolute entropies*, even though entropy (like energy) cannot be absolute. We made use of the tabulated value of the entropy of mercury in Eq. (7.41). In our derivation of the Sackur-Tetrode equation, it was the entropy *difference* between liquid and vapor that entered the calculation.

Simon version

Is there a version of the third law that recognizes the *actual* behavior of matter at low temperatures? Francis Simon (a student of Nernst) formulated two refinements:[42]

> The contribution of the entropy due to each factor within the system which is in internal equilibrium becomes zero as $T \to 0$;

and

> At absolute zero the entropy differences vanish between all those states of a system between which a reversible transition is possible *in principle* even at the lowest temperatures.

Let's parse these statements, particularly the concept of internal equilibrium.

A system is said to be in *internal equilibrium* if the state is not necessarily one of *absolute equilibrium* but is instead one of *local equilibrium* where other configurations of the system are *less* stable than the state under consideration.[8] There may be internal equilibrium with respect to one factor, but not with respect to another. A solid comprised of different isotopes may be in internal equilibrium with respect to thermal vibrations, but not with respect to the spatial distribution of isotopes. The Simon statement applies to processes not involving factors that are not in internal equilibrium, i.e., the third law applies to *processes during which internal constraints remain in effect*. A process between an initial metastable state of constrained equilibrium and a final state of unconstrained equilibrium (which sets in after the removal of constraints) is not of this kind. If a particular *set* of constraints stays in effect during otherwise reversible processes, the third law remains applicable. One may envision several samples of a substance, each kept at a different temperature T and each in a state of unconstrained equilibrium characteristic of that temperature. Now cool the samples rapidly (a *quench*) to a common low temperature T'. It may happen that the samples (of the same substance) develop different internal constraints. The entropy differences between them will not tend to zero as $T' \to 0$. However, a reversible process performed on *each sample* will produce no change in entropy as $T' \to 0$, provided that its particular constraints remain intact. With these considerations, Fowler and Guggenheim have stated the third law:[10, p246]

> For any isothermal process involving only phases in internal equilibrium or, alternatively, if any phase is in frozen metastable equilibrium, provided that the process does not disturb this frozen equilibrium, $\lim_{T \to 0} \Delta S = 0$.

Note that such refinements of the Nernst heat theorem refer to *processes* rather than absolute entropy.

---

[8]Systems with many states of local equilibrium became an active field of research known as spin-glass theory starting in the 1970s which has continued until the present.[43]

What is a law of nature?

The Simon statement is that the Nernst hypothesis is valid for those degrees of freedom among which equilibrium has been established. But then we're faced with the question of how we know a system is in equilibrium. The Simon version is not in the form of a *prediction*. The Planck version would seemingly be a prediction, except that it's not correct if it's intended as a prediction of the *actual* (rather than the *expected*) behavior of matter at low temperature. For these reasons, it's held by some that what's known as the third law of thermodynamics is not a law. At the very least, it's a different kind of law.

8.4 CONSEQUENCES OF THE THIRD LAW

Vanishing heat capacity

The heat capacity holding quantity X fixed is

$$C_X \equiv \left(\frac{\mathrm{d}Q}{\mathrm{d}T} \right)_X = T \left(\frac{\partial S}{\partial T} \right)_X . \tag{8.20}$$

By the third law, therefore, heat capacities vanish as $T \to 0$, because as $T \to 0$, $\Delta S \to 0$. In solids the heat capacity of phonons (quantized lattice vibrations) vanishes as T^3 at low temperature, and the heat capacity of conduction electrons vanishes linearly with T. The heat capacity of the photon gas $C_V \sim T^3$ for all temperatures, Eq. (5.12).

Vanishing slope of coexistence curve

The slope of the coexistence curve (Clausius-Clapeyron equation, Eq. (6.22)) involves the difference in entropy between the phases, $\mathrm{d}P/\mathrm{d}T = \Delta S/\Delta V$. The third law thus predicts that the slope of the coexistence curve vanishes as $T \to 0$,

$$\lim_{T \to 0} \left(\frac{\mathrm{d}P}{\mathrm{d}T} \right)_{\text{coexist}} = 0 . \tag{8.21}$$

For most substances there is only one phase at low temperatures. Liquid helium is an exception, which is a superfluid at low temperatures and pressures up to approximately 2.5 MPa. Studies of the melting curve of ^{4}He show that for $T \lesssim 1.4$ K, $\mathrm{d}P/\mathrm{d}T \propto T^8$ so that the slope of the coexistence curve vanishes as $T \to 0$.[44, p29]

Expansivity

By combining Eq. (1.20) with a Maxwell relation, the coefficient of thermal expansivity is related to a derivative of the entropy

$$\alpha = -\frac{1}{V} \left(\frac{\partial S}{\partial P} \right)_T .$$

The third law thus predicts that

$$\lim_{T \to 0} \alpha(T) = 0 . \tag{8.22}$$

The expansivity of solids vanish as $T \to 0$, which is why we stated in Section 8.1 that $P\mathrm{d}V$ work is not an efficient means of cooling at low temperature.

8.5 UNATTAINABILITY OF ABSOLUTE ZERO TEMPERATURE

The Nernst heat theorem is difficult to apply unless supplemented with microscopic information. Can the third law be formulated in a way that does not involve *non-macroscopic* information? The laws of thermodynamics are given in the form of *impossibility* statements, what you can't do.[9] In 1912, Nernst formulated the third law in the form of an impossibility statement:

> It is impossible by any procedure to reduce the temperature of a system to absolute zero in a finite number of operations.

We thus arrive at the beginning of this chapter—how *are* low temperatures produced? Let S depend on some thermodynamic property, X, as well as T: $S = S(T, X)$ or equivalently, $T = T(S, X)$. Figure 8.2 shows two possibilities for how T could depend on S for given values of X.

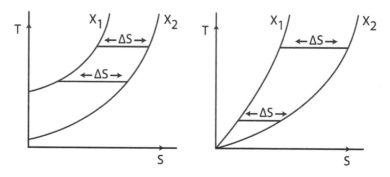

Figure 8.2 Hypothetical entropy curves

The left diagram in Fig. 8.2 is incompatible with the Nernst heat theorem: ΔS does not tend to zero as $T \to 0$. The diagram on the right, however, *is* compatible with the third law: $\Delta S \to 0$ as $T \to 0$. The right figure has been drawn so that $S(X) \to 0$ as $T \to 0$, but that's irrelevant; we could just as well have taken $S \to S_0$. Comparing Figs. 8.2 and 8.1, the entropy curves in Fig. 8.1 indicate that $T = 0$ *would* be attainable in a finite number of steps and thus are invalid (incompatible with the third law).

The principle of unattainability is synonymous with the third law. Consider a reversible adiabatic process from states $A \to B$, with A at temperature T' and B at T''. Assuming that temperatures are sufficiently low that a phase transition does not occur (so that we can ignore latent heats), we have for an isentropic process $(S_A(T') = S_B(T''))$

$$S_A(0) + \int_0^{T'} \frac{C}{T} \mathrm{d}T = S_B(0) + \int_0^{T''} \frac{C}{T} \mathrm{d}T \,, \tag{8.23}$$

where C is the heat capacity. The integrals in Eq. (8.23) exist because the heat capacities vanish as $T \to 0$. If the system is to be cooled to absolute zero temperature, set $T'' = 0$. From Eq. (8.23), therefore,

$$S_B(0) - S_A(0) = \int_0^{T'} \frac{C}{T} \mathrm{d}T \,. \tag{8.24}$$

By the stability requirement $C > 0$ (Section 3.10). From Eq. (8.24) therefore

$$S_B(0) - S_A(0) > 0 \,. \tag{8.25}$$

---

[9]Some of the major laws of physics can be given as impossibility statements. Consider: 1) the impossibility of distinguishing by means of local experiments an absolute standard of rest (principle of special relativity); 2) the impossibility of distinguishing by means of local experiments the acceleration of the frame of reference from the local effects of gravity (equivalence principle, foundation of the general theory of relativity); 3) the impossibility of assigning exact values to certain pairs of simultaneously measured quantities (Heisenberg uncertainty principle). The first law of thermodynamics can be formulated as an impossibility statement: It's impossible to construct a device working in a cycle that produces no effect other than the performance of work, i.e., if $\Delta U = 0$ (cycle), and $Q = 0$ ("no effect" clause), then it cannot be true that $W \neq 0$.

However, if the unattainability postulate is valid such a process cannot occur ($T'' = 0$) and Eq. (8.25) cannot be true. Thus, if unattainability holds,

$$S_B(0) - S_A(0) \leq 0. \tag{8.26}$$

By making the argument in the other order, $B \to A$, where A is to be at absolute zero, we would conclude

$$S_A(0) - S_B(0) \leq 0. \tag{8.27}$$

Consistency between Eq. (8.26) and Eq. (8.27) requires

$$S_A(0) = S_B(0). \tag{8.28}$$

The entropy *difference* for a reversible process ($A \rightleftharpoons B$) at $T = 0$ thus vanishes, in agreement with Nernst heat theorem. Unattainability implies the third law.

To show the converse, start with the validity of the third law, in which case the left side of Eq. (8.24) is zero. In that case no $T' > 0$ could be found that satisfies the equation (because $C > 0$). Therefore no initial temperature T' exists from which the system could be cooled to absolute zero. The third law implies unattainability.

8.6   RESIDUAL ENTROPY OF ICE

Ice is a substance having a nonzero low-temperature entropy. In the H_2O molecule, a pair of H atoms are covalently bonded to O in which the H-O-H angle is about $104°$ and the O-H distance is 0.096 nm. While the H_2O molecule is electrically neutral, the distribution of charge is not uniform. The O atom exerts a stronger attractive force on electrons than does the H nucleus, creating an electric dipole with negative charge being more concentrated at the O side of the molecule. In the liquid phase the partially-positive H atom on one water molecule is attracted to the partially-negative oxygen on a neighboring molecule, creating a *hydrogen bond*. The solid phase of H_2O, as determined from X-ray scattering, is one in which O atoms are arranged in a tetrahedrally-coordinated structure with each O atom surrounded by four other O atoms, held together through hydrogen bonds with the O-O length 0.276 nm, and with the H atoms such that one H atom is placed between each pair of O atoms (see Fig. 8.3). The H atoms do not have a regular arrangement; the lack of ordering among H atoms leads to a nonzero entropy at low temperature.

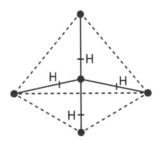

Figure 8.3   Tetrahedral bonding in ice. Oxygen atoms are the solid circles.

We can estimate the residual entropy through an argument due to Linus Pauling.[45] For N water molecules we have N oxygen atoms. In the tetrahedrally-coordinated solid phase there are $2N$ O-O bonds (divide $4N$ by two to avoid overcounting). For each O-O bond the H atom can be situated at two inequivalent positions between the O atoms—one at a distance of 0.1 nm and the other at a distance of 0.176 nm. There are thus $2^{2N} = 4^N$ "microstates" (arrangements) of the H atoms in the network of O-O bonds. There are 16 ways to arrange four H atoms around an O atom (see Fig. 8.4); of these only six are consistent with the *ice condition* (the arrangement that occurs in ice) that each O atom has two H atoms near to it and two that are far from it. The number of

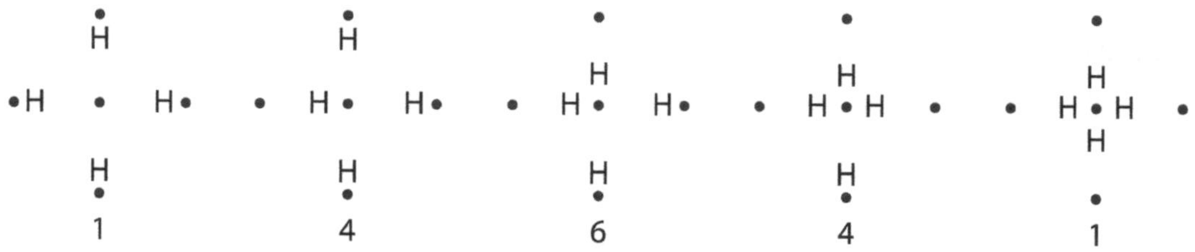

Figure 8.4   16 ways to arrange four H atoms. Only six are consistent with the ice condition.

arrangements of $2N$ H atoms compatible with the ice condition is thus

$$W = 4^N \left(\frac{6}{16}\right)^N = \left(\frac{3}{2}\right)^N .$$

Using Eq. (7.2), we obtain the residual entropy, per mole, $S = R\ln(1.5) = 0.405R = 0.806$ cal K^{-1}, which is quite close to the measured value $S = 0.82 \pm 0.15$ cal K^{-1}.[46]

CHAPTER SUMMARY

This chapter introduced the third law of thermodynamics which concerns the properties of systems near $T \approx 0$, and which to a large degree is about the behavior of entropy as $T \to 0$.

- The third law comes in two basic flavors (which are equivalent):
  - Absolute zero temperature cannot be reached in a finite number of steps;
  - As $T \to 0$, $\Delta S \to 0$.
- By the third law, anything related to a derivative of S vanishes as $T \to 0$:
  - Heat capacities vanish;
  - The slope of the coexistence curve for superfluid helium vanishes;
  - The thermal expansivity α vanishes.

EXERCISES

8.1 The Maxwell relation associated with the magnetic Gibbs energy is given in Eq. (8.5). Derive the Maxwell relations associated with the first law in the form of Eq. (8.3), the enthalpy, and the Helmholtz free energy. Show that:

$$\left(\frac{\partial H}{\partial S}\right)_M = \left(\frac{\partial T}{\partial M}\right)_S \qquad \left(\frac{\partial T}{\partial H}\right)_S = -\left(\frac{\partial M}{\partial S}\right)_H \qquad \left(\frac{\partial S}{\partial M}\right)_T = -\left(\frac{\partial H}{\partial T}\right)_M .$$

8.2 Show that $\dfrac{\partial(H, M)}{\partial(S, T)} = 1$. By the rules of calculus $dHdM = dSdT$.

8.3 Take the first law for a magnetic system to be $dU = TdS - PdV + HdM$. (We ignored the PdV term in Section 8.1, but let's keep it now.) Define the Helmholtz energy as usual, $F \equiv U - TS$, so that $F = F(T, V, M)$. Derive the three Maxwell relations associated with F. Show that

$$\left(\frac{\partial H}{\partial T}\right)_{V,M} = -\left(\frac{\partial S}{\partial M}\right)_{T,V} \qquad \left(\frac{\partial H}{\partial V}\right)_{T,M} = -\left(\frac{\partial P}{\partial M}\right)_{T,V} \qquad \left(\frac{\partial P}{\partial T}\right)_{V,M} = \left(\frac{\partial S}{\partial V}\right)_{T,M}.$$

8.4 Start with the first law in the form $dU = TdS - PdV + HdM$. Take U and S to be functions of T, V, and M, and derive the relation

$$\left(\frac{\partial U}{\partial M}\right)_{V,T} = H - T\left(\frac{\partial H}{\partial T}\right)_{V,M}.$$

Hint: Use a Maxwell relation derived in the previous exercise. This formula is the magnetic analog of Eq. (1.28).[10]

8.5 a. Show, using the first law in the form of Eq. (8.3), that dQ can be written in two ways:

$$dQ = \left(\frac{\partial U}{\partial T}\right)_M dT + \left[\left(\frac{\partial U}{\partial M}\right)_T - H\right] dM$$

and

$$dQ = \left[\left(\frac{\partial U}{\partial T}\right)_H - H\left(\frac{\partial M}{\partial T}\right)_M\right] dT + \left[\left(\frac{\partial U}{\partial H}\right)_T - H\left(\frac{\partial M}{\partial H}\right)_T\right] dH.$$

Assume in the first case that $U = U(T, M)$, and in the second, $U = U(T, H)$ and $M = M(T, H)$. These equations are the magnetic analogs of the results in Exercise 1.8.

  b. Show that you now have the expression for the heat capacity at constant H, Eq. (8.10). Show that the expression for the heat capacity at constant M is given by

$$C_M = \left(\frac{\partial U}{\partial T}\right)_M.$$

8.6 a. Show that the internal energy of an ideal paramagnet is independent of M (the analog of the energy of the ideal gas being independent of V). Start with the first law in the form

$$dS = \frac{1}{T}dU - \frac{H}{T}dM.$$

Assume that $U = U(T, M)$ and $S = S(T, M)$. Invoke the integrability condition on S:

$$\left[\frac{\partial}{\partial M}\left(\frac{\partial S}{\partial T}\right)_M\right]_T = \left[\frac{\partial}{\partial T}\left(\frac{\partial S}{\partial M}\right)_T\right]_M.$$

From the integrability condition, show that

$$\left(\frac{\partial U}{\partial M}\right)_T = -T^2 \frac{\partial}{\partial T}\left(\frac{H}{T}\right)_M.$$

Argue, for the equation of state given by Curie's law, that $(\partial U/\partial M)_T = 0$.

---

[10]Thermodynamic formulas pertaining to magnetic systems can sometimes (but not always) be found from their counterparts describing fluids through the substitutions $V \to M$ and $P \to -H$; such is true of formulas derived from the first law. Formulas that make use of the equation of state, however, do not have this property. The equation of state of the ideal gas is that $V \propto T/P$; the equation of state of the ideal paramagnet is not of this form. Using that M, an extensive quantity, corresponds to V, and H, an intensive quantity, corresponds to P, Curie's law effectively has the form $V \propto P/T$.

b. Now repeat, assuming $U = U(H,T)$, $S = S(H,T)$, and $M = M(H,T)$. Show that

$$\left(\frac{\partial U}{\partial H}\right)_T = H\left(\frac{\partial M}{\partial H}\right)_T + T\left(\frac{\partial M}{\partial T}\right)_H.$$

These formulas are the magnetic analogs of the results derived in Exercise 4.6. Show for substances obeying Curie's law that $(\partial U/\partial H)_T = 0$ (the analog of the internal energy of the ideal gas being independent of P). The internal energy of the ideal paramagnet thus depends *only* on T, $U = U(T)$; a magnetic analog of Joule's law.

c. Show, assuming $U = U(T,M)$ and $S = S(T,M)$, that $dU = C_M dT$. Hint: Start with $dU = TdS + HdM$ and use a Maxwell relation. We show in another exercise that C_M is independent of M, implying as well that U depends only on T.

8.7 a. Derive the following formulas (the analogs of the results obtained in Exercise 4.7):

$$\left(\frac{\partial C_H}{\partial H}\right)_T = T\left(\frac{\partial^2 M}{\partial T^2}\right)_H$$

and

$$\left(\frac{\partial C_M}{\partial M}\right)_T = -\frac{\partial}{\partial T}\left(T^2 \frac{\partial}{\partial T}(H/T)_M\right)_M.$$

Hint: Use the results of Exercise 8.6.

b. For the ideal paramagnet, show that:
   i. C_M is independent of M.
   ii. C_H is a function of H, $C_H = C_{H=0} + C(H/T)^2$.

c. Argue, because for a paramagnet $C_{H=0}$ is the same as $C_{M=0}$, and C_M is independent of M (ideal paramagnet), that

$$C_H = C_M + C\left(\frac{H}{T}\right)^2.$$

Thus, $C_H > C_M$. The heat capacity at constant M is measured without work performed on the system, whereas in measuring C_H work would be done, requiring a greater input of heat for the same temperature change.

d. Alternatively, argue that, because the internal energy for a paramagnet depends only on T (Exercise 8.6), we can replace $(\partial U/\partial T)_H$ in Eq. (8.10) with C_M to reach the same conclusion:

$$C_H = C_M - H\left(\frac{\partial M}{\partial T}\right)_H = C_M + C\left(\frac{H}{T}\right)^2.$$

8.8 a. Show, using the results of Exercise 8.5, that for an adiabatic process

$$\left(\frac{dM}{dH}\right)_S = \frac{C_M}{C_H}\frac{\left(\frac{\partial U}{\partial H}\right)_T - H\left(\frac{\partial M}{\partial H}\right)_T}{\left(\frac{\partial U}{\partial M}\right)_T - H}.$$

b. This formula simplifies for the ideal paramagnet. Show that:

$$\left(\frac{dM}{dH}\right)_S = \frac{C_M}{C_H}\left(\frac{\partial M}{\partial H}\right)_T.$$

c. It turns out that this equation, while derived for the ideal paramagnet, holds *generally*. Starting from the result derived in Exercise 8.8a., however, that is most likely not apparent. Here is a case where Jacobians simplify matters. Show that

$$\left(\frac{\partial M}{\partial H}\right)_S = \frac{\partial(M,S)}{\partial(H,S)} = \frac{\partial(M,S)}{\partial(M,T)}\frac{\partial(M,T)}{\partial(H,T)}\frac{\partial(H,T)}{\partial(H,S)} = \frac{C_M}{C_H}\left(\frac{\partial M}{\partial H}\right)_T.$$

d. Thus, we have the general result

$$\left(\frac{\partial M}{\partial H}\right)_S = \frac{C_M}{C_H}\left(\frac{\partial M}{\partial H}\right)_T.$$

Show that we have the analog of the result shown in Exercise 1.9,

$$\left(\frac{\partial H}{\partial M}\right)_S > \left(\frac{\partial H}{\partial M}\right)_T,$$

that on an H-M diagram, adiabats have a steeper slope than isotherms.

8.9 a. Sketch the Carnot cycle for a paramagnetic substance on a H-M diagram (let M be the abscissa and H the ordinate). The isotherms are simple to draw, given the equation of state, Eq. (8.6). To draw the adiabats, however, we require an equation describing adiabatic processes. Show that, along an adiabat, $H = H(M)$ satisfies the differential equation

$$\frac{dH}{dM} = \left(1 + \frac{M^2}{C_M C}\right)\frac{H}{M}.$$

Make use of the results derived in Exercises 8.7 and 8.8.

b. Show that the solution of this differential equation is in the form (assuming C_M to be constant)

$$H = AM\exp(M^2/(2C_M C)),$$

where A is an integration constant.

c. An alternate way of describing adiabatic processes is to start with the first law. Show that, for the ideal paramagnet:

$$C_M dT = HdM = \frac{MT}{C}dM,$$

the solution of which is (assuming C_M constant)

$$\ln T = \frac{1}{2CC_M}M^2 + \text{constant}.$$

Thus, for two points on the same adiabat,

$$\ln(T_1/T_2) = \frac{1}{2CC_M}\left(M_1^2 - M_2^2\right).$$

    d. Derive an expression for the efficiency of this Carnot cycle. You already know what the answer is, but show it.

8.10 The magnetic susceptibility (a response function) is defined as $\chi \equiv (\partial M / \partial H)_T$. Show that the third law predicts

$$\lim_{T \to 0} \left(\frac{\partial \chi}{\partial T} \right)_H = 0 ,$$

that is, χ becomes independent of temperature as $T \to 0$. Hint: Use a Maxwell relation.

8.11 The isothermal bulk modulus B, the inverse of the isothermal compressibility (see Eq. (1.20)), is defined as $B \equiv -V (\partial P / \partial V)_T$. Using the third law show that

$$\lim_{T \to 0} \left(\frac{\partial B}{\partial T} \right)_V = 0 ,$$

i.e., B becomes independent of temperature as $T \to 0$. Hint: Reach for a Maxwell relation.

8.12 Suppose that the structure of ice was such that the hydrogen atoms were exactly half way between oxygen atoms. What would the residual entropy be then?

CHAPTER 9

To this point

$\mathbf{M}$ ACROSCOPIC phenomena are subject to the laws of thermodynamics; an education in the physical sciences must therefore include a serious drubbing in the subject. Part I of this book is intended as a *précis* that emphasizes the structure of the theory, in preparation for a study of statistical mechanics. Before moving on to additional, specialized topics (Part II), let's recap the theory of thermodynamics, now that we've laid the groundwork.

Variables of thermodynamics: Intensive and extensive

A highly useful distinction is between extensive and intensive variables (Section 1.2). Understanding just this concept nearly comprises the entire subject!

- Extensive quantities scale with the size of the system, such as U, V, and N; they're additive over subsystems of a composite system. Internal energy scales with the amount of matter. The extensivity of S, however, is not as obvious because it's not tied to a microscopic property of matter (like mass or energy); it's a property of the equilibrium state. In axiomatic formulations of thermodynamics the extensivity of S is taken as a postulate. The first law can be written in the form

$$dS = \frac{1}{T}dU + \frac{P}{T}dV - \frac{\mu}{T}dN \,. \tag{3.15}$$

Infinitesimal changes in S are thus related to differentials of extensive quantities.

The emphasis on extensive variables arises from the division of energy into work and heat, with work being related to changes in macroscopically observable (extensible) quantities (Section 1.9). Heat (energy transfers to microscopic degrees of freedom) can also be given as a change in an extensive quantity, the entropy, with $dQ = TdS$.

- Intensive variables have the same value throughout a system, such as T, P, and μ; they occur in the theory as partial derivatives between extensive variables. From Eq. (3.30),

$$T = \left(\frac{\partial U}{\partial S}\right)_{V,N} \qquad P = -\left(\frac{\partial U}{\partial V}\right)_{S,N} \qquad \mu = \left(\frac{\partial U}{\partial N}\right)_{S,V} \,. \tag{3.30}$$

The occurrence of T in thermodynamic formulas almost always refers to the absolute temperature (defined in Section 2.4). T and S are both consequences of Carnot's theorem (Section 2.3). Absolute temperature provides meaning to the notion of zero temperature. There are exceptional systems characterized by negative absolute temperature (see Chapter 11).

While T and P are familiar quantities, the chemical potential μ is not. The quantity μ is the energy required to add a particle to the system at fixed S and V, Section 3.5 (equivalently at fixed T and P, Section 4.6). The chemical potential is not simply, as one might expect, a measure of the interactions between particles. The ideal gas ignores inter-particle interactions, yet its chemical potential is nonzero and involves fundamental constants. For the ideal gas,

$$\mu = -k_B T \ln \left(\frac{V}{N\lambda_T^3} \right), \qquad \text{(Exercise 7.9)}$$

where $\lambda_T = h/\sqrt{2\pi m k_B T}$ is the thermal wavelength, Eq. (7.51) (ostensibly the de Broglie wavelength). The chemical potential is generally a negative quantity, although there are exceptions (Section 3.5). That μ is nonzero for the ideal gas underscores that it involves entropy, a characteristic of the equilibrium state; even for the simplest macroscopic system (ideal gas), entropy is nontrivial and involves Planck's constant, Eq. (7.52). For cavity radiation, however (photons in equilibrium with matter), $\mu = 0$; for this system entropy scales linearly with average photon number (Section 5.2)—it's not possible to change N keeping S fixed.

Convex functions (Section 4.2) can be characterized in two equivalent ways: by their pointwise values or in terms of their tangents at a point through the Legendre transformation. The duality between points and tangents of convex functions is reflected in thermodynamics as a duality between extensive and intensive variables. The internal energy function $U = U(S, V, N)$ is convex (Exercise 3.5.) Intensive variables, which occur as the derivatives (slopes) of extensive quantities (as in Eq. (3.15)), can be used as independent state variables whenever experimental or theoretical considerations warrant. The thermodynamic potentials $H = U + PV$, $F = U - TS$, and $G = U - TS + PV$ are the three Legendre transformations of U (known as thermodynamic potentials) that can be formed from the products (having the dimension of energy) TS and PV; Section 4.3. Instead of $U = U(S, V)$, characterized by the two extensive quantities S and V, the Legendre transforms involve intensive quantities, $H = H(S, P)$ (enthalpy), $F = F(T, V)$ (Helmholtz free energy), $G = G(T, P)$ (Gibbs free energy). The thermodynamic potentials are not simply functions having convenient mathematical properties (through which Maxwell relations are established, Table 4.2), but represent potential energies stored in the system under controlled conditions, Table 4.1.

The first law expressed in terms of intensive variables is the Gibbs-Duhem equation

$$N\mathrm{d}\mu = -S\mathrm{d}T + V\mathrm{d}P. \qquad (3.32)$$

The "bridge" between intensive and extensive quantities is provided by Eq. (3.31), the Euler relation $U = TS - PV + \mu N$, which relies on the extensivity of S.

Equilibrium, the timeless state that thermodynamics describes

- Thermodynamic equilibrium is the macroscopically quiescent state where "nothing is happening." Taken literally, this definition is empty. *All* systems change in time, it being a matter of time scales, and time scales appreciably larger than that required to complete experiments render definitions of equilibrium problematic. All we can say is that a system that *appears* to be in equilibrium, *is* in equilibrium.

- A state of equilibrium is specified by the values of a finite set of quantities (state variables, what we can measure). State space is a mathematical space of the values of state variables; each point of the space represents a possible state of equilibrium. State variables are time independent and do not depend on the history of the system. The number of possible histories between equilibrium states is unlimited, but the number of state variables is quite limited. History-dependent substances are beyond the scope of thermodynamics. Because state variables are history independent, changes in state variables must be described in a way that's independent of the manner by which change is brought about. That's ensured by requiring differentials of state variables to be exact (Section 1.3).

- Equilibrium is transitive (zeroth law of thermodynamics), which implies 1) the existence of temperature as a state variable and 2) the existence of equations of state, functional relations among the values of state variables (Section 1.5).

- There are two equivalent criteria for stable equilibrium in isolated systems: (Section 4.1)

    - For systems of fixed energy, entropy is a maximum
    - For systems of fixed entropy, energy is a minimum

- Treating a system and its environment as a composite closed system, the equilibrium state characterized by $\Delta S = 0$ implies the equality of intensive variables between system and environment, T, P, and μ (Section 3.10). The values of intensive variables are "set" by the environment. Equality between system and environmental intensive variables holds for those conjugate to conserved quantities: U, V, and N. Photons are not conserved, and the theory requires $\mu = 0$ for cavity radiation (Section 5.2).

- The equilibrium state (characterized by maximum entropy) must be such that S as a function of state variables has negative curvature, $\Delta^2 S < 0$ (the *stability condition*). There are no macroscopic physical processes that drive $\Delta S < 0$ for isolated systems in equilibrium, yet we require mathematically that fluctuations about equilibrium lead to $\Delta S < 0$. Fluctuations in the theory of thermodynamics (a theory of equilibrium) are treated as virtual variations in the state of the system in which the condition of isolation is conceptually relaxed, akin to virtual displacements in analytical mechanics, which are conceived to occur at a fixed time.[1] The stability condition puts restrictions on the sign of response functions: $C_V > 0$ (heat capacity), $\beta > 0$ (isothermal compressibility), and $\partial \mu / \partial N > 0$ (Section 3.10). From the stability requirements, matter flows from regions of high to low chemical potential and entropy flows from high to low temperature (Section 3.11).

Internal energy: work, heat, and boundaries (first law of thermodynamics)

- The behavior of thermodynamic systems is determined by the nature of their boundaries. Three types of boundaries are distinguished: adiabatic, diathermic, and permeable (Table 1.1). Adiabatically isolated systems (enclosed by adiabatic boundaries) can interact with the environment through mechanical means only (no heat flow, no flow of particles); closed systems (enclosed by diathermic boundaries) restrict the flow of particles; open systems (enclosed by permeable boundaries) allow the flow of matter and energy. An additional type of system is distinguished, that of an isolated system with no interactions with the environment.

- Work W_{ad} performed on adiabatically isolated systems (adiabatic work) has the property that the transition it produces, $i \to f$ (between reproducible equilibrium states i and f), depends only on the pair (i, f) and not on the details of how W_{ad} is performed. From this discovery (path independence for the absorption of adiabatic work) we infer the existence of a state variable that couples to adiabatic work, the internal energy U such that $\Delta U = U_f - U_i = W_{ad}$. The energy of adiabatic work can be retrieved by letting the system perform adiabatic work on the environment, and thus is conserved.

- For work W performed under less restrictive, nonadiabatic conditions, it's found, for the same change of state $i \to f$ produced by W_{ad}, that $W \neq \Delta U$. The energy of mechanical work is thus not conserved in systems with diathermic boundaries; it's not all stored in the internal

---

[1] In statistical mechanics, fluctuations are treated as dynamical processes producing *correlations* between fluctuations at points separated in space and time.

energy of the system. Energy conservation is restored by recognizing new forms of energy—heat. The heat transferred to the system is the difference in work done on the system with the two types of boundaries that effect the same change in state, $Q \equiv \Delta U - W$. Internal energy is accounted for by the statement $\Delta U = W + Q$, the first law of thermodynamics. Work and heat are interconvertible forms of energy. Note the sign convention employed in this book.

- For small energy transfers the first law is written $dU = đQ + đW$. The notation dU indicates that the differential of U is exact (implied by the path independence of $\Delta U = W_{ad}$), but that the infinitesimal energy transfers $đQ$ and $đW$ are inexact. The number of possible combinations of $đQ$ and $đW$ that effect the same change dU is unlimited. While dU represents a small change in a physical quantity U, there do not exist substances Q and W of which $đQ$ and $đW$ represent small changes; $đQ$ and $đW$ are process-dependent modes of energy transfer.

- Infinitesimal, reversible heat transfers $(đQ)_{rev}$ comprise an exact differential when divided by the absolute temperature T at which the heat transfer takes place, $dS = (đQ)_{rev}/T$. The first law is then written completely in terms of exact differentials, $dU = T dS - P dV$. While such an expression was derived from considerations of reversible processes, as it entails state variables it can be applied to any process so long as a reversible path can be found connecting the end states of the process.

Irreversibility, time's implicit appearance in the theory

Irreversibility is the central message of thermodynamics; the sooner that recognition is made, the better. Several definitions underlie the concept of irreversibility.

- *Quasistatic processes* are: 1) performed sufficiently slowly that the system doesn't appreciably deviate from equilibrium, with state variables having values given by the equation of state, such that 2) work done on the system is only from forces holding the system in equilibrium, i.e., forces of friction have been eliminated.

- *Reversible processes* can be exactly reversed by making infinitesimal changes in the environment: System *and* environment are restored to their original conditions in a reversible process. Reversible processes are quasistatic, but not all quasistatic processes are reversible (Section 1.7). The concept of reversibility is problematic because it conceives of the reversal of processes in *time*. Strictly speaking, reversible processes are idealizations that do not exist: Equilibrium states do not involve change (time), yet reversible processes are taken to proceed through a sequence of equilibrium states. Any real change in thermodynamic state must occur at a finite rate, so the intermediate states cannot strictly be equilibrium states. Practically speaking, there will be some small yet finite rate at which processes occur such that disequilibrium has no observable consequences. Reversible processes can be plotted in state space.

- Not-reversible processes are termed irreversible. That "un-reversible" (irreversible) processes *exist* is central to thermodynamics. The second law of thermodynamics recognizes the existence of irreversibility and codifies our experience of it. Processes not reversible in time of necessity imply a *direction in time* for the occurrence of states linked by irreversible processes. Irreversibility simply *is*; we don't have to account for it.[2] Instead of worrying why macroscopic laws are time irreversible, the converse should be pondered why microscopic laws

---

[2]Recognizing the primary existence of irreversibility (without having to explain it) can be compared to the recognition that the property of matter known as inertia exists, without having to explain it. That inertial motion (free particles move without acceleration) simply *is*, is a foundation of the theory of relativity.

are time-reversal invariant.[3] We tend to think microscopic laws are more "real" than macroscopic; we're accustomed to breaking things apart in the attempt to say that the whole can be explained by the sum of the parts. The laws of thermodynamics codify our experience of the macroscopic world—surely a working concept of "reality." Sometimes macro-descriptions follow as a result of micro-descriptions, but not always. Temperature and entropy are quantities characteristic of the equilibrium state that do not derive from the microscopic properties of matter. The holy grail of physics is to succeed in ever-unifying the laws of physics. Until that happens, laws have their domains of validity. The second law is not a consequence of microscopic laws of motion; it's a separate principle.

- Irreversibility reveals the existence of states *inaccessible* from each other (see Chapter 10). As discussed in Section 1.7, take a gas enclosed by adiabatic boundaries at pressure and volume (P_0, V), and perform work on it in the form of stirring, holding V fixed. As a result, the pressure increases to a value $P_1 > P_0$. This process cannot be reversed (with $P_1 \to P_0$) by means of adiabatic work. Stirring the "other way" only increases the pressure further. With V fixed, there is no way to lower the pressure of an adiabatically isolated system by means of adiabatic work; the process is irreversible. In this example, the state K_0 characterized by the values of its state variables (P_0, V) cannot be attained from any of its states K ($P > P_0, V$) by means of adiabatic work. Such states are not equivalently related. All states K are accessible from K_0, $K_0 \to K$, but K_0 is not accessible from any state K, $K \nrightarrow K_0$. For a state K_0 inaccessible from K, if the system is in K_0 at time t, and K at another time t', then $t' > t$. States *time ordered* by irreversible processes could be labeled with the times at which such states occur, but they can also be labeled by the values of the entropy function, which stand in correspondence with the time order established by irreversibility; the entropy of isolated systems only increases, never decreases, Eq. (3.9).

Second law of thermodynamics

There have been nearly as many formulations of the second law as there have been discussions of it.—P.W. Bridgman, 1941[47, p116]

The mere phrase "second law of thermodynamics" is often enough to produce in students (and others) a certain state of unease. The second law has diverse, interrelated implications, yet there's seemingly no one, all-encompassing way of stating it.[4] Can one ever be sure that one has grasped its essence? Students (and others) often leap to "entropy" as the second law. While not incorrect, it misses the mark (in my opinion). I think it best to keep "intellectual distance" between entropy as a *consequence* of the second law, and the second law itself as the recognition of irreversibility. It's remarkable how much physics follows from the observation that heat does not spontaneously flow from cold to hot. Among the nexus of concepts linked by the second law, we have:

- *Clausius form of second law*: Heat cannot of itself flow from cold to hot. Note that merely having the word "flow" in this statement implies a direction in time.

- *Kelvin form of second law*: Heat can't be converted entirely into work at a single temperature. Waste heat must be involved; the efficiency of heat engines cannot be 100%. The Kelvin and Clausius statements are equivalent (Section 2.2).

---

[3]At the *really* microscopic level, time-reversal symmetry *is* broken. By the CPT-theorem of quantum field theory, any Lorentz-invariant local theory (commutation relations obey the spin-statistics connection) is invariant under the combined operations of C (charge conjugation, interchange particles for their antiparticles), P (parity, inversion of space axes through the origin), and T (invert time, $t \to -t$). The discovery in 1964 of CP violations in weak decays implies (if CPT invariance is sacred) that time-reversal symmetry is broken at the sub-nuclear level for processes involving the weak interaction. At the *super* macroscopic level, we have an expanding universe, which provides a sense for the direction of time.

[4]We could say then that the second law itself is entropic, because it has so many equivalent ways of understanding it.

- *Carathéodory principle*: Wait, there's more! The Carathéodory formulation of the second law, fully equivalent to the Kelvin form, is covered in Chapter 10.

- *Carnot's theorem*: The efficiency of a heat engine cannot exceed that of a reversible engine operating between the same reservoirs. Carnot's theorem is the source of the *inequalities* associated with the second law. The three statements—Clausius, Kelvin, Carnot—are impossibility statements. Many of the major laws of physics can be formulated as negative statements, what you can't do (Section 8.5).

- *Absolute temperature*: The universality of Carnot's theorem implies that temperature can be formulated universally, independent of the thermometric properties of substances. The ratio of absolute temperatures is defined as the ratio of the absolute values of the heats reversibly exchanged with reservoirs, Eq. (2.11). Absolute temperature T is measured on the Kelvin scale, specified by Eq. (2.14).

- *Clausius inequality*: Carnot's theorem implies, for a heat engine operating between reservoirs at absolute temperatures T_h and T_c with $T_h > T_c$, the inequality $|Q_c| \geq Q_h \left(T_c / T_h \right)$, where Q_c is the heat expelled to the reservoir at T_c, and Q_h is the heat absorbed from the reservoir at T_h, with equality holding for reversible engines. The greater the inefficiency, the greater is the heat expelled at the lower temperature. This basic inequality generalizes to an arbitrary cycle as an inequality on the integral $\oint \mathrm{d}Q/T \leq 0$, where $\mathrm{d}Q$ is the infinitesimal heat transfer (positive or negative) at the absolute temperature T. Overall the integral is non-positive because expelled heat is treated as negative, and the magnitude of the heat expelled at the lower temperature is never less than the heat absorbed at a higher temperature.

- *Entropy*: For reversible cycles $\oint \left(\mathrm{d}Q \right)_{\mathrm{rev}} /T = 0$, implying $\mathrm{d}S \equiv \left(\mathrm{d}Q \right)_{\mathrm{rev}} /T$ is an exact differential. The Clausius definition tells us what the differential of S is; it doesn't directly define entropy. The Boltzmann-Planck formula does that with $S = k_B \ln W$, where W is the number of microstates compatible with a given state of equilibrium (subject to constraints). The Clausius definition applies to closed systems (fixed amount of matter). The extension of entropy to open systems requires that it be extensive, a feature built into the statistical entropy. (The non-extensive entropy obtained from the Clausius definition is the root of the confusion in the so-called Gibbs's paradox.) Entropy and absolute temperature are both consequences of Carnot's theorem, fraternal twins born from the same mother.

- *Entropy creation*: The Clausius inequality in differential form is $\mathrm{d}S \geq \mathrm{d}Q/T$, where equality holds for reversible heat transfers. Thus, $\mathrm{d}Q/T$ does not account for all contributions to $\mathrm{d}S$. The entropy change associated with irreversibility, $\mathrm{d}S_i \equiv \mathrm{d}S - \mathrm{d}Q/T$, closes the gap between $\mathrm{d}Q/T$ and $\mathrm{d}S$, where, by the Clausius inequality, $\mathrm{d}S_i \geq 0$. Introducing $\mathrm{d}S_i$ exemplifies the same logic as the first law, in that whereas heat transferred to the system is the difference between the change in internal energy and the work done, $Q = \Delta U - W$, irreversible entropy production is the difference between the total entropy change and that due to heat transfers, $\Delta S_i = \Delta S - \int \mathrm{d}Q/T$. Heat transfers with the environment, $\mathrm{d}Q/T$, can be positive, negative, or zero. Only when heat is added reversibly to a system is the entropy change given solely in terms of heat transfers. Otherwise, $\Delta S_i > 0$ is entropy created by irreversibility and is always positive. We show in Chapter 14 that the *rate* of entropy production due to irreversibility is non-negative, Eq. (14.47).

- *Entropy only increases in isolated systems*: For adiabatically isolated systems $\mathrm{d}S \geq 0$. The entropy of isolated systems can only increase, never decrease, perhaps the most far-reaching result of thermodynamics. In approaching equilibrium, the entropy of an isolated system achieves a maximum value subject to constraints on the system. The mere fact that there are processes, such as the free expansion, where neither energy nor matter is transferred to the system, yet there is a change between equilibrium states implies there must be another state variable characterizing the system. Entropy is that new state variable.

- *Energy dissipation and free energy*: Not all of the energy change ΔU in thermodynamic processes can be converted into work when $\Delta S > 0$, a phenomenon known as energy dissipation. Energy is not lost, its ability to produce work is diminished through the increase in entropy. *Free energy* is the energy available for work. The Helmholtz free energy $F \equiv U - TS$ provides the maximum obtainable work at constant T: $[\Delta F]_T = \Delta U - T\Delta S = W_{\max}$. The Gibbs free energy is the maximum available "other work" W' (work other than PdV work) for processes at constant T and P: $[\Delta G]_{T,P} = \Delta U - T\Delta S + P\Delta V = W'_{\max}$. The energy *dissipated* in a process (energy *not* available for work) is $T_0\Delta S$, where T_0 is the coldest temperature obtained in the process.

Thermodynamic potentials

- From the first law $dU = TdS - PdV$, and thus we infer that U is naturally a function of S and V, $U = U(S, V)$. Hence, $[\Delta U]_S = W$. Internal energy is the storehouse of adiabatic work. (Reversible processes with $dQ = 0 \Rightarrow dS = 0$.)

- Entropy is not accessible to experimental control (can you buy an entropy meter?), other than isolating the system adiabatically so that $dS = 0$. Because $U(S, V)$ is a convex function of its arguments (Exercise 3.5), we can use Legendre transformations to develop equivalent state variables (having the dimension of energy) that are functions of quantities more amenable to experimental control. Absolute temperature is the value of the tangent to the energy surface associated with changes in S, $T = (\partial U/\partial S)_{V,N}$ and $-P$ is the tangent to the energy surface associated with changes in V. Legendre transformations of U are $F = U - TS$, $H = U + PV$, and $G = U - TS + PV$. From the differentials of these functions, Eq. (4.2), we have that $F = F(T, V)$, $H = H(S, P)$ and $G = G(T, P)$.

- One theoretical advantage of introducing Legendre transformations of $U(S, V)$ is that they provide a rich source of relations among certain partial derivatives of state variables that are not otherwise apparent. The differentials of each of the functions U, F, H, G are exact, and the integrability condition Eq. (1.1) implies a set of thermodynamic identifies known as Maxwell relations, Table 4.2.

- The quantities U, F, H, G each represent a type of potential energy stored in the system under experimentally controlled conditions: $[\Delta U]_S = W_{\mathrm{ad}}$, $[\Delta H]_{S,P} = W'_{\mathrm{ad}}$, $[\Delta F]_{T,V} = W'$, $[\Delta G]_{T,P} = W'$, where W' is "other" work (see Table 4.1). Furthermore, $[\Delta F]_T = W$ and $[\Delta H]_P = Q$. Just as a knowledge of the types of potential energy helps one solve problems in mechanics, knowing the interpretations of the potentials can prove highly advantageous in efficiently working problems in thermodynamics. Under the conditions indicated, we have $[\Delta U]_{S,V} \leq 0$, $[\Delta H]_{S,P} \leq 0$, $[\Delta F]_{T,V} \leq 0$, and $[\Delta G]_{T,P} \leq 0$, where equality holds for reversible processes. The quantities U, H, F, and G are each a minimum in equilibrium. A further result is that the chemical potential is the Gibbs energy per particle, $G = N\mu$.

Chemical thermodynamics

Every equation of thermodynamics applies to a well-defined chemical substance; the connection between chemistry and thermodynamics is strong. Much as physicists might want to minimize it, applications of thermodynamics require at least a limited knowledge of chemical concepts.

- For a substance distributed among several phases coexisting in equilibrium, its chemical potential is independent of phase, $\mu^I = \mu^{II} = \mu^{III} = \cdots$, Eq. (6.19).

- For a system having k chemical species in π phases, the Gibbs phase rule $f = 2+k-\pi$ is the number of intensive variables that can be varied without disturbing the number of phases in equilibrium, Eq. (6.20). A result of this generality should be appreciated by physics students; it follows directly from the condition that in equilibrium the Gibbs energy is a minimum.

- Between two phases involving a single substance ($f = 1$), only one intensive variable can be varied along the coexistence curve. The slope of the coexistence curve is given by the Clausius-Clapeyron equation, Eq. (6.22). For a single substance in three phases ($f = 0$), no variations are possible: The triple point occurs at a unique combination of T and P.

- The law of mass action, Eq. (6.42), derived from the theory of chemical equilibrium should be known to physicists as it plays a major role in the theory of carrier densities in semiconductors.

- Electrochemical cells are used to measure changes in free energy (either F or G) which play a pivotal role in formulating the third law of thermodynamics. If you want to understand the third law, you need to understand how the experiments are conducted, and thus at least a cursory understanding of electrochemical cells.

Sackur-Tetrode equation: It's a good thing

The Sackur-Tetrode formula for the entropy of the ideal gas (Section 7.4),

$$S = Nk_B \left[\frac{5}{2} + \ln \left(\frac{V}{N\lambda_T^3} \right) \right] , \tag{9.1}$$

is a significant achievement. Let us count the ways.

- All parts of it have been tested against experiment. It's consistent with the heat capacity ($C_V = T(\partial S/\partial T)_{V,N} = 3Nk_B/2$) and the equation of state ($P = T(\partial S/\partial V)_{T,N} = Nk_BT/V$); these "tests" however would be met by an expression of the form $S = Nk_B \ln(T^{3/2}V)$, what was derived in Eq. (3.21) before imposing extensivity. The strongest test of Eq. (9.1) is that all terms appearing in it are required to reproduce the entropy of a gas in contact with its liquid phase (Section 7.4), which does not involve taking logarithmic derivatives; constants matter!

- Equation (9.1) embodies extensivity (N scales with V for fixed density). While that shouldn't come as a surprise (extensivity is built into the statistical entropy), having an experimentally-tested, extensive expression for S is worthy. That alone should lay to rest Gibbs's paradox (Section 7.7). As discussed in Section 3.7, extensivity must be imposed on the thermodynamic entropy as an additional requirement; the Clausius definition is incomplete and does not pertain to open systems.

- It involves Planck's constant, perhaps the first association of h with material particles. The statistical entropy, $S = k_B \ln W$, is a connection between microscopic and macroscopic, with W being the number of microstates per macrostate. The Sackur-Tetrode formula provides experimental confirmation—from thermodynamic measurements—of the quantum-mechanical picture that each particle carries with it a fundamental volume of phase space, h^3. There is a granularity of phase space (in units of h^3) that provides a natural means by which to count microstates. Equation (9.1) can be derived when W is given as in Eq. (7.57).

- We *better* be able to calculate the entropy of the ideal gas, the simplest macroscopic system. In doing so, however, we arrive at the end of thermodynamics and the beginning of statistical physics—we *can't* calculate S for the ideal gas without Planck's constant. Equation (3.27) is as far we can get with thermodynamics alone (even after making it extensive); we noted there the need for a characteristic quantity having the dimension of action.

Ideal gas vs. photon gas

We let pass without further comment the remark in Chapter 1 that Eq. (1.28) is one of the most important equations in thermodynamics,

$$\left(\frac{\partial U}{\partial V}\right)_T = T \left(\frac{\partial P}{\partial T}\right)_V - P . \tag{1.28}$$

Equation (1.28) follows from Eq. (3.12) and a Maxwell relation (Exercise 4.6), but the larger point is that it cannot be derived unless we have entropy as a state variable. Equation (1.28) also occurs in the Carathéodory formulation (Section 10.1) as related to the integrating factor for đQ and the absolute temperature, which is an alternative path to the existence of entropy as a state variable. Dimensionally, Eq. (1.28) relates energy density to pressure, itself an energy density.[5] It's of interest to compare the ideal gas (noninteracting particles in thermal equilibrium) with a system of photons in thermal equilibrium (cavity radiation).

- Using the ideal gas equation of state, we find from Eq. (1.28)

$$\left(\frac{\partial U}{\partial V}\right)_T = 0 . \qquad \text{(ideal gas)}$$

  By ignoring inter-particle interactions, the internal energy of the gas is independent of volume and is a function of temperature only, $U(T, V) = U(T)$.

- For the photon gas, the energy *density* $u = U/V$ is independent of volume under isothermal conditions, Eq. (5.4),

$$\left(\frac{\partial u}{\partial V}\right)_T = 0 . \qquad \text{(photon gas)}$$

  Equation (5.4) does not follow from Eq. (1.28), it follows from the second law (cavity radiation is isotropic and independent of the size and shape of the cavity, and the emissivity of a black body depends only on the energy density). Thus, $U(T, V) = Vu(T)$. Equation (5.4) applies only for isothermal conditions; the energy density is *not* independent of volume for adiabatic processes, $V (\partial u/\partial V)_S = -4u/3$, Eq. (5.32). The temperature of the photon gas changes in adiabatic processes, such as in the Cosmic Microwave Background. The pressure of cavity radiation can be calculated using classical methods, with the result $P = \frac{1}{3}u$, Eq. (5.6). (Compare with $P = \frac{2}{3}u$ for the ideal gas.) By combining the equation of state Eq. (5.6) with Eq. (1.28), we obtain the energy density $u(T) = aT^4$, Eq. (5.8), where a is the radiation constant. The entropy of the photon gas then follows from thermodynamics, $S = \frac{4}{3}aVT^3$, Eq. (5.13). In comparison with the complexity of the Sackur-Tetrode formula, Eq. (9.1), the entropy of cavity radiation is quite simply related to the energy, $S = \frac{4}{3}U/T$.

- In the isothermal expansion of an ideal gas $\Delta U = 0$ because the energy is a function of T only. To maintain the gas at constant temperature, heat is absorbed from a reservoir and is converted entirely into work (not a violation of the second law—there's no cycle). The entropy change $\Delta S = Nk_B \ln(V_f/V_i)$. No chemical work is performed because d$N = 0$, even though $\mu \neq 0$.

- In the isothermal expansion of a photon gas, there's no chemical work because $\mu = 0$, even though d$N \neq 0$ (because $\Delta S \neq 0$). The work done at constant T is, from the Helmholtz free energy, $W = \Delta U - T\Delta S = -\frac{1}{3}aT^4\Delta V = -P\Delta V$. The heat absorbed, $T\Delta S = \frac{4}{3}aT^4\Delta V = \frac{4}{3}\Delta U$, exceeds the change in energy by $\frac{1}{3}\Delta U$; heat is not all converted into work.

---

[5]Pressure has the dimension of energy density. Note from the ideal gas law, $PV = Nk_BT$, that energy equals energy.

The heat absorbed comes from the energy of added photons. From Section 5.2, $S = 3.6k_B N$, and thus $\Delta S = 3.6k_B \Delta N$. The entropy change of the photon gas in an isothermal expansion can be written $\Delta S = 3.6k_B N \, (\Delta V/V)$, where N is the initial number of photons (Exercise 5.5). The entropy change of the ideal gas for the same process is $\Delta S = Nk_B \, (\Delta V/V)$. The change in entropy of the photon gas is thus considerably larger than that of an ideal gas with the same number of particles.

- For the photon gas, the isothermal change in S is due to the change in the number of photons, photons that are created at all frequencies in proportion to the Planck distribution at absolute temperature T. The number of photons between frequency ν and $\nu + d\nu$ in a cavity of volume V is ($\beta \equiv (k_B T)^{-1}$)

$$N(\nu)d\nu = \frac{8\pi V}{c^3} \frac{\nu^2}{e^{\beta h\nu} - 1} d\nu \, ,$$

where $N = \int_0^\infty N(\nu)d\nu$. Clearly $\Delta N(\nu) \propto \Delta V$ at a fixed temperature. Without the photon picture, the increase in S is difficult to understand.

- In adiabatic processes for the photon gas $VT^3 = \text{constant}$. For $V \propto R^3$, where R is a characteristic length, $T \propto R^{-1}$. Even though the temperature changes in such a process, U/T is a constant. In reversible adiabatic processes, cavity radiation retains its black-body spectral distribution (Section 5.3).

Entropy, the constant stranger

> ... it is to be feared that we shall have to be taught thermodynamics for several generations before we can expect beginners to receive as axiomatic the theory of entropy.
> —J. C. Maxwell, 1878[2, p668]

Humans have known about entropy for 150 years, with its meaning having perplexed physicists such as Gibbs, Maxwell, Boltzmann, Planck, and Schrödinger. To what extent can we say that we understand it? Were Maxwell to wake up today would he think our attempts at entropy education a success, or would he say that we still have several generations to go? Chapter 1 introduced the zeroth and first laws of thermodynamics, the concepts of temperature and conservation of energy. If that's all there was to the subject, this might be a relatively thin book. As it is, however, thermodynamics simply *is* about entropy: Practically everything from Chapter 2 on has been about entropy in some way. Entropy, like quantum mechanics and relativity, is outside our everyday experience of the physical world, and must be learned through concerted reflection.[6] Offered, for your consideration, some musings on entropy.

- Entropy and absolute temperature are two sides of the same coin. Both arise as consequences of the Carnot theorem. The product TS has the dimension of energy. Because we have artificially given absolute temperature the unit Kelvin, entropy is forced to have the units Joule/Kelvin. Absolute temperature more naturally should have the dimension of energy and entropy should be dimensionless (Sections 2.4, 3.2). Entropy as a dimensionless quantity lends credence to the statistical picture that entropy is related to a number, $S/k_B = \ln W$. We'll see in Chapter 12 that entropy can be considered a special (but important!) case of a more general quantity, *information*, which is dimensionless.

- How did its inventor Rudolf Clausius view entropy? He wrote in 1865:[15, p327]

> The whole mechanical theory of heat rests on two fundamental theorems: that of the equivalence of heat and work, and that of the equivalence of transformations.

---

[6]Einstein's words are apt here: "Common sense is the collection of prejudices acquired by age eighteen."

Clausius called the second law the *Law of the Equivalence of Transformations*. Before coining the term entropy, he referred to Q/T as the *equivalence value* of a transformation. A *transformation* for Clausius is a conversion of heat into work or work into heat, or a transfer of heat between bodies at one temperature to another temperature. He wanted to assign a mathematical magnitude to transformations, and found that Q/T served that purpose. The equivalence value for a given transformation can be positive or negative. He discovered that $\oint (đQ)_{\mathrm{rev}} /T = 0$ for reversible cycles, and thus all reversible cycles have the same equivalence value, but more importantly that the positive and negative values of the transformations comprising a reversible cycle cancel. For non-reversible cyclical processes, the positive and negative transformations are not equal and they can differ only in such a way that the negative transformations predominate, $\oint đQ/T \leq 0$, Eq. (3.2). Clausius considered S the *transformational content* of a body, which is the genesis of the word entropy, adapted from the Greek $\tau\rho o\pi\acute{\eta}$ ("in transformation").

- Clausius would separately write the first and second laws as

$$dQ = dU - dW \qquad\qquad dQ = TdS .$$

To quote Clausius:[15, p366]

> But in order to be able to bring the equations into conjunction, we will suppose that they relate to one and the same reversible change of a body. In this case the thermal element dQ is the same in both equations, hence we can eliminate it from the equations, ...

$$T dS = dU - dW . \tag{9.2}$$

Gibbs would take Eq. (9.2) as the starting point for his mathematical characterization of the equilibrium state of a fixed mass of material.

- Walther Nernst referred to the second law as the *Law of the Transmutability of Energy*, which he summarized as follows:[48, p2]

> According to the second principle there is, for isothermal variations of a system, a function F which has, for such variations, the same properties as U; $F_2 - F_1$ expresses the maximal external work which can be obtained in the change considered, and this quantity is likewise independent of the nature of the method by which the maximum work considered is obtained.

Nernst viewed the second law *not in terms of entropy*, but in terms of free energies. It appears that he *avoided the concept of entropy altogether*; one can formally eliminate S in favor of thermodynamic potentials: $-TS = F - U = G - H$. In his work on the third law (Section 8.2) Nernst measured the differences $\Delta G - \Delta H$ and $\Delta U - \Delta F$ which are related to ΔS through the Gibbs-Helmholtz equations, Eq. (6.52).

- Ilya Prigogine, winner of the 1977 Nobel Prize in Chemistry, referred to the second law as the *Principle of the Creation of Entropy* to emphasize the role played by irreversibility; entropy is created in irreversible changes in state.[49, p32]

- Maxwell wrote that "The second law of thermodynamics has the same degree of truth as the statement that if you throw a tumblerful of water into the sea, you cannot get the same tumblerful of water out again." Quoted in [50]. The second law captures the propensity of macroscopic systems to *spread out*—if there are more ways to be over there than over here, then over there is where you'll find the system, most of the time.

So, what is thermodynamics?

In reading (and writing) about thermodynamics, the *wordiness* of the subject is always apparent. Compared with other branches of physics, thermodynamics seems to be alone in the amount of verbosity required to impart its ideas. In preparing this book, I've continually kept in mind the following question: What is the essence of the subject and why is it so difficult to convey succinctly? Why thermodynamics requires so much exposition is perhaps due to the simultaneous lack of precision of its concepts (and hence their astonishing generality) and yet their great utility, seemingly in spite of it all. Einstein wrote, concerning thermodynamics:

> A theory is the more impressive the greater the simplicity of its premises is, the more different kinds of things it relates, and the more extended is its area of applicability. Therefore the deep impression which classical thermodynamics made upon me. It is the only physical theory of universal content concerning which I am convinced that, within the framework of the applicability of its basic concepts, it will never be overthrown ....[51, p33]

With that said, having now delved into the subject with considerable depth, can we say what thermodynamics is, standing on one foot? If you had to come up with a "tagline" for thermodynamics, what would it be? A theme set forth in Chapter 1 is that thermodynamics concerns equilibrium, internal energy, and irreversibility, concepts that line up with the three main laws of thermodynamics, the zeroth, first, and second. Yet, as we've emphasized, irreversibility is the chief feature of thermodynamics, that constraints once removed cannot be re-established without performing work on the system. A definition of thermodynamics might be:[7]

> Thermodynamics is the study of the equilibrium properties of macroscopic systems for which the release of constraints cannot be undone without the expenditure of work.

---

[7]Consider the definition of thermodynamics offered by H.B. Callen, which is a nod to the more microscopic interpretation of thermodynamics: "Thermodynamics is the study of the macroscopic consequences of myriads of atomic coordinates, which, by virtue of the statistical averaging, do not appear explicitly in a macroscopic description of a system."[1, p7]

II

Additional Topics in Thermodynamics

II

Additional Topics in Thermodynamics

CHAPTER 10

Carathéodory formulation

ENTROPY was discovered by a somewhat circuitous path through the efficiency of heat engines,[1] a finding that in hindsight could appear serendipitous. Were we just *lucky* to have discovered something so fundamental in this way? Can it be seen *directly* that entropy as a state variable is contained in the structure of thermodynamics, without the baggage of heat engines? It can, as shown by Constantin Carathéodory in 1909. That Carathéodory's work is not more widely known is due to the unfamiliarity of the mathematics—the existence of integrating factors, a topic we develop in this chapter. For Carathéodory's treatment to be without reference to heat engines it must necessarily be more abstract than the traditional approach. We start by examining, in detail, the condition for integrability introduced in Section 1.3.

10.1   INTEGRABILITY CONDITIONS AND THERMODYNAMICS

Scalar and vector fields

A *scalar field* is a function that assigns a number to every point of its domain. The temperature distribution in physical space, for example, $T = T(x, y, z)$, is a scalar field. State variables can be considered scalar fields in *state space*. For example, the equation of state $T = T(P, V)$ is a scalar field in the space spanned by P and V. The same is true of any state function in the appropriate space of variables, e.g., $U = U(S, V, N)$. In the following, we denote a generic state function Φ as a scalar field in a space of n state variables, $\Phi = \Phi(x_1, \cdots, x_n)$.

The change in Φ between nearby points in its domain is, from calculus,

$$d\Phi = \sum_{i=1}^{n} \frac{\partial \Phi}{\partial x_i} dx_i \equiv \sum_{i=1}^{n} F_i(x_1, \cdots, x_n) dx_i \equiv \boldsymbol{F} \cdot d\boldsymbol{r} , \tag{10.1}$$

where *we treat the derivatives* $\partial \Phi / \partial x_i$ *as components of a vector field.*[2] Linear differential expressions $\sum_{i=1}^{n} F_i dx_i$ are known as Pfaffian differential forms in n variables.[3] Thus, there's a connection between vector fields and Pfaffians.

Scalar fields Φ gives rise to vector fields through their gradients, $\nabla\Phi$. Vector fields feature prominently in physics.[4] It's not clear, however, that they should find any use in thermodynamics. As we'll see, treating thermodynamic quantities as components of vector fields leads to an increased understanding of the mathematical foundation of the second law. A central question we must address

---

[1]Entropy emerges as a special case of the Clausius inequality, which is a consequence of Carnot's theorem.

[2]A vector field is an assignment of a vector to every point of the appropriate space.

[3]A Pfaffian differential form is, in modern language, called a 1-form. That begs the question: Are there 2-forms, 3-forms? Yes, but those are concepts not needed in this book. When a Pfaffian form is equated with zero, $\sum_i F_i dx_i = 0$, it's referred to as a Pfaffian differential equation.

[4]For example, in electromagnetism, fluid mechanics, and general relativity.

is that whereas a vector field can always be obtained from a scalar field, to what extent is the converse true: *Does a given vector field F imply the existence of a scalar field?* That is, while $\Phi \Rightarrow \nabla\Phi$, does $F \Rightarrow \Phi$? The prototypical example is the relationship between the electric field and the electrostatic potential function, $E = -\nabla V$, with V derived from E, $V(r) = -\int E \cdot dr$. We posed a similar question in Section 1.3, that while the differential of a function generates a linear differential form, does an arbitrary Pfaffian represent the differential of an integrable function?

Changes in state variables Φ are independent of the means by which change is effected (Section 1.2). Thus, $\int_1^2 d\Phi = \Phi(2) - \Phi(1)$ is independent of the integration path between the endpoints $(1, 2)$. Under what conditions is the integral $\int_C F \cdot dr$ independent of the curve C joining the endpoints such that[5]

$$\int_C F \cdot dr = \Phi(2) - \Phi(1) ? \tag{10.2}$$

A necessary and sufficient condition for Eq. (10.2) to hold is the equality of the derivatives,[9, p357]

$$\frac{\partial F_j}{\partial x_i} = \frac{\partial F_i}{\partial x_j} . \qquad (i, j = 1 \cdots n) \tag{10.3}$$

There are $\frac{1}{2}n(n-1)$ independent relations implied by Eq. (10.3), the *integrability conditions*, the same conditions for the n-variable Pfaffian $\sum_{i=1}^n F_i dx_i$ to be an exact differential.[6]

Rotational and irrotational vector fields

It's convenient to adopt a special notation for the integrability conditions. Let[7]

$$T_{ij} \equiv \frac{\partial F_j}{\partial x_i} - \frac{\partial F_i}{\partial x_j} . \qquad (i, j = 1 \cdots n) \tag{10.4}$$

If Eq. (10.3) is satisfied, then $T_{ij} = 0$. Thus, we have a way of classifying vector fields: either *all* the terms T_{ij} are zero, or not. In the special case of $n = 3$, the terms in Eq. (10.4) are the components of the curl $C \equiv \nabla \times F$, with $C_i = \sum_{jk} \varepsilon_{ijk}\partial_j F_k$, where ε_{ijk} is the Levi-Civita symbol for a three-dimensional space,[8] with $C_1 = T_{23}$, $C_2 = T_{31}$, and $C_3 = T_{12}$. A three-dimensional vector field F is therefore integrable when it has zero curl, $\nabla \times F = 0$. Because $F = \nabla\Phi$ satisfies Eq. (10.2), $\{T_{ij} = 0\}$ implies (for $n = 3$) the result from vector calculus, $\nabla \times \nabla\Phi = 0$.

Integrable vector fields are thus curl free, right? Basically, but not quite; it's a question of terminology. *The curl exists only in three dimensions.* The vector cross product is *not defined in spaces of dimension $n \geq 4$*. The "trick" of associating the components of a vector C_i with the integrability conditions T_{jk} only works when $n = 3$, which is the non-trivial solution of $n = \frac{1}{2}n(n-1)$. The quantities $\{T_{ij}\}$ *are the generalization of the curl in higher-dimensional spaces,* and $\{T_{ij} = 0\}$ is the generalization of[9] the classic result from vector analysis, curl(grad) $= 0$. Vector fields F satisfying Eq. (10.3) (in any dimension) are said to be *irrotational*. Vector fields not satisfying Eq. (10.3) are called *rotational* and the associated Pfaffian is inexact. These definitions are summarized in Table 10.1.

---

[5]To be clear, dr does not denote an infinitesimal vector in real space, but in state space; just consider that $F \cdot dr$ is shorthand for $\sum_i F_i dx_i$.

[6]Equation (10.3) generalizes Eq. (1.1) to n variables; we also recognize Eq. (10.3) as that from which the Maxwell relations are generated, Section 4.5.

[7]The quantities T_{ij} in Eq. (10.4) are the elements of an antisymmetric, second rank tensor, $T_{ij} = -T_{ji}$.

[8]Learned comment: The isomorphism between the elements of an antisymmetric tensor T_{ij} and the components of a vector C_i is an instance of *Hodge duality*. Only in three dimensions is the dual of a second-rank antisymmetric tensor a vector. In four dimensions, for example, the dual of a vector is a third-rank antisymmetric tensor.

[9]In electromagnetism expressed in four-dimensional spacetime, the two relations $B = \nabla \times A$ and $E = -\nabla\phi - \partial A/\partial t$ are "packaged" into the four-dimensional generalization of the curl (antisymmetric tensor) $F_{\mu\nu} \equiv \partial_\mu A_\nu - \partial_\nu A_\mu$, where the potential four-vector $A_\mu \equiv (-\phi/c, A)$, $\mu = 0, 1, 2, 3$, with 0 the time index, $\partial_\mu \equiv (\partial_0, \nabla)$, and $x^0 \equiv ct$.

Table 10.1 Rotational character of vector fields and exactness of associated Pfaffian

Integrability condition	Vector field, $\boldsymbol{F}$	Pfaffian, $\sum_i F_i dx_i$
all $T_{ij} = 0$	Irrotational	Exact
some $T_{ij} \neq 0$	Rotational	Inexact

Rotational vector fields and integrating factors

Differentials of state functions Φ are exact and hence integrable: The integral $\int_C d\Phi$ establishes a unique value of Φ at any point up to a constant. Associated with Φ is an irrotational vector field, $\boldsymbol{\nabla}\Phi$. Surfaces of constant Φ are referred to as *equipotential surfaces*;[10] they're defined by displacements $d\boldsymbol{r}$ such that $\boldsymbol{\nabla}\Phi \cdot d\boldsymbol{r} = 0$. Said differently, $\boldsymbol{\nabla}\Phi$ *is everywhere orthogonal to equipotential surfaces.* Irrotational vector fields imply scalar fields, and vice versa.

Rotational vector fields, however, *do not imply the existence of scalar fields*—unique numbers cannot be assigned to points because $\int_C \boldsymbol{F} \cdot d\boldsymbol{r}$ in general depends on the integration path C. Nevertheless, there *may* be a way in which rotational vector fields can be associated with equipotential surfaces, if *special curves* $\widetilde{C}$ can be found that are everywhere orthogonal to $\boldsymbol{F}$, so that

$$\int_{\widetilde{C}} \boldsymbol{F} \cdot d\boldsymbol{r} = 0 \, . \qquad (\boldsymbol{F} \text{ rotational, special curve } \widetilde{C}, \text{ independent of endpoints})$$

If the integrand vanishes at all points along $\widetilde{C}$, the integral is obviously independent of its endpoints. Contrast with the line integral of an irrotational vector field,

$$\int_C \boldsymbol{F} \cdot d\boldsymbol{r} = \Delta\Phi \, . \qquad (\boldsymbol{F} \text{ irrotational, arbitrary curves } C, \text{ depends on endpoints})$$

Curves $\widetilde{C}$ orthogonal to $\boldsymbol{F}$ *do not necessarily exist.* When such curves exist, they mesh together to form an equipotential surface if it turns out they're also orthogonal to an *irrotational* vector field (call it $\boldsymbol{G}$) that's locally *colinear* with $\boldsymbol{F}$. In that case $\boldsymbol{F}$ can be *scaled* so that $\lambda(x,y)\boldsymbol{F}$ is locally coincident with $\boldsymbol{G}$, i.e., $\boldsymbol{G} = \lambda(x,y)\boldsymbol{F}$. The scaling factor λ is called the integrating factor. The idea is schematically illustrated in Figure 10.1. *Rotational vector fields can thus be divided into two classes: those that do and do not possess integrating factors.*

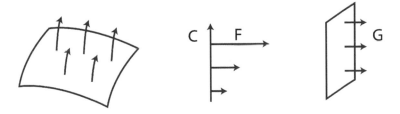

Figure 10.1 Left: Equipotential surface is orthogonal to an irrotational vector field. Middle: Path $\widetilde{C}$ is orthogonal to a rotational vector field, $\boldsymbol{F}$. Right: $\lambda(x,y)\boldsymbol{F} = \boldsymbol{G}$ is locally coincident with an irrotational vector field.

---

[10]The locus of points in state space of constant temperature is referred to as an isotherm; the locus of points of constant entropy, an isentrope; constant pressure, an isobar; and constant volume, an isochore. For want of a better word, we'll refer to the locus of points with constant values of the generic state function Φ as an equipotential surface. In mathematics, the set of points where a function $f(x_1, \cdots, x_n) = $ constant is referred to as its *level set*.

In what follows we consider the conditions under which vector fields possess integrating factors. It turns out that Pfaffians in two variables are fundamentally different from those with three or more variables; we treat separately the cases of $n = 2$ and $n \geq 3$.

Integrating factors for Pfaffians in two variables always exist

For two independent variables, $\boldsymbol{F} \cdot \mathrm{d}\boldsymbol{r} = 0$ is equivalent to

$$\mathrm{d}Q \equiv M(x, y)\mathrm{d}x + N(x, y)\mathrm{d}y = 0 . \tag{10.5}$$

We're changing notation here to make the presentation more concrete: $(x_1, x_2) \equiv (x, y)$ and $(F_1, F_2) \equiv (M, N)$. (We're also giving the Pfaffian a suggestive name, $\mathrm{d}Q$.) Equation (10.5) implies the first-order ordinary differential equation

$$\frac{\mathrm{d}y}{\mathrm{d}x} = -\frac{M(x, y)}{N(x, y)} \equiv f(x, y) . \tag{10.6}$$

Solutions to Eq. (10.6) exist under very general conditions, *regardless of whether the Pfaffian $\mathrm{d}Q$ is exact or not*.[11] The solutions to Eq. (10.6) are a one-parameter family of non-intersecting curves

$$\psi(x, y) = k , \tag{10.7}$$

where k is a constant. Through every regular point[12] in the plane there passes exactly one curve that's a solution to Eq. (10.6).[14, p22]

By differentiating Eq. (10.7), we have

$$\mathrm{d}\psi = \frac{\partial \psi}{\partial x}\mathrm{d}x + \frac{\partial \psi}{\partial y}\mathrm{d}y = 0 . \tag{10.8}$$

Because Eq. (10.8) describes the same curves as do the solutions of Eq. (10.5),

$$\frac{\mathrm{d}y}{\mathrm{d}x} = -\frac{M(x, y)}{N(x, y)} = -\frac{\partial \psi/\partial x}{\partial \psi/\partial y} \quad \Longrightarrow \quad \frac{1}{M}\frac{\partial \psi}{\partial x} = \frac{1}{N}\frac{\partial \psi}{\partial y} . \tag{10.9}$$

Equation (10.9) implies the existence of a function, call it $\lambda(x, y)$. Thus,

$$\frac{\partial \psi}{\partial x} = \lambda(x, y)M(x, y) \qquad \frac{\partial \psi}{\partial y} = \lambda(x, y)N(x, y) . \tag{10.10}$$

The rabbit is in the hat. Combining Eqs. (10.10) and (10.8), and using Eq. (10.5),

$$\mathrm{d}\psi = \lambda(x, y)\left(M(x, y)\mathrm{d}x + N(x, y)\mathrm{d}y\right) = \lambda(x, y)\mathrm{d}Q .$$

Curves for which $\mathrm{d}Q = 0$ are therefore the same as $\mathrm{d}\psi = 0$ ("equipotential"), implying the existence of an integrating factor.[13] *A Pfaffian differential equation in two variables always possesses an integrating factor.*

What have we accomplished geometrically? We have a vector field $\boldsymbol{F}$ that does not satisfy the integrability condition, Eq. (10.3), and we have constructed a "stretched" vector field $\boldsymbol{G} \equiv \lambda(x, y)\boldsymbol{F}$ that does, where the scaling factor $\lambda(x, y)$ is different at every point in the plane (see Fig. 10.2). $\boldsymbol{G}$ is everywhere colinear with $\boldsymbol{F}$, and both are orthogonal to the curves $\psi(x, y) = k$. Along curves such that $\mathrm{d}Q = 0$, $\psi(x, y)$ has a constant value so that $\mathrm{d}\psi = 0$. Note that in this case $\boldsymbol{F} \cdot \nabla \times \boldsymbol{F} = 0$, even though $\nabla \times \boldsymbol{F} \neq 0$, because $\nabla \times \boldsymbol{F}$ is orthogonal to the plane. We also have $\boldsymbol{G} \cdot \nabla \times \boldsymbol{G} = 0$ because $\nabla \times \boldsymbol{G} = 0$. The significance of $\boldsymbol{F} \cdot \nabla \times \boldsymbol{F} = 0$ is discussed in Eq. (10.12).

---

[11]That's why Pfaffian differential equations in two variables always possess an integrating factor.
[12]At a *regular point*, $M(x, y) \neq 0$ and $N(x, y) \neq 0$.
[13]Equipotential curves were obtained for the ideal gas in the form of $TV^{3/2} = k =$ constant, Exercise 1.13.

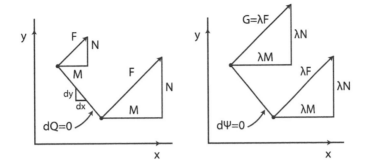

Figure 10.2 Stretched vector field produced by the integrating factor, $\lambda(x, y)$. F is rotational; $G = \lambda F$ is irrotational. The line such that $F \cdot dr = 0$ is embedded in the equipotential surface $d\psi = 0$ where $G = \nabla\psi$.

Integrating factor for Pfaffians in $n \geq 3$ variables

Consider a three-dimensional vector field such that $\nabla \times F \neq 0$ (rotational). We seek a function $\lambda(x_1, x_2, x_3)$ that "stretches" F into an irrotational field, so that $\nabla \times (\lambda F) = 0$, or

$$\lambda \nabla \times F + \nabla \lambda \times F = 0 . \tag{10.11}$$

Take the inner product of Eq. (10.11) with F: $\lambda F \cdot \nabla \times F + F \cdot (\nabla \lambda \times F) = \lambda F \cdot \nabla \times F = 0$. A necessary condition for the existence of an integrating factor is thus

$$F \cdot \nabla \times F = 0 . \qquad (n = 3) \tag{10.12}$$

Equation (10.12) is necessary because it presumes that λ exists; Eq. (10.12) must be satisfied irrespective of the form of λ. Three-dimensional vector fields, however, are not *necessarily* orthogonal to their curl (as is the case for $n = 2$). *Integrating factors don't have to exist for $n = 3$.* Ditto for $n \geq 4$.

Example. Does the Pfaffian $x_2 dx_1 + dx_2 + dx_3$ possess an integrating factor? In this case, $F \cdot \nabla \times F = -1$; it does not have an integrating factor.

For there to be an integrating factor, the irrotational vector G has n components λF_i, where $\lambda = \lambda(x_1, \cdots, x_n)$, for which the integrability conditions apply,

$$\frac{\partial(\lambda F_j)}{\partial x_i} - \frac{\partial(\lambda F_i)}{\partial x_j} = 0 . \tag{10.13}$$

As with Eq. (10.3), there are $n(n-1)/2$ independent conditions implied by Eq. (10.13). Equation (10.13) implies the set of equations

$$\omega_{ij} \equiv \lambda \left(\frac{\partial F_j}{\partial x_i} - \frac{\partial F_i}{\partial x_j} \right) + \left[F_j \frac{\partial \lambda}{\partial x_i} - F_i \frac{\partial \lambda}{\partial x_j} \right] = 0 . \tag{10.14}$$

Equation (10.14) is the analog of Eq. (10.11) for $n \geq 3$. To find the conditions analogous to Eq. (10.12), proceed as follows. For (i, j) in Eq. (10.14), pick $k \neq i$ or j. Multiply Eq. (10.14) by F_k ($k \neq i, j$). Cyclically permute k, i, j and add the equations. For example, for $(i, j, k) = (1, 2, 3)$,

$F_3\omega_{12} + F_1\omega_{23} + F_2\omega_{31} = 0$. When this is done, the terms in square brackets in Eq. (10.14) cancel, leaving us with the condition that

$$\lambda \boldsymbol{F} \cdot \boldsymbol{\nabla} \times \boldsymbol{F} = 0 \,, \qquad \text{(in } x_i, x_j, x_k \text{ space)}$$

where the curl is taken in the three-dimensional space spanned by x_i, x_j, x_k. There are "n choose 3" ways of choosing three-dimensional subspaces from an n-dimensional space (see Eq. (7.4))

$$\binom{n}{3} = \frac{1}{6}n(n-1)(n-2) \,, \qquad (n \geq 3) \qquad (10.15)$$

and *each provides a necessary condition for* $\boldsymbol{F}$ *to have an integrating factor.* For $n = 2$, no special conditions need be placed on the vector field; integrating factors always exist in that case. The higher the dimension of the space (for $n \geq 3$), the more the number of conditions that must be satisfied by the components of $\boldsymbol{F}$ for an integrating factor to exist.

Sufficient condition for integrability

Equation (10.12) is also a *sufficient* condition. To show this, we must demonstrate the existence of an integrating factor as a consequence of Eq. (10.12). Suppose momentarily that one of the coordinates in the differential equation $\boldsymbol{F} \cdot \mathrm{d}\boldsymbol{r} = \sum_{i=1}^{3} F_i(x_1, x_2, x_3)\mathrm{d}x_i = 0$ is held fixed, say x_3. The expression $\sum_{i=1}^{2} F_i(x_1, x_2, x_3)\mathrm{d}x_i = 0$ would then involve two independent variables and we can invoke the result that there's a scalar function $U(x_1, x_2)$ such that $F_i = \lambda \partial U/\partial x_i$ for $i = 1, 2$. However, because the F_i are functions of all three coordinates, we have that $U = U(x_1, x_2, x_3)$. Substituting into the original Pfaffian we have the intermediate result that

$$\boldsymbol{F} \cdot \mathrm{d}\boldsymbol{r} = \lambda\frac{\partial U}{\partial x_1}\mathrm{d}x_1 + \lambda\frac{\partial U}{\partial x_2}\mathrm{d}x_2 + F_3\mathrm{d}x_3 = 0 \,. \qquad (10.16)$$

Equation (10.16) is equivalent to $\boldsymbol{F} \cdot \mathrm{d}\boldsymbol{r} = \sum_{i=1}^{3} \frac{\partial U}{\partial x_i}\mathrm{d}x_i + \left[\frac{1}{\lambda}F_3 - \frac{\partial U}{\partial x_3}\right]\mathrm{d}x_3 = 0$, or

$$\boldsymbol{F} \cdot \mathrm{d}\boldsymbol{r} = \boldsymbol{\nabla} U \cdot \mathrm{d}\boldsymbol{r} + \boldsymbol{K} \cdot \mathrm{d}\boldsymbol{r} = 0 \,, \qquad (10.17)$$

where $\boldsymbol{K} \equiv (0, 0, K)$ with $K \equiv \lambda^{-1}F_3 - \partial U/\partial x_3$. From Eq. (10.17) we infer that $\boldsymbol{F} = \boldsymbol{\nabla} U + \boldsymbol{K}$. Clearly, $\boldsymbol{\nabla} \times \boldsymbol{F} = \boldsymbol{\nabla} \times \boldsymbol{K} \neq 0$, because $\boldsymbol{F}$ is rotational by assumption. Evaluating $\boldsymbol{F} \cdot \boldsymbol{\nabla} \times \boldsymbol{F}$,

$$\boldsymbol{F} \cdot \boldsymbol{\nabla} \times \boldsymbol{F} = \frac{\partial U}{\partial x_1}\frac{\partial K}{\partial x_2} - \frac{\partial U}{\partial x_2}\frac{\partial K}{\partial x_1} \equiv \frac{\partial(U, K)}{\partial(x_1, x_2)} = 0 \,, \qquad (10.18)$$

where we assume from Eq. (10.12) that $\boldsymbol{F} \cdot \boldsymbol{\nabla} \times \boldsymbol{F} = 0$. The vanishing Jacobian in Eq. (10.18) implies that a relation exists between U and K not explicitly involving x_1 or x_2 (Exercise 10.1). Thus, $K = K(U, x_3)$. Equation (10.17) is then equivalent to $\mathrm{d}U + K(U, x_3)\mathrm{d}x_3 = 0$, a Pfaffian in two variables. Hence there's a family of scalar functions, $\Phi(U, x_3) = k$. Because $U = U(x_1, x_2, x_3)$, we infer the existence of scalar functions $\psi(x_1, x_2, x_3) = k$. There is therefore an equipotential surface associated with the original Pfaffian, which must possess an integrating factor.

Integrating factors are unique up to a multiplicative constant

Let $\sum_{i=1}^{n} F_i\mathrm{d}x_i = 0$ possess an integrating factor $\lambda(x_1, \cdots, x_n)$ so that

$$\lambda F_i = \frac{\partial \phi}{\partial x_i} \,, \qquad (i = 1, \cdots, n) \qquad (10.19)$$

i.e., for each i, λF_i is the ith-component of the gradient of a function ϕ. Equation (10.19) implies that $0 = \sum_i \lambda F_i \mathrm{d}x_i = \sum_i (\partial \phi / \partial x_i) \mathrm{d}x_i = \mathrm{d}\phi$. If λ exists, the solution curve of $\boldsymbol{F} \cdot \mathrm{d}\boldsymbol{r} = 0$ lies in the surface $\phi = c$ where c is a constant. Consider an arbitrary function of ϕ, $V(\phi)$. Then, using Eq. (10.19),

$$\mathrm{d}V = \frac{\mathrm{d}V}{\mathrm{d}\phi}\mathrm{d}\phi = \frac{\mathrm{d}V}{\mathrm{d}\phi}\sum_i \frac{\partial \phi}{\partial x_i}\mathrm{d}x_i = \lambda \frac{\mathrm{d}V}{\mathrm{d}\phi}\sum_i F_i \mathrm{d}x_i = 0 \ .$$

If λ is an integrating factor, so is $\lambda(\mathrm{d}V/\mathrm{d}\phi)$; λ is unique up to a multiplicative constant.

Application to thermodynamics

Two degrees of freedom

The first law for a simple fluid is $\mathrm{d}Q = \mathrm{d}U + P\mathrm{d}V$. Taking $U = U(P,V)$,

$$\mathrm{d}Q = \left(\frac{\partial U}{\partial P}\right)_V \mathrm{d}P + \left[P + \left(\frac{\partial U}{\partial V}\right)_P\right]\mathrm{d}V \ .$$

Is $\mathrm{d}Q$ exact? Is the integrability condition, Eq. (10.3), satisfied? No (check it!). Is there an integrating factor, λ? We've just shown that one is guaranteed to *exist*—but can we find it? Form the Pfaffian,

$$\lambda \left(\frac{\partial U}{\partial P}\right)_V \mathrm{d}P + \lambda \left[P + \left(\frac{\partial U}{\partial V}\right)_P\right]\mathrm{d}V \ ,$$

where $\lambda = \lambda(P,V)$. We require that the integrability condition be satisfied:

$$\frac{\partial}{\partial V}\left[\lambda \left(\frac{\partial U}{\partial P}\right)_V\right]_P = \frac{\partial}{\partial P}\left(\lambda \left[P + \left(\frac{\partial U}{\partial V}\right)_P\right]\right)_V \ .$$

Take the derivatives indicated; λ must be such that

$$\left(\frac{\partial U}{\partial P}\right)_V \left(\frac{\partial \lambda}{\partial V}\right)_P = \lambda + \left(\frac{\partial \lambda}{\partial P}\right)_V \left[P + \left(\frac{\partial U}{\partial V}\right)_P\right] \ .$$

This equation can be simplified. Divide through by $(\partial \lambda / \partial P)_V$ and rearrange:

$$\left(\frac{\partial U}{\partial P}\right)_V \left(\frac{\partial \lambda}{\partial V}\right)_P \left(\frac{\partial P}{\partial \lambda}\right)_V - \left(\frac{\partial U}{\partial V}\right)_P = \lambda \left(\frac{\partial P}{\partial \lambda}\right)_V + P \ . \tag{10.20}$$

Use the cyclic relation to equate

$$\left(\frac{\partial \lambda}{\partial V}\right)_P \left(\frac{\partial P}{\partial \lambda}\right)_V = -\left(\frac{\partial P}{\partial V}\right)_\lambda \ . \tag{10.21}$$

Combine Eqs. (10.21) and (10.20):

$$-\left[\left(\frac{\partial U}{\partial P}\right)_V \left(\frac{\partial P}{\partial V}\right)_\lambda + \left(\frac{\partial U}{\partial V}\right)_P\right] = \lambda \left(\frac{\partial P}{\partial \lambda}\right)_V + P \ . \tag{10.22}$$

Use Eq. (3.48) to equate the terms in square brackets in Eq. (10.22) with the derivative

$$\left(\frac{\partial U}{\partial V}\right)_\lambda = \left(\frac{\partial U}{\partial V}\right)_P + \left(\frac{\partial U}{\partial P}\right)_V \left(\frac{\partial P}{\partial V}\right)_\lambda \ .$$

Thus we arrive at the *starting line* of what we require of λ:

$$\left(\frac{\partial U}{\partial V}\right)_\lambda + P = -\lambda \left(\frac{\partial P}{\partial \lambda}\right)_V \equiv \lambda^{-1}\left(\frac{\partial P}{\partial \lambda^{-1}}\right)_V \ , \tag{10.23}$$

where we've written $\lambda(\partial/\partial\lambda) = -\lambda^{-1}(\partial/\partial\lambda^{-1})$ to put Eq. (10.23) into a suggestive form, namely that of Eq. (1.28). Equation (10.23) is a challenging differential equation to solve for λ. However, it becomes *identical* to Eq. (1.28) (a valid equation of thermodynamics) if we choose $\lambda = T^{-1}$, the integrating factor for dQ obtained through the second law.

Three degrees of freedom

Consider fluids A and B in thermal contact. By the zeroth law there's a connection between the variables P_A, V_A, P_B, V_B provided by the empirical temperature functions $\phi_A(P_A, V_A) = \phi_B(P_B, V_B) \equiv T$. Take as the three independent variables V_A, V_B, and T, with $U = U(V, T)$. The first law can then be written

$$dQ = dU_A + P_A dV_A + dU_B + P_B dV_B$$
$$= \left(P_A + \frac{\partial U_A}{\partial V_A}\right) dV_A + \left(P_B + \frac{\partial U_B}{\partial V_B}\right) dV_B + \left(\frac{\partial U_A}{\partial T} + \frac{\partial U_B}{\partial T}\right) dT$$
$$\equiv F_1 dx_1 + F_2 dx_2 + F_3 dx_3 .$$

Is dQ exact? Answer: Evaluate the integrability conditions T_{ij} from Eq. (10.3),

$$T_{12} = 0 \qquad T_{13} = -\frac{\partial P_A}{\partial T} \qquad T_{23} = -\frac{\partial P_B}{\partial T} . \tag{10.24}$$

The vector field F is rotational because some $T_{ij} \neq 0$. Does it possess an integrating factor? Equation (10.12), $F \cdot \nabla \times F = 0$ (the *one* condition for F to have an integrating factor for $n = 3$) implies that

$$F \cdot \nabla \times F = F_1 T_{23} + F_2 T_{31} + F_3 T_{12} = 0 . \tag{10.25}$$

Because $T_{12} = 0$, we require from Eq. (10.25) $\frac{T_{13}}{F_1} = \frac{T_{23}}{F_2}$, or

$$\left[\frac{(\partial P/\partial T)_V}{P + (\partial U/\partial V)_T}\right]_B = \left[\frac{(\partial P/\partial T)_V}{P + (\partial U/\partial V)_T}\right]_A . \tag{10.26}$$

The term on the left is independent of A, while that on the right is independent of B. Because they are equal, however, *neither* depends on A or B! Thus,

$$\frac{(\partial P/\partial T)_V}{P + (\partial U/\partial V)_T} \equiv f(T) , \tag{10.27}$$

is a *universal function*. Even though each quantity $[(\partial P/\partial T)_V, (\partial U/\partial V)_T]$ depends separately on the physical system, Eq. (10.26) asserts that the combination does not. The terms in Eq. (10.27) define an *absolute* measure of temperature in that its numerical value is system independent. *An integrating factor for dQ implies the existence of an absolute temperature.* Using Eq. (1.28) in Eq. (10.27), we find that $f(T) = T^{-1}$. Because Eq. (1.28) ultimately relies on entropy as a state variable, the "T" in $f(T) = T^{-1}$ is the absolute temperature implied by Carnot's theorem.

What's the integrating factor? From Eq. (10.14) we require

$$F_i \frac{\partial \ln \lambda}{\partial x_j} - F_j \frac{\partial \ln \lambda}{\partial x_i} = \frac{\partial F_j}{\partial x_i} - \frac{\partial F_i}{\partial x_j} \equiv T_{ij} , \tag{10.28}$$

a set of $n(n-1)/2$ partial differential equations for $\ln \lambda(x_1, \cdots, x_n)$. For $n = 3$, Eq. (10.28) implies the three equations

$$F_1 \frac{\partial \ln \lambda}{\partial x_2} - F_2 \frac{\partial \ln \lambda}{\partial x_1} = T_{12} \quad F_1 \frac{\partial \ln \lambda}{\partial x_3} - F_3 \frac{\partial \ln \lambda}{\partial x_1} = T_{13} \quad F_2 \frac{\partial \ln \lambda}{\partial x_3} - F_3 \frac{\partial \ln \lambda}{\partial x_2} = T_{23} . \tag{10.29}$$

From Eq. (10.24), $T_{12} = 0$, so the first equation in Eq. (10.29) can be satisfied by taking λ to be independent of x_1 and x_2. The other two equations then reduce to

$$\frac{\partial \ln \lambda}{\partial x_3} = \frac{T_{13}}{F_1} = \frac{T_{23}}{F_2} = -f(T) , \qquad (10.30)$$

with $f(T)$ given by Eq. (10.27). Equation (10.30) implies that

$$\lambda = C \exp \left(- \int f(T) \mathrm{d}T \right) , \qquad (10.31)$$

where C is a constant. From Eq. (1.28), $f(T) = T^{-1}$, and Eq. (10.31) gives $\lambda = C T^{-1}$.

10.2 CARATHÉODORY THEOREM

A Pfaffian in two variables always possesses an integrating factor. The situation is not as clear for three or more variables. *When* an integrating factor exists, the solution curves to $\mathrm{d}Q = 0$ (adiabats) are *unique*: through any point in state space there is only one adiabat. More than one adiabat through a point entails a violation of the second law (Section 2.3). *The second law requires uniqueness of adiabats*. Points on the same adiabat are said to be *accessible* by an adiabatic process; points on different adiabats are *inaccessible* (by adiabatic processes). Points 1 and 2 in Fig. 10.3 are connected by an adiabat (accessible). Point 3 cannot be reached from 1 or 2 by adiabatic processes; it's inaccessible.

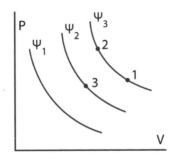

Figure 10.3   Point 2 is accessible from point 1; point 3 is inaccessible from point 1

Carathéodory's theorem consists of the equivalence of two statements, $A \Leftrightarrow B$. The proof consists of showing the implications $A \Rightarrow B$ and $B \Rightarrow A$. The statements are:

A: The vector field $\boldsymbol{F}$ associated with Pfaffian $\mathrm{d}Q = \boldsymbol{F} \cdot \mathrm{d}\boldsymbol{r}$ is integrable.

B: In the neighborhood, however small, of a point P in the vector field, there are points P' inaccessible from P by adiabatic paths, the solution curves of the Pfaffian differential equation $\mathrm{d}Q = 0$.

$A \Rightarrow B$

If $\boldsymbol{F}$ is integrable, there exist nonintersecting surfaces $\psi(x_1, \cdots, x_n) = k$, each with the property that it contains curves such that $\mathrm{d}Q = \boldsymbol{F} \cdot \mathrm{d}\boldsymbol{r} = 0$. Because the surfaces are nonintersecting, there are points in the space (even infinitesimally nearby) inaccessible by adiabatic paths. Thus, $A \Rightarrow B$.

$B \Rightarrow A$

We follow a proof for $n = 3$ devised by Max Born.[52, p145] Proofs for arbitrary n are available.[53] In the neighborhood of any point P there is (by assumption) an inaccessible point M (see Fig. 10.4). Construct through P a line L which is not a solution of đ$Q = 0$. Pass a plane

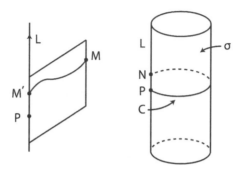

Figure 10.4   The solution curve C must be closed

through M and L. On this plane there will be a curve passing through M that satisfies đ$Q = 0$. Let the curve intersect L at M'. M' is inaccessible to P, but can be made to lie as close to P as we want by choosing M sufficiently close to P. Suppose that L is moved parallel to itself to generate a closed cylnder, σ; the precise shape of σ is unimportant. On σ draw a curve C passing through P that's a solution to đ$Q = 0$. Let C intersect L at point N. It follows that N and P must coincide. By continuously deforming the cylinder σ we can make N move along a segment of L surrounding P. In this way we would develop a band of accessible points in the vicinity of P. But this is contrary to the assumption that arbitrarily close to P there exist points on L (such as M') that are inaccessible from P. Thus, N must coincide with P.

There is thus a closed curve C everywhere orthogonal to $\boldsymbol{F}$, $\oint_C \boldsymbol{F} \cdot \mathrm{d}\boldsymbol{r} = 0$ (the Pfaffian đ$Q \equiv \boldsymbol{F} \cdot \mathrm{d}\boldsymbol{r} = 0$). By Stokes's theorem, $\oint_C \boldsymbol{F} \cdot \mathrm{d}\boldsymbol{r} = \int_{S(C)} \boldsymbol{\nabla} \times \boldsymbol{F} \cdot \mathrm{d}\boldsymbol{S}$, where $S(C)$ is the surface bounded by C. In the limit of C an infinitesimal closed curve, we can replace the integral with $\hat{\boldsymbol{n}} \cdot (\boldsymbol{\nabla} \times \boldsymbol{F}) \Delta S = 0$, where $\hat{\boldsymbol{n}}$ is the direction normal to $S(C)$ and parallel to $\boldsymbol{F}$. Thus, $\boldsymbol{F} \cdot \boldsymbol{\nabla} \times \boldsymbol{F} = 0$, a necessary and sufficient condition for $\boldsymbol{F}$ to be integrable. $B \Rightarrow A$.

We then have Carathéodory's theorem:

> The existence of points in every arbitrarily small neighborhood of a given point in a vector field inaccessible by paths that are the solution to the Pfaffian differential equation đ$Q = 0$, is necessary and sufficient for the existence of an integrating factor of the associated Pfaffian đ$Q = \boldsymbol{F} \cdot \mathrm{d}\boldsymbol{r}$.

The theorem can be summarized schematically:

> Points inaccessible by đ$Q = 0 \iff$ Integrating factor for đQ.

10.3   CARATHÉODORY'S PRINCIPLE AND THE SECOND LAW

A Pfaffian in two variables is guaranteed to have an integrating factor. For a fluid described by P and V, we found T^{-1} to be the integrating factor for đQ because in that case Eq. (10.23) reduces to Eq. (1.28),

$$\frac{(\partial P/\partial T)_V}{(\partial U/\partial V)_T + P} = \frac{1}{T} .$$

But where did Eq. (1.28) come from? It was derived *after* entropy had been established as a state variable (Exercise 4.6), i.e., after T^{-1} as an integrating factor for đQ had been discovered through the second law. What then has been "gained" by this foray into the mathematics of integrating factors? Perhaps nothing *in this case*. Yet we have learned that for systems described by two thermodynamic variables, the *existence of entropy may be inferred independently of the second law.*

For a system with three degrees of freedom (two fluids in thermal contact), for đQ to have an integrating factor we found in Eq. (10.27) that the identical group of terms

$$f \equiv \frac{(\partial P/\partial T)_V}{(\partial U/\partial V)_T + P}$$

must be a universal quantity akin to the absolute temperature. (A magnetic system with three degrees of freedom is analyzed in Exercise 10.2, where again Eq. (1.28) plays a role.) Recall that heats Q_1, Q_2 exchanged reversibly with reservoirs have the property $Q_1/Q_2 = \phi(\theta_1)/\phi(\theta_2)$, where ϕ is a universal function, Eq. (2.10). Absolute temperature is *defined* as ϕ, $T \equiv \phi(\theta)$. With T so defined, the machinery of thermodynamics gives $f = T^{-1}$; f itself, however, could serve as an absolute temperature, what systems in equilibrium have in common.[14] The second law, absolute temperature, and integrating factors are interrelated, as illustrated in Fig. 10.5. The existence of an integrating

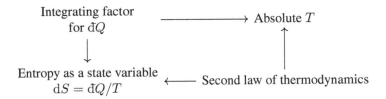

Figure 10.5 Integrating factor for đQ implies entropy as well as absolute temperature

factor for đQ not only allows a new state variable to be defined, *it implies the existence of absolute temperature.* The second law accomplishes the same tasks: Entropy and absolute temperature are consequences of the second law. Seen this way, the existence of integrating factors is on the same footing as the second law.

Yet integrating factors do not *necessarily* exist for Pfaffians in $n \geq 3$ variables. Moreover, the number of conditions placed on a Pfaffian for it to possess an integrating factor increases dramatically with n, Eq. (10.15). Thus, we can't simply *require* đQ to have an integrating factor. In contrast, the second law leads to the integrating factor T^{-1} for *any* physical system, regardless of the number of state variables n.

Carathéodory was therefore led to *postulate* the existence of inaccessible states. The *Carathéodory principle* is an alternate way of formulating the second law:

> In the neighborhood of any point in state space there are points inaccessible by adiabatic processes.

In that way, all thermodynamic systems possess an integrating factor for đQ (by the Carathéodory theorem), and hence entropy as a state variable, as summarized in Fig. 10.6.

---

[14]It might be thought that Eq. (10.26) only confirms the zeroth law, that the quantity in Eq. (10.27) defines an empirical, not absolute temperature. The terms in Eq. (10.26), however, have *the same functional form for each material*. Even though the individual terms in Eq. (10.27) are separately material specific, the combination is *material-independent*. We're faced with a similar situation in Eq. (2.10), what defines the absolute temperature, that the ratio of heats Q_1/Q_2 reversibly expelled to and absorbed from reservoirs equals the ratio of a universal function of the empirical temperatures $\phi(\theta_1)/\phi(\theta_2)$, *no matter what empirical temperature is used for each reservoir.*

Existence of states
inaccessible by   ⟷   Existence of entropy
adiabatic processes   as a state variable

Figure 10.6   Equivalence between entropy and states inaccessible by adiabatic paths

Is the Carathéodory principle equivalent to the Kelvin and Clausius statements of the second law, or is it implied by them? It's in the form of an *impossibility statement*;[15] that much it shares with the other laws of thermodynamics. One can show that the negation of the Carathéodory principle implies the negation of the Kelvin form of the second law, and thus the Kelvin statement implies Carathéodory's principle.[54] It can also be shown that the inference goes the other way, Carathéodory's principle implies the Kelvin statement.[55] Thus, Carathéodory's principle is an equivalent statement of the second law.

CHAPTER SUMMARY

In this chapter we introduced the Carathéodory formulation of the second law, which is that entropy exists as a state variable if there are states inaccessible from a given state by adiabatic paths. The Carathéodory formulation is equivalent to the Kelvin statement of the second law.

- The first law of thermodynamics is in the form of a Pfaffian differential form, $đQ = \sum_i F_i(x_j) \mathrm{d}x_i$, where we can consider the thermodynamic variables F_i as comprising the components of a vector field $\boldsymbol{F}$ in the space of state variables $x_1, \cdots, x_n$. With the coordinate differentials packaged into an infinitesimal displacement vector $\mathrm{d}\boldsymbol{r}$ we can write $đQ = \boldsymbol{F} \cdot \mathrm{d}\boldsymbol{r}$.

- When the Pfaffian is exact (inexact), $\boldsymbol{F}$ is termed an irrotational (rotational) vector field. For $n = 3$ the distinction between rotational and irrotational is whether $\nabla \times \boldsymbol{F} \neq 0$ or $\nabla \times \boldsymbol{F} = 0$. For $n > 3$ the distinction is whether $T_{ij} \neq 0$ or not, where T_{ij} is defined in Eq. (10.4).

- Irrotational vector fields give rise to a scalar field Φ because $T_{ij} = 0$ is necessary and sufficient for $\int_C \boldsymbol{F} \cdot \mathrm{d}\boldsymbol{r} = \Delta\Phi$ to be independent of the path C. Irrotational vector fields are the gradients of a scalar field, $\boldsymbol{F} = \nabla\Phi$, which are orthogonal to the equipotential surfaces of Φ.

- Rotational vector fields may possess special curves $\widetilde{C}$ for which $\int_{\widetilde{C}} \boldsymbol{F} \cdot \mathrm{d}\boldsymbol{r} = 0$, the solution curves of the differential equation $đQ = 0$. Along such paths it may be possible to scale $\boldsymbol{F}$ so that $\boldsymbol{F}$ is colinear with an irrotational vector field $\boldsymbol{G}$, $\lambda(x_j)\boldsymbol{F} \equiv \boldsymbol{G}$. The quantity λ (when it exists) is an integrating factor. In such cases the solution curve $đQ = 0$ is embedded in an equipotential surface of the scalar field associated with $\boldsymbol{G}$. For $n = 3$, $\nabla \times (\lambda\boldsymbol{F}) = 0$.

- A Pfaffian in two variables always possesses an integrating factor. Integrating factors do not necessarily exist for Pfaffians in three or more variables ($n \geq 3$).

- Inaccessible points are points P' in the neighborhood of a given point P that cannot be reached along the solution curves to $đQ = 0$. Carathéodory's theorem is that the existence of inaccessible points is necessary and sufficient for the existence of an integrating factor.

- Carathéodory's principle is that in the neighborhood of any equilibrium state there are points in state space inaccessible by adiabatic processes. In that case, by Carathéodory's theorem an integrating factor for $đQ$ exists, implying the existence of a state variable, entropy. Carathéodory's principle is equivalent to the Kelvin statement of the second law.

---

[15]It is impossible to reach nearby points in state space by adiabatic processes.

EXERCISES

10.1 Given functions $u(x, y)$ and $v(x, y)$, show that for a relation $F(u, v) = 0$ to exist between u and v *not explicitly involving* x and y, the Jacobian between (u, v) and (x, y) must identically vanish,

$$\frac{\partial(u, v)}{\partial(x, y)} = 0.$$

Hint: Differentiate F with respect to x and then differentiate F with respect to y. The vanishing Jacobian is a necessary condition for there to be a relation between u and v not involving x or y. It's sufficient as well.

10.2 Consider a magnetic system with the first law given by

$$đQ = dU + PdV - HdM.$$

Take U, P, H to be functions of the independent variables T, V, M, so that

$$đQ = \left(\frac{\partial U}{\partial T}\right)_{V,M} dT + \left[P + \left(\frac{\partial U}{\partial V}\right)_{T,M}\right] dV + \left[\left(\frac{\partial U}{\partial M}\right)_{T,V} - H\right] dM.$$

a. Is $đQ$ integrable? Evaluate the integrability conditions Eq. (10.4). Show that

$$T_{12} = \left(\frac{\partial P}{\partial T}\right)_{V,M} \qquad T_{23} = 0 \qquad T_{31} = \left(\frac{\partial H}{\partial T}\right)_{V,M}.$$

$T_{23} = 0$ follows from a Maxwell relation, Exercise 8.3.

b. For an integrating factor to exist, Eq. (10.12) must hold. Show that Eq. (10.12) requires the equality

$$-\frac{T_{31}}{F_3} = \frac{-(\partial H/\partial T)_{V,M}}{(\partial U/\partial M)_{T,V} - H} = \frac{1}{T} = \frac{(\partial P/\partial T)_{V,M}}{(\partial U/\partial V)_{T,M} + P} = \frac{T_{12}}{F_2}.$$

Both sides are equal to T^{-1}, as can be seen from Eq. (1.28) and the result of Exercise 8.4. Of course, we can only make such conclusions when we know that $dS = đQ/T$. Note that the equality holds under the substitutions $P \leftrightarrow -H$ and $V \leftrightarrow M$ one obtains in comparing $-PdV \leftrightarrow HdM$.

c. Show from an application of Eq. (10.28) that we can take λ as independent of V and M, and that the differential equation for λ is

$$\frac{\partial \ln \lambda}{\partial T} = -\frac{T_{12}}{F_2} = \frac{T_{31}}{F_3} = -\frac{1}{T},$$

the same equality obtained above. Thus, $\lambda = C/T$.

10.3 Show that Eq. (10.23) can be written $\left(\frac{\partial U}{\partial V}\right)_\lambda = -\left(\frac{\partial(\lambda P)}{\partial \lambda}\right)_V$. Use the cyclic relation to show that Eq. (10.23) is equivalent to:

$$\left(\frac{\partial(\lambda P)}{\partial U}\right)_V = \left(\frac{\partial \lambda}{\partial V}\right)_U.$$

This equation is what one obtains in seeking an integrating factor for the first law written as $đQ = dU + PdV$. Comparing with the result of Exercise 3.11 (which presumes that $dS = đQ/T$ is an exact differential), we find once again that $\lambda = C/T$.

CHAPTER 11

Negative absolute temperature

C AN absolute temperature be *negative*? Several considerations would indicate, *not*. Yet, starting in the 1950s,[56] and continuing to the present,[57] experiments have been reported that are consistent with the finding of negative absolute temperature. In this chapter we discuss the conditions under which absolute temperature can be a negative quantity.

11.1 IS NEGATIVE ABSOLUTE TEMPERATURE POSSIBLE?

Negative absolute temperature does not mean colder than absolute zero

The universality of Carnot's theorem implies that absolute temperature can be formulated independently of the thermometric properties of materials (see Section 2.4). The ratio of absolute temperatures T and T_0 equals the ratio of the absolute value of heats Q and Q_0 reversibly exchanged with reservoirs at those temperatures, Eq. (2.11):

$$\frac{T}{T_0} = \frac{Q}{Q_0} \, . \tag{11.1}$$

If we take $T_0 = 273.16$ and Q_0 to be the heat exchanged with a reservoir at the triple point of H_2O, we have the Kelvin scale. We're free, however, to take T_0 to be any number we want, positive *or negative*. For T_0 negative, *all* absolute temperatures would be negative. Yet regardless of the sign of T_0, the smallest value Q can have is zero: Zero is the least amount of heat that can be transferred. The smallest-magnitude absolute temperature is therefore *zero*, whether approached through a series of positive or negative temperatures (if $T_0 < 0$). Whatever is the sense that negative absolute temperature has meaning, *it does not mean temperature colder than absolute zero*.

Absolute temperature—the distribution of states available to a system

What aspect of physical systems does absolute temperature characterize?[1] It's certainly related to the *motion* of atoms[2]—the internal energy of the ideal gas, which ignores inter-particle interactions, is $U = \frac{3}{2}Nk_BT$. Negative temperature would then be paradoxical if absolute temperature represented *just* kinetic energy.[3] Absolute temperature also represents *potential energy*, the energy stored in

---

[1] We've posed that question about entropy several times in this book; now we ask the same about temperature.

[2] Yet what kind of motion? Does the temperature of a can of gas increase if it's placed on a moving train car?

[3] To be clear, we're talking about *absolute* temperature—positive or negative—not empirical temperature. Nothing rules out negative empirical temperature. When we refer to negative temperature, negative absolute temperature is implied.

the internal degrees of freedom of the molecules of a gas. The *equipartition theorem* (not shown in this book) states that $\frac{1}{2}k_BT$ is the *average* energy[4] associated with each *quadratic degree of freedom*, those for which energy depends on the square of some quantity.[5] Absolute temperature thus characterizes the total energy of a system.[6] We can't rely, however, too heavily on the equipartition theorem, as it breaks down at low temperature,[7] and moreover not all types of energy are associated with quadratic degrees of freedom as provided by the classical picture. The energy of the photon gas is given by $U = aVT^4$, Eq. (5.11), and (as it turns out) the energy of a collection of independent electrons varies as T^2 at low temperature. While the total energy of a system depends on absolute temperature, the form of the relationship is system dependent, and not always linear in T.

We can't therefore *simply* relate absolute temperature to internal energy (rather energy is related to temperature). Temperature is defined as that which systems in equilibrium have in common (Section 1.5). The meaning of temperature should emerge from an understanding of the basis for equilibrium. A body in equilibrium has the maximum entropy possible, subject to constraints. Repeating a familiar argument (Section 3.10), for two systems A and B separated by a rigid, diathermic boundary (and otherwise isolated), the entropy of the combined system is a maximum in equilibrium, $\delta(S_A + S_B) = 0$, and its internal energy is conserved, $\delta(U_A + U_B) = 0$. Equilibrium is therefore characterized by the equality

$$\frac{\delta S_A}{\delta U_A} = \frac{\delta S_B}{\delta U_B} .$$

The ratio $\delta S/\delta U$ is something that systems in equilibrium have in common, and serves as a proxy for temperature. From Eq. (3.30), absolute temperature is related to $\delta S/\delta U$ by

$$\frac{1}{T} = \left(\frac{\partial S}{\partial U}\right)_{V,N} . \tag{11.2}$$

Equation (11.2) is a more abstract characterization of absolute temperature than the attempt to relate it to the energy of specific systems: The variation of entropy with energy *is* the inverse absolute temperature.[8] Reaching for Eq. (7.2), we have[9]

$$\frac{1}{k_BT} = \frac{1}{W}\left(\frac{\partial W}{\partial U}\right)_{V,N} . \tag{11.3}$$

The *inverse* absolute temperature probes the fractional change in W (number of microstates per macrostate) with changes in internal energy U. Temperature therefore characterizes the *system*, through the dependence of W on U, and is not a microscopic property of matter.[10] Equation (11.2) is sufficiently general that it can accommodate negative temperature.

---

[4]Energy averages are calculated from the probability distribution function of energy states (which depends on the absolute temperature), such as the Maxwell-Boltzmann distribution, Eq. (7.28).

[5]The Hamiltonian of a classical harmonic oscillator is $H = p^2/(2m) + \frac{1}{2}m\omega^2x^2$; the kinetic and potential energy terms are quadratic degrees of freedom.

[6]One should see that absolute temperature characterizes a system *as a whole*. It embodies the average properties of a system, averages that are constructed from the probability of occurrence of *all states available to a system*. In saying that the average kinetic energy of the molecules of a gas is $\langle\frac{1}{2}mv^2\rangle = \frac{3}{2}k_BT$, it implies there are atoms moving faster than the average and those that are moving slower than the average, all in the same system at absolute temperature T.

[7]At sufficiently low temperatures *it's no longer true that the average energy associated with degrees of freedom is proportional to* T. The law of Dulong and Petit, for example (the heat capacity of solids is $3nR$, page 21), which can be accounted for using the equipartition theorem, breaks down at low temperatures, as required by the third law of thermodynamics, and requires the machinery of quantum mechanics to understand.

[8]There's a change of emphasis here, away from $\partial S/\partial U = T^{-1}$, the variation of S with respect to U is controlled by T^{-1} to $T^{-1} = \partial S/\partial U$; T^{-1} is $\partial S/\partial U$. Consider from electrostatics $\nabla \cdot E = \rho/\epsilon_0$. We normally read this as the divergence of the E-field is controlled by the charge density. Turn it around, $\rho = \epsilon_0\nabla \cdot E$; charge density *is* the divergence of the field.

[9]Note that Eq. (11.2) is a result of thermodynamics, whereas Eq. (11.3) is from the statistical theory of entropy.

[10]Can we speak of the temperature of a single particle?

$T < 0 \iff$ Well-isolated subsystems

Intensive thermodynamic quantities have the same value at all spatial locations of the system, and for a system truly in equilibrium we can speak of *the* temperature. Systems, however, can consist of subsystems *internally* in equilibrium (Section 8.3), but not mutually in equilibrium.[11] The Purcell-Pound experiment of 1951 made use of the nuclear spin of ^{7}Li in a LiF crystal.[56] Spins, situated at lattice sites of the crystal, form a subsystem magnetically interacting among themselves, and only weakly interacting with the lattice. The *spin-lattice relaxation time*[12] can range from minutes[58] to hours.[59] For times shorter than the spin-lattice relaxation time, a system of spins can be treated as an essentially isolated system having its own *spin temperature* which can differ from that of the lattice.[60]

Most systems cannot have a negative temperature because increasing the energy of the system increases the entropy (as depicted in the left portion of Fig. 11.1), and by Eq. (11.2), $T > 0$. For most systems, S is a *strictly increasing* function of internal energy U. There are exceptional systems, however, where $S = S(U)$ is not a monotonically increasing function, such as shown in the right portion of Fig. 11.1. Spin systems are of this type. The energy of a system of spins *saturates*: Spins,

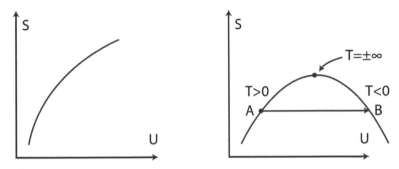

Figure 11.1 Left: $S = S(U)$ is an increasing function for systems with an unbounded energy spectrum, what's usually the case. Right: For systems with a bounded spectrum, $S = S(U)$ is not monotonic, allowing for the possibility of negative temperature.

tiny magnets, can only all line up with an imposed magnetic field. Energy maximizes in such a system, which is associated with an ordered, i.e., low-entropy state.[13]

Consider a system of N spin-$\frac{1}{2}$ particles[14] in an applied magnetic field $\boldsymbol{H}$. Each particle can have two energy values corresponding to the spin either aligned or anti-aligned with the field; there are thus 2^N spin configurations. The energies of the spins can range from $U = -N\mu H$, where μ is the magnetic moment of one spin, with all spins aligned with the field, to $U = N\mu H$, with all spins anti-aligned with the field. A configuration of n "down" spins (anti-aligned with the field) has energy $U(n) = -(N - 2n)\mu H$, $n = 0, \cdots, N$. For a given energy state (labeled by n), there are

$$\binom{N}{n} = \frac{N!}{n!(N-n)!}$$

---

[11]In plasmas, for example (ionized gases), electrons and ions can have different temperatures. Because of the disparity in their masses, electrons come to equilibrium amongst themselves faster than they come to equilibrium with ions or neutral atoms. In a given plasma, electrons can have temperatures as high as several tens of thousand Kelvin, whereas the ion temperature might be in the range 500–1000 K.

[12]The time over which the spin system comes to equilibrium with the lattice (spin-lattice relaxation time), and the time over which spins come to equilibrium amongst themselves (spin-spin relaxation time) can be measured with nuclear magnetic resonance (NMR).

[13]Negative temperatures are not possible without quantum effects, e.g., the discreteness of spin angular momentum.

[14]The ^{7}Li nucleus has spin $\frac{3}{2}$, which complicates the argument, but does not qualitatively change the conclusion.

ways of arranging the n down spins. The maximum number of ways of distributing down spins is for $n = N/2$ (show this). Let $n = N/2 + m$, where $m = -N/2, \cdots, N/2$. The energies can thus be written $U(m) = 2m\mu H$. By the same analysis leading to Eq. (7.17), the entropy of the spin system has the form depicted in the right portion of Fig. 11.1:

$$S/k_B = \ln \left[\frac{N!}{(N/2 + m)!(N/2 - m)!} \right] = N \ln 2 - \frac{2}{N}m^2 + \cdots . \tag{11.4}$$

Systems possessing a subsystem with an entropy function as in Eq. (11.4) can be prepared in a negative temperature state by quickly reversing the magnetic field,[15] as indicated schematically in Fig. 11.1 with the transition $A \to B$. We can see this by combining Eq. (11.4) with $U = 2m\mu H$:

$$S(U, H) = k_B \ln \left[\frac{N!}{\left(\dfrac{N}{2} + \dfrac{U}{2\mu H}\right)! \left(\dfrac{N}{2} - \dfrac{U}{2\mu H}\right)!} \right] . \tag{11.5}$$

The entropy therefore has the symmetry $S(U, H) = S(-U, -H)$, implying from Eq. (11.2) that for such a system, $T(H) = -T(-H)$.

11.2   NEGATIVE ABSOLUTE IS HOTTER THAN POSITIVE ABSOLUTE

We derived in Eq. (7.28) the probability $P(E_j)$ that a system in equilibrium at absolute temperature T has energy E_j occupied, with

$$P(E_j) = Z^{-1}e^{-E_j/k_B T} , \tag{11.6}$$

where Z is the normalization factor on the probability distribution.[16] Equation (11.6) implies that the number of particles N_j having energy E_j, relative to the number N_0 in the ground state E_0, is

$$\frac{N_j}{N_0} = \exp\left(-(E_j - E_0)/k_B T\right) . \tag{11.7}$$

As T is increased (through positive values), the *population* of higher-energy states increases, but in such a way that $N_j < N_0$, for $j = 1, 2, 3, \cdots$. If we could let $T \to \infty$, we would have from Eq. (11.7) that all states would be *equally* populated, with $N_j = N_0$ for all $j \geq 1$. Such a situation would require an *infinite* amount of energy if the energy spectrum is *unbounded*.[17] By letting T become negative in Eq. (11.7), we would have $N_j > N_0$ for all $j \geq 1$, where higher-energy states are *more* populated than lower-energy states—a *population inversion*.[18] To have a population inversion with every state E_j more populated than E_{j-1} would require "more" than an infinite amount of energy, a nonsensical result. *A system with an unbounded energy spectrum cannot attain negative temperatures.*

Such a conclusion does not apply to systems having a *bounded* energy spectrum, such as a finite system of spins in a magnetic field. The entropy function for such a system achieves a maximum

---

[15]The proviso of changing the field rapidly (faster than the spin-spin relaxation time) is so that the spins do not have time to follow the change in the direction of the field.

[16]The energy appearing in Eq. (7.28) refers to the kinetic energy of a gas particle (a continuous quantity), which has been made discrete through our approximation of coarse-graining phase space. The energy E_j appearing in Eq. (11.6) is taken to be discrete, in accord with the spin system under consideration here. The *form* of Eq. (7.28), the "Boltzmann factor," is more general than the result obtained through the specific derivation given in Section 7.3.

[17]For example, a quantum harmonic oscillator of angular frequency ω has an unbounded energy spectrum described by $E_n = \left(n + \frac{1}{2}\right)\hbar\omega$, for $n = 0, 1, 2, \cdots$.

[18]Lasers operate in a medium that can achieve population inversion among a *few* energy levels, not *all* energy levels required for negative temperature. A laser is not in thermal equilibrium, but is maintained in a non-equilibrium state.

value, as seen in Eq. (11.4), and indicated in the right portion of Fig. 11.1. Note that the top of the entropy curve, where $\partial S/\partial U = 0$, is consistent with *either* $T = \infty$ *or* $T = -\infty$. We have the surprising result, for a system with a bounded energy spectrum, that a state of $T = \infty$ is the *same* as one for which $T = -\infty$. If temperature is conceived as a measure of hotness, we must conclude that *negative temperatures are hotter than positive temperatures*, in particular that *negative temperatures are hotter than infinite temperature*. To avoid such a non-intuitive conclusion would require us to come up with a replacement for the Kelvin temperature scale. If we were to replace $T \to -1/T$, we would have a linear progression from cold to hot, as shown in Fig. 11.2.

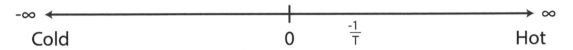

Figure 11.2 With $(-T)^{-1}$ as temperature there is a linear progression from cold to hot

Measuring negative temperature

Our interpretation of negative temperature is that of a population inversion. How to detect a population inversion? A beam of electromagnetic energy (of suitably chosen frequency), when passed through a system with a *normally* populated set of energy levels, will absorb energy—more lower-energy states are populated than higher-energy states; there are more absorbers than emitters and the beam is attenuated. Conversely, a beam of electromagnetic energy passed though a system in which higher-energy states are more populated than lower-energy states, will *increase* the intensity of the beam through *stimulated emission*, the process that underlies the operation of lasers.

11.3 NEGATIVE-TEMPERATURE THERMODYNAMICS

The hottest negative temperature an object can have (as measured on the Kelvin scale) is -0, while the coldest negative temperature an object can have is $-\infty$ (see Fig. 11.2). Of two objects A and B, each at negative temperature, object A is hotter than B if $|T_A| < |T_B|$. Consider passing heat Q from $A \to B$. The change in entropy would be given by

$$\Delta S = \frac{-Q}{-|T_A|} + \frac{Q}{-|T_B|} = Q\left(\frac{1}{|T_A|} - \frac{1}{|T_B|}\right) > 0 \,. \tag{11.8}$$

Thus, heat flows spontaneously from hot to cold, as required by the second law of thermodynamics. The same conclusion holds if object B has a positive temperature. In that case passing heat Q from $A \to B$ entails the entropy change $\Delta S = Q\left(1/|T_A| + 1/T_B\right) > 0$.

We can pretty much do thermodynamics with objects characterized by negative temperature as for objects at positive temperature. The only exception would appear to be the process indicated in Fig. 11.3. A nominal "refrigerator" operating between reservoirs at negative temperature would require that work be done *on* the environment. Such a process violates the second law, however. See Exercise 11.2.

CHAPTER SUMMARY

In this chapter we considered exceptional systems that can temporarily exist in states of negative absolute temperature. Ordinarily, absolute temperature is positive.

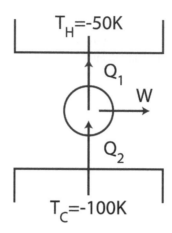

Figure 11.3 Process conserves energy, but violates the second law

- To produce negative absolute temperatures, a system must contain a subsystem that: 1) Attains equilibrium with the rest of the system relatively slowly; 2) Comes to internal equilibrium relatively quickly; and 3) Is finite in extent such that its energy spectrum is bounded.

- Negative absolute temperatures are hotter than positive temperatures. In particular, negative temperatures are hotter than infinite positive temperature.

- Systems with negative temperature are characterized by a population inversion: Higher-lying energy states are more populated than lower-energy states.

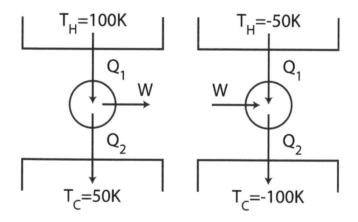

Figure 11.4 Left: Positive-temperature reservoirs. Right: Negative-temperature reservoirs

EXERCISES

11.1 a. The left portion of Fig. 11.4 depicts the operation of a reversible heat engine with positive-temperature reservoirs. Show using Eq. (11.1) that $Q_2 = \frac{1}{2}Q_1$, and thus, to conserve energy, there must be work done on the environment. What is the magnitude of W? Note that in Eq. (11.1), it's the absolute values of the heats that are involved.

b. The right portion of Fig. 11.4 depicts a "heat engine" operating between hot and cold reservoirs having negative temperatures. Show that $Q_2 = 2Q_1$, and therefore work must be done *on* the device to make it function as depicted. What is the magnitude of W? Note that we are using the work to "amplify" the amount of heat delivered to a lower temperature. Heat spontaneously flows from hot to cold without the expenditure of work, see Eq. (11.8).

11.2 Show that the cycle in Fig. 11.3 is not possible, even though it conserves energy. Hint: Allow heat Q_1 to flow back to the cold reservoir by a direct conduction path. Show that you have a compound engine violating the Kelvin form of the second law.

CHAPTER 12

Thermodynamics of information

It is to states of systems . . . incompletely defined that the problems of thermodynamics relate.—J.W. Gibbs[3, p166]

12.1   ENTROPY AS MISSING INFORMATION

THERMODYNAMICS pertains to systems *incompletely specified*. The quantities we can control experimentally are limited to those that affect changes in internal energy through *work* as the mode of energy transfer. Macroscopic systems are comprised of enormous numbers of microscopic components whose coordinates are not known—hence the term incompletely specified. Energy transfers to microscopic degrees of freedom are known as *heat*; heat couples to the large number of degrees of freedom not explicitly accessible to macroscopic observation. Heat is an extensive quantity—heat capacities scale with the size of the system—yet we know it's meaningless to speak of the heat contained in a system because heat transfers are process-dependent; heat is not a state variable. Entropy, however, *is* a state variable, closely related to heat transfers, that's defined by the state of the system. Entropy is a measure of the microstates that are consistent with the macrostate specified by the other state variables (Chapter 7).

Entropy can therefore be said to represent the *missing information* about a system that can't be accounted for in macroscopic descriptions—what we don't know; the larger the entropy, the more the missing information.[1] The concept that entropy—which represents degrees of freedom not subject to our control—also represents "missing information," what could *potentially* be known about a system, is an appealing idea, but can it be made precise? As the term is customarily used, *information* connotes subjective aspects of facts learned in everyday life, yet we're suggesting a connection between something nominally qualitative (information) with something *physical* (entropy). *Can information be quantified* in a way suitable for use in physical theories? *Is information physical*? The connection between entropy and information is a significant development in our understanding of the second law, the gift that seems to keep on giving.

The link between entropy and information has a curious origin in *Maxwell's demon*, a *thought experiment* to beat the second law of thermodynamics.[2] The second law codifies our experience that heat spontaneously flows from hot to cold, and that the reverse is never observed. If we were to discover exceptional circumstances in which the second law fails, it would be a different *kind* of

---

[1]Equivalently, the *less* information (more missing information) we have about a system, the greater is the entropy.
[2]There is a long history of thought experiments in physics, from Galileo to Einstein to Schrödinger.

physical law from those that hold without exception.[3] As early as 1867, Maxwell argued that the second law could be beaten by a hypothetical being with the ability to discern the quickly-moving atoms of a gas from the slowly-moving, and to *separate them without performing work*, thereby transferring heat from cold to hot with no other effect. The hypothetical being became known as Maxwell's demon. "The demon," once having been born on the stage of physics, took on a life of its own through attempts to explain how it does or does not violate the second law, and has led to advances in understanding the connections between thermodynamics, information, and computing. This chapter is an introduction to the rich web of concepts spawned by Maxwell's demon,[4] from a hypothetical way of beating the second law, to the physical foundations of information processing.

12.2   MAXWELL'S DEMON: A WAY TO BEAT THE SECOND LAW?

In Chapter 7 we discussed some of the subtleties associated with entropy, including the possibility of its subjectivity. Maxwell weighed in on the topic, using energy dissipation as an entrée:

> It follows . . . that the idea of dissipation of energy depends on the extent of our knowledge. Available energy is the energy which we can direct into any desired channel. Dissipated energy is energy which we cannot lay hold of and direct at our pleasure, such as the energy of the confused agitation of molecules which we call heat. Now, confusion, like the correlative term order, is not a property of material things themselves, but only in relation to the mind which perceives them. . . . the notion of dissipated energy would not occur to a being who could not turn any of the energies of nature to his own account, or to one who could trace the motion of every molecule and seize it at the right moment. It is only to a being in the intermediate stage, who can lay hold of some forms of energy while others elude his grasp, that energy appears to be passing inevitably from the available to the dissipated state.[2, p646]

That passage refers to what would become known as the demon. Maxwell elaborated in his *Theory of Heat*, and it's useful to quote at length from that work:

> One of the best established facts in thermodynamics is that it is impossible in a system enclosed in an envelope which permits neither change of volume nor passage of heat, and in which both the temperature and the pressure are everywhere the same, to produce any inequality of temperature or of pressure without the expenditure of work. This is the second law of thermodynamics, and it is undoubtedly true as long as we can deal with bodies only in mass, and have no power of perceiving or handling the separate molecules. But if we conceive of a being whose faculties are so sharpened that he can follow every molecule in its course, such a being, whose attributes are as essentially finite as our own, would be able to do what is impossible to us. For we have seen that molecules in a vessel full of air at uniform temperature are moving with velocities by no means uniform, though the mean velocity of any great number of them, arbitrarily selected, is almost exactly uniform. Now let us suppose that such a vessel is divided into two portions, A and B, by a division in which there is a small hole, and that a being, who can see the individual molecules, opens and closes this hole, so as to allow only the swifter molecules to pass from A to B, and only the slower molecules to pass from B to A. He will thus, without expenditure of work,[5] raise the temperature of B and lower that of A, in contradiction to the second law of thermodynamics.[62, p328]

---

[3]There are certainly exceptions to Newtonian dynamics—the fast and the small—yet within its domain of validity, Newton's laws always apply. The classical statements of the second law of thermodynamics contain words such as "never" and "impossible," yet Maxwell provided a hypothetical scenario by which it could possibly fail.

[4]We can only scratch the surface. The book by Leff and Rex is a valuable resource on Maxwell's demon.[61]

[5]Note the "sole result" clause implicitly being invoked here; see Chapter 2.

Maxwell did not call the mythical doorkeeper a demon (typically depicted as in Fig. 12.1)—that would come from Kelvin, who referred to "Maxwell's intelligent demons," and the name has stuck. Conceived 150 years ago and pronounced dead several times in the intervening years, the demon continues to thrive as a conceptual framework for the connection between information and thermodynamics.[63]

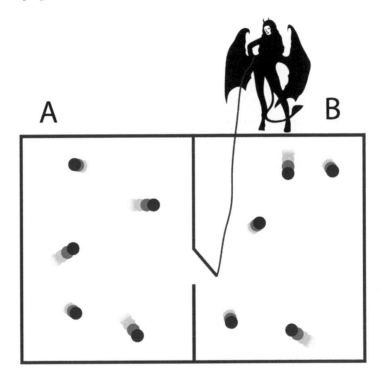

Figure 12.1   The demon separates cold atoms from hot so as to violate the second law

When Maxwell introduced the demon, the distribution of speeds of the atoms of a gas at a fixed temperature *had only recently been established* (the Maxwell-Boltzmann distribution, Eq. (7.26)).[6] Maxwell asked the natural question: Is it possible to separate the atoms of a gas ("lay hold of") into two populations, fast and slow, without the expenditure of work? His purpose in hypothesizing a creature to beat the second law was to illustrate that the second law has a *statistical* character and *cannot be reduced to dynamics*. He continues (in his *Theory of Heat*):

> This is only one of the instances in which conclusions which we have drawn from our experience of bodies consisting of an immense number of molecules may be found not to be applicable to the more delicate observations and experiments which we may suppose made by one who can perceive and handle the individual molecules which we deal with only in masses.

> In dealing with masses of matter, while we do not perceive the individual molecules, we are compelled to adopt what I have described as the statistical method of calculation, and to abandon the strict dynamical method, in which we follow every motion by the calculus.

---

[6]The absolute temperature T is related to the *average* kinetic energy of the particles of a gas, $\langle mv^2/2 \rangle = 3k_B T/2$; see Chapter 11. It's meaningful therefore to speak of hotter and cooler, more and less energetic (than the average), atoms *in the same gas at temperature T*.

It would be interesting to enquire how far those ideas about the nature and methods of science which have been derived from examples of scientific investigation in which the dynamical method is followed are applicable to our actual knowledge of concrete things, which, as we have seen, is of an essentially statistical nature, because no one has yet discovered any practical method of tracing the path of a molecule, or of identifying it at different times.

Indeed, it *would* be of interest to inquire how far we can take mechanics to describe all the atoms in macroscopic systems! Despite Maxwell's admonition that we are compelled to abandon the methods of analytic mechanics, it didn't stop others from making the attempt. The subject of statistical mechanics begins with the work of Maxwell and Boltzmann.

12.3 DEMISE OF THE DEMON: FLUCTUATIONS AND INFORMATION

Maxwell's intent in introducing the demon was not to challenge the second law, but rather to emphasize that *without* the ability to follow the course of individual atoms, our knowledge of collections of atoms is *limited* to their statistical properties.[7] Maxwell apparently was not concerned with whether the demon could actually exist, even in principle; merely conceiving it was sufficient for his purposes. It begs the question, however—*can* Maxwell's demon exist? The demon has been *hypothesized* to have certain abilities, but are those abilities within the realm of physical possibility?

Death by thermodynamics

The demon is tasked with violating the second law by transferring heat from cold to hot with no other effect. To that end, the system must be mechanically isolated: The demon can't have help from sources of work outside the system. The system must also be thermally isolated; lowering the entropy of the gas cannot come about because of heat transfers with the environment.[8] As we now discuss, thermodynamic isolation *prevents the demon from doing its job*, if it's subject to the laws of physics.[9]

---

[7]Reminiscent of quantum mechanics? Leon Rosenfeld described the two modes of description—dynamical and statistical—as *complementary* in the sense of the principle of complementarity in quantum mechanics:[64, p812] "Above all, it [the analysis of the two modes of description] is characterized by the mutual exclusiveness of the two descriptions: conditions allowing of a complete mechanical (and electrodynamical) description exclude the possibility of applying to the system any of the typical thermodynamical concepts; and conversely, the definition of the latter requires conditions of observation under which the mechanical parameters essentially escape our control." The subtlety of this statement can best be appreciated after we have some familiarity with statistical mechanics. To assign a definite temperature to a system, it must be able to exchange energy with its environment, Section 3.10, implying the impossibility of assigning a definite value to the energy. To have the energy well defined, one must isolate the system, implying the impossibility of assigning a definite value of the temperature. It can be shown that the product of the root mean squares of the fluctuations in internal energy and temperature are related by $\sqrt{\langle (\Delta U)^2 \rangle} \sqrt{\langle (\Delta T)^2 \rangle} = k_B T^2$—the same form as the Heisenberg uncertainty principle.

[8]The demon must therefore be *inside* the system, not as depicted in Fig. 12.1.

[9]We can't very well explain one miracle by invoking another—where does that stop? To keep the discussion within the realm of physical possibilities, the activities of the demon should be subject to the laws of physics, *including the second law of thermodynamics*. Should the demon succeed in violating the second law, one would conclude that the second law contains the seeds of its own limitations. It's not uncommon for laws of physics to indicate their domains of validity. The Newtonian theory of gravitation, for example, has no natural length scale associated with it—it's meant to apply over *any* distance. The general theory of relativity, which supersedes Newtonian gravity, features an intrinsic length associated with a mass M, the Schwarzschild radius $r_S \propto GM/c^2$. Quantum mechanics predicts another fundamental length, the Compton wavelength $\lambda_C = h/(Mc)$. The two lengths become equal at the *Planck length*, $L_P \equiv \sqrt{hG/c^3}$. We would not expect general relativity (a classical theory), or the Standard Model of particle physics (which excludes gravity), to apply at the Planck length. Incidentally, Planck defined what are known as the Planck units in his *Theory of Heat Radiation*.[27, p173]

The demon heats up so much it can't function

The entropy of isolated systems never decreases, $\Delta S \geq 0$. To whatever extent the demon can lower the entropy of the gas, *its entropy must increase even more in the process.* Unless its heat capacity C is infinite, the temperature of the demon will rise with increasing entropy, $dT = (T/C)\,dS$. In 1912, the physicist Marian Smoluchowski noted that fluctuations known as *Brownian motion*[10] would become increasingly amplified with rising temperature[11] to the point where the demon would be unable to function. *A demon that sets out to violate the second law is rendered ineffective by the second law!* To quote Richard Feynman: "It [the demon] has but a finite number of internal gears and wheels, so it cannot get rid of the extra heat that it gets from observing the molecules. Soon it is shaking from Brownian motion so much that it cannot tell whether it is coming or going, much less whether the molecules are coming or going, so it does not work."[36, p46-5] A computer simulation reported in 1992 found that, indeed, a finite demon becomes sufficiently thermalized that it can't operate as intended.[65] It's ironic that Brownian motion, a topic outside the purview of thermodynamics,[12] comes to the rescue of the second law.

How does the demon see?

The demon operates in an enclosed, isolated system in thermal equilibrium. In order to sort the molecules, *the demon must see them.* Yet, located within such a system, the demon is immersed in cavity radiation where *it's impossible to see anything;* Exercise 5.6. That point seems to have gone unnoticed in the physics literature until the 1940s. Leon Brillouin wrote in 1951: "The demon cannot see the molecules, hence, he cannot operate the trap door and is unable to violate the second principle."[66]

In order to see the molecules, the demon must be equipped with *a light source having a temperature greater than that of the gas* (noted by Brillouin[66] and independently by Dennis Gabor[13][67]). Let T_l denote the temperature of the light source, where $T_l > T$, the temperature of the gas, and of the demon. If the source emits energy E which is absorbed by the gas, there is an increase in entropy associated with the energy transfer, $\Delta S_E = E\left(T^{-1} - T_l^{-1}\right) > 0$. The demon must detect at least one scattered photon of frequency ν, where $h\nu > k_B T$. The entropy change of the demon in absorbing the photon is $\Delta S_d = (h\nu/T) > 0$. If the demon lets a molecule through the trap door, the number of microstates of the gas is *diminished,* so that its entropy decreases by an amount (see Eq. (7.65)) $\Delta S_g = k_B \ln(1 - |\Delta W|/W) \approx -k_B |\Delta W|/W$, where $|\Delta W|/W \ll 1$. The change in entropy of just the demon in detecting light scattered from the gas, and of the gas in its atoms being let through the trap door is therefore

$$\Delta S_d + \Delta S_g = k_B \left[\frac{h\nu}{k_B T} - \frac{|\Delta W|}{W}\right] > 0, \tag{12.1}$$

because $h\nu/(k_B T) > 1$ and $|\Delta W|/W \ll 1$. The entropy *increase* of the demon exceeds the entropy *decrease* of the gas. Thus, while an experimental apparatus ("demon") might sort molecules,

---

[10]References to Smoluchowski's work are given in [61]. Brownian motion is a *stochastic process* in which a large particle is subject to random impacts from smaller particles. In this case the demon is the Brownian "particle" that's being bombarded by the molecules of the gas.

[11]The average size of fluctuations increase with temperature. For example, $\langle(\Delta T)^2\rangle = k_B T^2/C_V$.

[12]Fluctuations strictly speaking cannot be incorporated within the framework of thermodynamics. Thermodynamics is concerned with equilibrium, the quiescent state where state variables have fixed values. We introduced fluctuations in Section 3.10 through the device of *virtual variations* of the system in which the constraint of thermal isolation is conceptually relaxed. That's basically what's done in statistical mechanics through the theory of *ensembles,* a conceived large collection of identical copies of a system, each prepared consistent with known information about the system. Fluctuations could be motivated phenomenologically by relaxing the concept (see Section 1.2) that thermodynamic measurements are *the* values of state variables, and instead represent the average result of many measurements on a system.

[13]Gabor was awarded the 1971 Nobel Prize in Physics for the invention of holography.

there's no violation of the second law.[14] Note that implicit in Eq. (12.1) is the quantum nature of electromagnetic radiation—one cannot absorb a fraction of a photon. If light behaved as described by the classical theory, where the energy of light rays can be made arbitrarily small, there *could* be a violation of the second law. Another instance where the consistency of thermodynamics relies on the underlying quantum nature of the physical world.[15]

Does measurement affect entropy?

The demon must, because it operates in isolation, increase its entropy, causing its own demise. Brillouin and Gabor showed that observing the gas with light is one way by which the entropy of the demon increases. Other means by which the demon can detect the gas, not involving light, have been proposed,[61, p8] prompting the question: *Does measurement always entail an increase in entropy?* Does the demon's act of sorting, which requires it to discern fast from slow molecules, and hence which involves an interaction between demon and molecules—an act of measurement—does that *always* result in $\Delta S > 0$? *Is measurement inherently a source of irreversibility?* The answer, which emerged only in the second half of the 20th century, is *not necessarily* (see Section 12.5).

Szilard's engine: Role of information

If a mechanical demon can't beat the second law, could a demon function if it possessed *intelligence*, as Maxwell originally suggested? Maxwell apparently assumed that intelligence can function without energy cost. He wrote:

> Then the number of molecules from A to B are the same as at first, but the energy in A is increased and that in B diminished, that is, the hot system has got hotter and the cold colder and yet no work has been done, only the intelligence of a very observant and neat-fingered being has been employed.[61, p5]

The physicist Leo Szilard undertook a careful examination of the role played by the involvement of an observer. In his 1929 article,[16] "On the Decrease of Entropy in a Thermodynamic System by the Intervention of Intelligent Beings," he devised an idealized process—another thought experiment, now known as *Szilard's engine*—that's mostly mechanical, but which requires the participation of an intelligent being, and which shows that *a violation of the second law is prevented if entropy is associated with measurement*. The minimal "intelligence" we require of a demon is the ability to store information, if only briefly, i.e., it must have *memory*. An act of measurement is not complete until the result is recorded somehow.[17]

The Szilard engine consists of a closed volume V, maintained at a constant temperature T through contact with a heat reservoir, that contains a single particle in random thermal motion (see Fig. 12.2). A partition is introduced in the middle of the system (without work being performed), dividing it into two equally-sized chambers. At this point, the particle is either on one side of the

---

[14] In 2009, the operation of a Maxwell demon was reported that used a pair of laser beams tuned to atomic transitions and configured in such a way to create a potential barrier that atoms could cross in one direction, but not in the other, effecting a separation of cold and hot atoms.[68] The entropy increase of the "demon" (experimental arrangement that separates the atoms) exceeded the entropy decrease of the atoms. In 2016, a *photonic* Maxwell's demon was reported.[69]

[15] In 1941, P.W. Bridgman (who was awarded the 1946 Nobel Prize in Physics for work on high-pressure physics) wrote:[47, p156] "If the Maxwell demon had been invented yesterday instead of in the last century I believe he would not have caused as much consternation." While game seemingly over for the demon, it still had lessons to teach us concerning the role of information in the performance of its job.

[16] An English translation is reprinted in Leff and Rex.[61, p124] Szilard's article is also reprinted in Wheeler and Zurek.[70, p539] Measurement, which Szilard identified as key to understanding the demon, is fundamental to the theory of quantum mechanics. The title of Szilard's article is misleading, which seems to imply that an intelligent demon could violate the second law, yet the article argues the opposite, that no being, intelligent or otherwise, can do so.

[17] The demon could therefore be a computer connected to suitable detectors for discerning the molecules of the gas. The demon need not possess human intelligence.

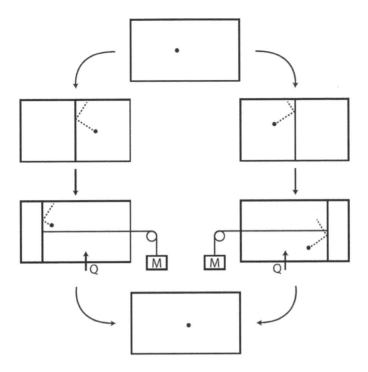

Figure 12.2 Szilard engine violates the second law unless entropy is associated with the observer

partition or the other, with equal probability. Whichever side it's on, an intelligent observer (the demon) introduces a "load" connected to the partition—which is free to slide without friction—against which the particle does work in its random collisions with the partition. The numerous collisions with the partition produce an average pressure, and we may consider the movement of the partition against the load a reversible, quasistatic process. The partition is removed when it reaches the end of the chamber, restoring the system to its original configuration.[18] In the process work $|W|$ has been done on the environment at the expense of the heat absorbed from the reservoir, with $Q = |W|$. The first law of thermodynamics is satisfied.

The second law, however, is violated: Heat has been completely converted into work—an engine with 100% efficiency.[19] As a result, the entropy of the reservoir decreases.[20] Assuming we can use the ideal gas law (with $N = 1(!)$), work $|W| = k_B T \ln 2$ has been expended. The entropy change of the reservoir is $\Delta S = -Q/T = -|W|/T = -k_B \ln 2$. The entropy change of the mechanical system, however, which has been returned to its original state, is zero.[21] Is there a way to save the second law here? Yes, if we enlarge our concept of what the *system* is. Szilard argued that the system obeys the second law if *the entropy of the observer increases as a result of the measurement it made*

[18]Better descriptions of Szilard's engine are given in Leff and Rex,[61] in Brillouin,[71] and in the article by Charles Bennett.[72] Bennett describes detectors of his own invention for determining which side of the partition the particle is on.

[19]The disorganized energy of heat has been entirely converted into the "organized" energy of work, seemingly with no other effect—what's prohibited by the second law.

[20]For engines operating between two reservoirs, the decrease in entropy of the hot reservoir is compensated by the entropy increase of the colder reservoir. With the Szilard engine, there's no "waste heat" (until we bring the observer into account), leading to an entropy decrease in the one and only one reservoir.

[21]The movement of the partition is presumed to have been done reversibly. Szilard's engine is thus (before we invoke the observer) a reversible cyclic process for which $\Delta S < 0$.

in determining which side of the partition the particle was on. Szilard's engine, *by design*, consists of a mechanical system *plus* an observer that must make a measurement. He wrote:

> One may reasonably assume that a measurement procedure is fundamentally associated with a certain definite average entropy production, and that this restores concordance with the second law. The amount of entropy generated by the measurement may, of course, always be greater than this fundamental amount, but not smaller.[61, p127]

The "fundamental" entropy increase is $k_B \ln 2$.

Szilard thus made a generic connection between entropy and measurement, or more generally *information*. Without knowing it, he discovered the "bit" of information[22] associated with two equally likely outcomes: Before we detect the particle, it can be on either side of the partition with equal probability. The fundamental role of measurement in quantum mechanics had been recognized at the time of Szilard's work, but the significance of *information* in physics had yet to be appreciated. Szilard did not specify *how* detecting the particle brings about an entropy increase; he "merely" noted the necessity of recognizing the role played by information in interpreting the second law. Thus, he did not *prove* that $\Delta S > 0$ because of measurement (as is sometimes stated), but his tremendous insight argued for such a connection. Szilard recognized what would only later emerge as the essential aspects we require of an intelligent demon: measurement, information, and especially memory.

It's not important that Szilard's engine operates with only one molecule and appears hopelessly contrived and impractical. What you should ask is: Is it *impossible*? In 2010, a group reported an experimental realization of a Szilard engine involving a colloidal particle in a fluid bath at uniform temperature.[73] By acting on information about fluctuations in the location of the particle (as observed using a CCD camera) they could demonstrate conversion of *information* about the particle to extract work against the forces acting on the particle. In 2014, a group reported the experimental realization of a Szilard engine that uses a single electron.[74]

12.4   IS ENTROPY INFORMATION?

In 1948 the mathematician Claude Shannon published a landmark paper, "A Mathematical Theory of Communication,"[75] which was soon thereafter reprinted in book form.[76] Shannon is considered the founder of *information theory*, what has wide-ranging applications in mathematics, statistics, computer science, physics, electrical engineering, economics, and the biological sciences. Shannon's innovation—what led to its use in so many disciplines—is a way to *quantify* information. As we'll see, his *measure* of information, Eq. (12.5), bears resemblance to entropy, with their commonality being that entropy represents "missing information," what's required to specify the microstates of a system given only the specification of the macrostate.[23]

Shannon defined information in a specific way, that at times can appear paradoxical, which takes into account the *surprise element* one experiences in learning of events whose occurrence was not previously certain. Consider the following two statements: 1) The sun rose this morning; 2) The sun exploded this morning. Which conveys more "news"? If we were repeatedly told the first statement, we'd quickly tune it out: No news there. The second statement, however, is real news: It's unexpected. The more *likely* an event, the *less* is the information conveyed in learning of its occurrence, and conversely. Information, as the term is used in information theory, is a measure of the *uncertainty* inherent in a system, what *could* be learned about a system, but which is not currently known.[24] Information, like entropy, represents the potentialities inherent in a system.

---

[22]The term *bit* wasn't coined until the 1940s. Beware of potential confusion: A bit as a *binary digit* (0 or 1) is not the same as a *bit of information*, what's defined in Section 12.4. Binary digits and bits of information are different.

[23]Information is thus missing information. Got it?

[24]Again, "information" is missing information. If you get confused, think *uncertainty*. That way, the parallel between information and entropy is more readily apparent—uncertainty of information associates with physical disorder. We note here that E.T. Jaynes used information theory as a way to formulate statistical mechanics.[77]

Invoking *likelihood* paves the way for bringing *probability* into information theory.[25] If for an event x having probability p_x for its occurrence, let I_x denote the information conveyed upon learning that x has happened. A fundamental tenet of information theory is that I_x can be obtained from a *function* of p_x, $I_x = I(p_x)$. One way to capture the "surprise" aspect of information would be to guess that I_x is simply $1/p_x$. That turns out not to be correct, but it's close.

The form of the information function $I(p)$ will be inferred from a set of axioms (see below). One simple requirement is that $I(p)$ be a non-negative function of p for[26] $0 < p \leq 1$. Denote by $p_{\overline{x}}$ the probability that x does *not* occur. Clearly,[27] $p_x + p_{\overline{x}} = 1$. For the non-occurrence of x, let $I_{\overline{x}}$ be the information conveyed that x did not happen, where it's assumed that $I_{\overline{x}}$ is obtained from the *same* information function, $I_{\overline{x}} = I(p_{\overline{x}})$. The *average* information in a system for which x can occur as the outcome of an experiment is given by the expression[28] $p_x I(p_x) + p_{\overline{x}} I(p_{\overline{x}})$. It might seem strange that we've included the probability of what *doesn't* happen. Consider tossing a coin, and let x denote the occurrence of "heads." Unless we have a one-sided coin (whatever that is), or a coin in which heads always occurs, we have to consider the other possibility, of not-heads, or tails. There would be no information conveyed in tossing a coin that always lands with heads showing—no news. *For a system to have non-zero information, there must be at least two outcomes for experiments performed on it.* For p the probability of heads showing, the average information is thus $pI(p) + (1-p)I(1-p)$. If $p \to 1$, to have zero information conveyed, we require $I(1) = 0$ and moreover that $\lim_{x \to 0} xI(x) = 0$. Note the symmetry in this expression between $I(p)$ and $I(1-p)$.

Requirements on the information function

We can now generalize to the case of mutually exclusive events $(x_1, \cdots, x_n)$ that occur with probabilities p_i such that $\sum_{i=1}^{n} p_i = 1$. Following tradition, we use the symbol[29] H to denote the information of a system characterized by a set of probabilities $\{p_i\}$:

$$H(p_1, \cdots, p_n) = \sum_{i=1}^{n} p_i I(p_i) . \tag{12.2}$$

The form of Eq. (12.2) emphasizes that information, what we're calling H, is a function of the probabilities $\{p_i\}$, *which are presumed known*. The function $I(p)$ is unknown at this point. Equation (12.2) serves as a "bootstrap": It motivates the concept that *information is associated with the probabilities p_i of the outcomes of experiments*. The actual form of $H(p_1, \cdots, p_n)$ is determined by the following set of requirements, which imply that H occurs in the form of Eq. (12.2).

We impose four requirements on $H(p_1, \cdots, p_n)$:

1. $H(p_1, \cdots, p_n)$ is a continuous function in each of its arguments, a reasonable request;

2. $H(p_1, \cdots, p_n)$ is a symmetric function of its arguments. The information associated with a sequence of outcomes is independent of the order in which the outcomes occur;

---

[25] As noted on page 97, a certain level of familiarity with the theory of probability is assumed.

[26] The case of $p = 0$ requires special handling; it can be incorporated into information theory, but not directly in the function $I(p)$.

[27] Behind this simple result lies several ideas from the rules of probability. Two events are said to be *mutually exclusive* if they both cannot occur at the same time. For x and y mutually exclusive events, the probability of both occurring is clearly zero. This is stated in the form that $p(x \text{ and } y) = 0$. If one asks for the probability that x *or* y occurs, the rule is that $p(x \text{ or } y) = p(x) + p(y)$. The event x and the event $\overline{x}$ (the non-occurrence of x) cannot both occur at the same time, and are mutually exclusive. The probability that x happened or not, is clearly unity.

[28] The *expectation value* of a quantity x that could take on a range of values $(x_1, \cdots, x_n)$, each occurring with probability p_i, is given by $\sum_{i=1}^{n} x_i p_i$ where the probability distribution is normalized, $\sum_{i=1}^{n} p_i = 1$. Consider calculating the average height of a group of people; you'd see this formula in action.

[29] Shannon[76, p51] chose H because it resembles the "H" in Boltzmann's H-theorem (mentioned on page 98).

3. If all probabilities p_i are equal, with $p_i = 1/n$, $H(1/n, \cdots, 1/n)$ is a monotonically increasing function of n. As n is increased, i.e., as more arguments are added to $H(p_1, \cdots, p_n)$, more outcomes become available to the system, which if all are equally likely, should imply more uncertainty, more information in the information-theoretic sense;

4. If some of the outcomes of a given experiment can be realized as the outcomes of "sub-experiments" (what we'll explain shortly), the value of H for the original experiment can be expressed as the weighted sum of the values of H for the sub-experiments.

To understand what's meant by requirement four—which is actually quite strong, it determines the form of $H(p_1, \cdots, p_n)$ without appealing to Eq. (12.2)—consider Fig. 12.3. Suppose there are

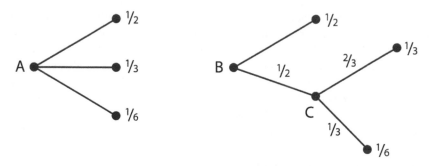

Figure 12.3   Information associated with experiment A is the same as that for the compound experiment BC

three outcomes of an experiment having probabilities $p_1 = \frac{1}{2}$, $p_2 = \frac{1}{3}$, and $p_3 = \frac{1}{6}$; call this experiment A. The information associated with A would be indicated $H\left(\frac{1}{2}, \frac{1}{3}, \frac{1}{6}\right)$. Suppose the outcomes of A can be realized by the outcomes of two successive experiments, call them B and C, where in experiment B two outcomes occur with equal probabilities $\frac{1}{2}$, followed by subsidiary experiment C, performed on one of the outcomes of B, that itself has two outcomes with probabilities $\frac{2}{3}$ and $\frac{1}{3}$. We require that the two experiments—A, and B followed by C—yield the same information:

$$H\left(\frac{1}{2}, \frac{1}{3}, \frac{1}{6}\right) = H\left(\frac{1}{2}, \frac{1}{2}\right) + \frac{1}{2}H\left(\frac{2}{3}, \frac{1}{3}\right), \tag{12.3}$$

where the coefficient of $\frac{1}{2}$ is the weighting factor because experiment C occurs only half the time. Requirement four can then be restated:

4. If $p_n = q_1 + q_2 > 0$, we require that

$$H\left(p_1, \cdots, p_{n-1}, q_1, q_2\right) = H\left(p_1, \cdots, p_n\right) + p_n H\left(\frac{q_1}{p_n}, \frac{q_2}{p_n}\right). \tag{12.4}$$

Requirement four generalizes. It can be shown by induction that

$$H(p_1, \cdots, p_{n-1}, q_1, \cdots, q_m) = H(p_1, \cdots, p_n) + p_n H\left(\frac{q_1}{p_n}, \cdots, \frac{q_m}{p_n}\right),$$

where $p_n = q_1 + \cdots + q_m > 0$. If every probability p_i can be expressed as a sum of "sub-probabilities," $p_i = \sum_{j=1}^{m_i} q_{i,j}$, $i = 1, \cdots, n$, then it can be shown that requirement four implies

$$H(q_{1,1}, \cdots, q_{1,m_1}; \cdots; q_{n,1}, \cdots, q_{n,m_n}) = H(p_1, \cdots, p_n) + \sum_{i=1}^{n} p_i H\left(\frac{q_{i,1}}{p_i}, \cdots, \frac{q_{i,m_i}}{p_i}\right).$$

Shannon measure of information

Shannon showed that the *only* function meeting these requirements has the form[76, p116]

$$H(p_1, \cdots, p_n) = -K \sum_{i=1}^{n} p_i \ln p_i , \qquad (12.5)$$

what's known as the *information entropy*, or the *Shannon entropy*, or, better, *Shannon's measure of information*,[30] where $K > 0$ is an arbitrary positive constant that sets the *scale* of information.[31] That K must be positive is so that requirement three is met. For $p_i = 1/n$, $i = 1, \cdots, n$, Eq. (12.5) implies that

$$H\left(\frac{1}{n}, \cdots, \frac{1}{n}\right) = K \ln n . \qquad (12.6)$$

The constant K must be positive so that $K \ln n$ is a monotonically increasing function of n. Starting from Eq. (12.5), it can be shown that $H(\frac{1}{n}, \cdots, \frac{1}{n})$ is the *maximum value* of $H(p_1, \cdots, p_n)$; see Exercise 12.6. Maximum information—maximum uncertainty—is for all outcomes equally likely. The imposition of *constraints*, therefore, can only serve to *decrease the information*. If a constraint is imposed, we have more knowledge about the system, and less uncertainty.

The choice of K is partially dictated by the *base* of the logarithm. In Eq. (12.5) we've used the natural logarithm, $\ln x \equiv \log_e x$. In general $\log_a x = \log_a b \log_b x$; thus, $\log_b x = \ln x / \ln b$. By changing the base of the logarithm, one affects the value of K. If one works with base-2 logarithms, then with $K = 1$, information is measured in *bits*. The basic information function $I(p)$ (comparing Eq. (12.5) with Eq. (12.2)) is therefore $I(p) = -K \ln p = K \ln(1/p)$, which captures the "surprise" aspect of information—the smaller the probability of an event occurring, the greater the information conveyed when it occurs.[32] Note that $I(1) = 0$ and $\lim_{x \to 0} x I(x) = 0$ as we noted previously was required of the function[33] $I(x)$.

Example. A string of letters from the English alphabet, together with blank spaces, would, from Eq. (12.6), have an information of $H = \log_2 27 \approx 4.75$ bits per character if they all occurred with equal frequency. If one uses the known probabilities with which these symbols occur in the English language, the information (obtained from Eq. (12.5)) is $H \approx 4.03$ bits per character.[71, p8] Note that H *goes down* when we impose the constraint of using the actual probabilities in which letters occur, versus assuming they all occur randomly. If one takes into account *correlations* between letters, such as the letter u almost always follows the letter q, H goes down even more, close to two bits per character.

Is information entropy?

Except for a scale factor, Eq. (12.5) is identical to Eq. (7.21), what was obtained from the Boltzmann entropy. One might wonder if information is the same as entropy, especially when Eq. (12.5) is referred to as the information entropy. Equation (7.21) does not represent the entropy of a physical system because it can't be connected to the parameters characterizing equilibrium, T, P, and μ, as required by thermodynamics. Despite the similarity of the formulas, information and entropy are logically distinct. Entropy is a property of systems in thermodynamic equilibrium. Information is

---

[30]Calling Eq. (12.5) an entropy causes confusion. Best to call it simply the measure of information.

[31]More thorough proofs that Eq. (12.5) is the only function satisfying the requirements on H are given in Khinchin.[78]

[32]In fact, the function $I(p) = -K \ln p$ is known as the *surprisal* associated with the event for which the probability of occurring is p. The information H is thus the *average surprisal* associated with the system!

[33]To show that $\lim_{x \to 0} x \ln x = 0$, start with $\lim_{x \to 0} \ln x / (1/x)$ and apply l'Hôpital's rule.

associated with any system for which we know the probability distribution of its experimental outcomes. Seen this way, *entropy is a special case of information.* Information applies to any system for which its probability distribution is known; entropy applies to systems for which the probability distribution characterizes the state of thermodynamic equilibrium. Entropy can be said to be information, but information does not necessarily imply entropy.[34]

Example. To illustrate that entropy is information, we can use Eq. (12.6) (the *maximum* information) to determine the entropy of the ideal gas (which, being in thermodynamic equilibrium, must be a maximum), $H = k_B \ln W$, where W is the number of distinct phase-space configurations (which we've assumed occur with equal a priori probabilities, what's required in the derivation of Eq. (12.6)) as given by Eq. (7.58) and we have set $K = k_B$. This expression leads to the Sackur-Tetrode formula when Stirling's approximation is used; see Exercise 7.8.

It's not surprising that information and entropy are similar in their mathematical features: They both spring from the use of probability. The major requirement imposed on H, Eq. (12.4), provides for how information combines for compound systems. Referring to Fig. 12.3, experiment A is equivalent to the compound experiment BC, where $H(BC) = H(B) + H_B(C)$, where $H_B(C)$ is the information in experiment C, given that B has already occurred. These are very nearly the same requirements imposed on the Boltzmann entropy , $S(W_1 W_2) = S(W_1) + S(W_2)$; see Section 7.1.

Does information help us understand Maxwell's demon?

After the partition is introduced in Szilard's engine, there is one bit of information associated with the two equally-likely possibilities: $H(\frac{1}{2}, \frac{1}{2}) = 1$ (Exercise 12.5). After the demon determines the location of the particle, there is zero information, $H(1, 0) = 0$ (Exercise 12.3). Information has *decreased* (in the sense of Shannon information); we know more about the system, namely which side of the partition the particle is on. *How does a loss of information about the gas translate into an increase in the entropy of the demon?* Szilard did not say. Brillouin, who had already *shown* how the demon's entropy increases—based on detecting the gas by means of light signals—*postulated* a general connection between information and entropy based on this one example.[66][71] Brillouin distinguished two *kinds* of information—*bound* and *free*. Free information is associated with knowledge, which has no physical significance, until it is transmitted by some physical means, when it becomes bound information. Bound information is that conveyed by physical means, information that can be related to the physical states of systems. Thus, Brillouin introduced the important concept that *information is physical.* He also defined a rather confusing concept, that of *negentropy*—negative entropy—as related to information. Brillouin proposed a generalization of the second law:[35] $\Delta(S - H) \geq 0$, or as he stated it, $\Delta(-S + H) \leq 0$. We won't elaborate on negentropy, because the emphasis on information acquisition proved to be something of a *distraction*: It diverted attention from the *other* aspect of an intelligent demon as identified by Szilard: memory. Szilard foresaw that memory was part of the mix in understanding the demon:

> When such beings make measurements, they make the system behave in a manner distinctly different from the way a mechanical system behaves when left to itself. We show that it is a sort of memory faculty, manifested by a system where measurements occur, that might cause a permanent decrease of entropy and thus a violation of the second

---

[34]There is the issue of the dimensionless scale factor of information, K, which is arbitrary except for being positive. Entropy conventionally has as a scale k_B, yet as we've argued (Section 3.2) entropy should naturally be considered a dimensionless quantity—the only reason that we have k_B as a scale for entropy is because we measure absolute temperature in an artificial way, in units of degrees Kelvin.

[35]Linking information with entropy requires that we identify the arbitrary constant K in the definition of information with Boltzmann's constant: $K = k_B$. If information is measured in bits, we should identify $K = k_B \ln 2$.

law of thermodynamics, were it not for the fact that the measurements themselves are necessarily accompanied by a production of entropy.[61, p124]

12.5 INFORMATION IS PHYSICAL

Information is inevitably inscribed in a physical medium. It is not an abstract entity. It can be denoted by a hole in a punched card, by the orientation of a nuclear spin, or by the pulses transmitted by a neuron. The quaint notion that information has an existence independent of its physical manifestation is still seriously advocated. This concept, very likely, has its roots in the fact that we were aware of mental information long before we realized that it, too, utilized real physical degrees of freedom.
—Rolf Landauer, 1999[79]

In Szilard's engine, the demon must *record* which side of the partition the particle is on before it can act: The demon can't make a decision on how to proceed unless the particle's location is *registered*, if only briefly. *A measurement is not a measurement until the result of measurement has been recorded.* At the end of the cycle, to restore Szilard's engine to its original state, the contents of the demon's memory must be *erased*. Erasing memory *increases* information (in the sense of information theory)—the number of possibilities available to memory registers increases upon erasure. More uncertainty, more entropy? Perhaps it's the *erasure* of information, not its acquisition, which causes the entropy of the demon to increase.

In 1961, the physicist Rolf Landauer published a seminal article,[80] "Irreversibility and Heat Generation in the Computing Process," proposing a link between information and the thermodynamics of its manifestation in physical systems, an article that engenders discussion to this day. Landauer's thesis, in brief, is that *information is physical*, that information is stored in physical media, is manipulated by physical means, and is a quantity subject to the laws of physics.

Logical irreversibility

Landauer introduced the concept of *logical irreversibility*, which pertains to information-processing operations for which *the outputs do not stand in one-to-one correspondence with the inputs*. For example, of the three elementary logic gates AND, OR, and NOT, only NOT entails a unique relationship between output and input. Inputs to the NOT gate having truth values $(0, 1)$ correspond to the output values $(1, 0)$. For the AND gate, however, only the input truth values $(1, 1)$ correspond to the output value 1. The other three combinations of input truth values $(0, 0)$, $(0, 1)$, and $(1, 0)$ each correspond to the output value 0. There is not therefore a unique relationship between output and input; one can't recover the input data from the output data, and is in that sense, irreversible. Likewise, for the OR gate, only the input truth values $(0, 0)$ correspond to the output value 0; the other three cases of input values each correspond to the output value 1. For the AND and OR gates, if each of the four types of inputs occurred randomly, there would be 2 bits of information on the input (Eq. (12.6), $\log_2 4 = 2$, where $K = 1$), yet only 1 bit of information on the output $(\log_2 2 = 1)$. Logically-irreversible gates therefore *destroy* information. The NOT gate conserves information, in the sense that 1 bit in, 1 bit out.[36] We can say that information-destroying operations "compress" logical possibilities, much as we speak of compressing a gas. Memory registers entail an even greater compression of information. Suppose an n-bit memory register—which represents 2^n possible states—is *reset*; after the operation the register, which had 2^n possibilities, is now in a definite state. The reset operation compresses many logical states into one; n bits of information are destroyed in the resetting process.

---

[36] Shannon information is not related to the truth value, or meaning of a statement.

Landauer's principle

Landauer's premise is that to each *logical* state of a computer there corresponds a *physical* state of the computer. Landauer proposed that for information destroyed in logically-irreversible operations, there is a corresponding reduction in the entropy of the physical degrees of freedom that carry the information—what are termed the *information-bearing degrees of freedom*. Basically, associated with the compression of the phase space spanned by the logical states of the system, is a compression of the phase space of the physical, information-bearing degrees of freedom. By the second law of thermodynamics, associated with a decrease in entropy of the information-bearing degrees of freedom, is a concomitant increase in entropy of the *non-information bearing degrees of freedom* of the computer's hardware.[37] This idea has come to be known as *Landauer's principle*. As a consequence, erasing information (clearing a memory register), a logically irreversible operation, results in entropy being created in the environment. In particular, Landauer argued that *there is a minimal heat $k_B T \ln 2$ generated by irreversible single-bit operations*—the same number noted by Szilard as required to save the second law in the operation of the Szilard engine. Given the smallness of this energy (3×10^{-21} J at room temperature), it would be difficult to put this prediction to an experimental test. Quite recently, however, several groups have announced experimental verification of the Landauer lower bound on the energy dissipated by the erasure of a single bit of information.[81][82][83]

Whither the demon?

Landauer's principle is not independent of the second law of thermodynamics, and thus it alone cannot be used to save the second law from the demon.[38] Coupled with another development, however, the demon can be decisively vanquished. Starting in 1973, Charles Bennett (and others) showed it's possible in principle to build computers that operate without energy losses.[84] The "trick" to making computation thermodynamically reversible is not to discard information! Unwanted by-products of the computation process can be eliminated by reversing the program, restoring the machine to its original state.[39]

The significance of reversible computing[40] is that the demon can acquire information about the gas, record that information, and sort the molecules in a way that does not dissipate energy. That alone invalidates Brillouin's assertion, who, based on one example—information acquisition by light, concluded that measurements are *necessarily* dissipative processes. In 1982, Bennett noted[41] that *the demon cannot violate the second law because it must erase information previously obtained about the gas*.[85] The demon dies an entropic death because it must discard information previously collected. A way "out" of that conclusion would be to postulate that the demon has an *infinite* amount of memory—no need to ever erase. As noted in Section 12.3, we could get out of the conclusion that fluctuations cause the demon's demise by letting it have an infinite heat capacity. Is an infinite amount of memory tantamount to infinite heat capacity? Nothing in this world is infinite, but it suggests a connection between information storage (or erasure) and the ability to absorb heat.

---

[37] Do you see the demon rearing its head? The entropy of information-bearing degrees of freedom is reduced at the expense of the increase of entropy of the non-information-bearing degrees of freedom.

[38] We cannot use the second law to prove the second law.

[39] Bennett's work spawned a field of research known as *reversible computation*. The subject is reviewed in Bennett[85], in Feynman[86], and in Nielsen and Chuang.[87] The history of reversible computation has been recounted by Bennett.[88]

[40] In addition to its role in vanquishing the demon, the concept of reversible computing using classical degrees of freedom—a heretofore unstated assumption—paved the way for the concept of a *quantum computer*.[86][87]

[41] This point was made earlier by Oliver Penrose (1970), which apparently went unnoticed.[89, p225–226]

CHAPTER SUMMARY

This chapter introduced the thermodynamics of information, a concept that finds wide use in physics and astrophysics, as we'll see in the next chapter. The link between physics and information began in 1867, with Maxwell's thought experiment of a creature capable of beating the second law of thermodynamics. That it took 100 years to finally understand why the demon could not operate as intended, strengthens our understanding of the second law. And it's not just an academic exercise. Thermodynamics, once considered an ancient subject with little to add to modern physics, is quite viable as a tool in determining the fundamental limits of computing technology and in shaping our ideas of the universe (Chapter 13).

- Maxwell hypothesized the demon as a way of illustrating the nature of the second law of thermodynamics. His point was not to challenge the second law, but to note that because we can't follow around the molecules of a gas, our knowledge is limited to its statistical properties. Maxwell, it seems, was not concerned with whether the demon could actually exist, even in principle. That didn't stop others, however, from investigating whether the demon is possible within the confines of the laws of physics.

- The demon must operate in an isolated system; it must operate inside the system it's tasked with lowering the entropy of. The entropy of isolated systems can only increase. If the demon is to lower the entropy of the gas, its entropy must rise in the process. Smoluchowski, in 1912, noted that if in the operation of the demon its entropy increases, thermal fluctuations would render it incapable of operating. Computer simulations reported in 1992 confirm that picture.[65]

- In 1929, Szilard introduced a thought experiment, now known as Szilard's engine. Szilard's engine is designed to convert heat absorbed from a single reservoir entirely into work, what would be a violation of the second law, if not for the intervention of an observer, whose entropy must increase as the result of a measurement it makes to effect the operation of the engine. Because the volume of the engine doubles in the process, it performs work on the environment, $|W| = k_B T \ln 2$. Szilard concluded that the entropy of the observer would have to increase by at least $k_B \ln 2$ to preserve the second law, the same result predicted by Landauer (1961), and what has recently been measured experimentally. Szilard's engine is what Landauer would characterize as logically irreversible, involving the destruction of a bit of information, an act that must be associated with an increase in entropy.

- The notion of bits of information—a phrase we use routinely today—emerged only after Shannon's development of information theory in 1948. Information is a generalization of entropy. Both represent the potentialities inherent in a system—what we don't know about a system, but could.

- In the 1950s, Brillouin, and independently Gabor, noted that the demon, being in an isolated system, is enveloped by cavity radiation, implying it would be unable to see the molecules of the gas, and hence would be unable to effect a violation of the second law. In order for the demon to see the gas, it must be equipped with a light source having a temperature greater than that of the gas. They showed that under these circumstances, the entropy increase of the demon exceeds the entropy decrease of the gas. Based on this one example (detecting the gas with light signals), it was concluded (erroneously) that all acts of measurement increase the entropy of the demon in a way that prevents it from violating the second law. This work proved to be something of a red herring. It was shown by Bennett and others starting in the 1970s that it's possible to build computers that operate reversibly, i.e., without dissipation. Thus, it's not true that measurement necessarily entails an entropy increase.

- Measurement implies memory. A measurement is not a measurement until the result of measurement is recorded. To bring Szilard's engine back to its starting condition, the memory of the demon, which had to record which side of a partition the molecule is on, must be erased. The demon, if it has a finite amount of memory, will suffer an entropic demise because it must erase its memory registers.

EXERCISES

12.1 The information associated with a coin-tossing experiment is given by the expression

$$H(p) = pI(p) + (1 - p)I(1 - p),$$

where p is the probability of a coin showing "heads" and $I(p)$ is the basic information function. Show that $H(p)$ is maximized for $p = \frac{1}{2}$, irrespective of the specific form of $I(p)$.

12.2 Verify Eq. (12.3) using Eq. (12.5).

12.3 a. Show that *in general* $H(1,0) = 0$ (no information is associated with a certain event) is contained in the axioms satisfied by H; don't use Eq. (12.5). Hint: Use Eq. (12.4) with $p_1 = \frac{1}{2}$ and $p_2 = \frac{1}{2} = \frac{1}{2} + 0$, i.e., $q_1 = \frac{1}{2}$ and $q_2 = 0$. Then use the symmetry requirement that $H(\frac{1}{2}, \frac{1}{2}, 0) = H(0, \frac{1}{2}, \frac{1}{2})$. Also, $H(0,1) = H(1,0)$.

   b. Show that in general $H(p_1, \cdots, p_n, 0) = H(p_1, \cdots, p_n)$, i.e., including an impossible event in the list of outcomes does not alter the information contained among a set of possible events. Hint: Use Eq. (12.4) with $p_n = q_1 > 0$, i.e., $q_2 = 0$, and what you've just shown, $H(1,0) = 0$.

12.4 Show that Eq. (12.5) implies Eq. (12.6).

12.5 Show for $K = 1$, and working with base-2 logarithms, that Eq. (12.5) implies $H(\frac{1}{2}, \frac{1}{2}) = 1$. Flipping a fair coin therefore conveys 1 bit of information. Argue that flipping N coins (or one coin N times) implies N bits of information.

12.6 Show that the maximum value of $H(p_1, \cdots, p_n)$, as specified by Eq. (12.5), is given by Eq. (12.6), i.e., the maximum uncertainty is for all outcomes to be equally likely. Hint: Follow the derivation in Section 7.3; introduce a Lagrange multiplier to enforce the normalization on the probability distribution.

CHAPTER 13

Black hole thermodynamics

I T was discovered in the 1970s that the equations describing black holes can be placed in correspondence with the laws of thermodynamics. At first considered a formality, the connection between black holes and thermodynamics has proven to be of deep significance, contributing to our ideas of the universe, and offering insights into a possible quantum theory of gravity—the frontier of theoretical physics. A modern presentation of thermodynamics should at least touch on this important development. To start from first principles, however, would require a background in general relativity and quantum field theory, topics outside the scope of this book. In this chapter we present the basics of black hole thermodynamics.

13.1 BLACK HOLES AND THERMODYNAMICS

A black hole is a region of spacetime having a gravitational field sufficiently strong that nothing can escape from it.[1] The proper description of such fields requires Einstein's general theory of relativity, the theory of gravitation that superseded the Newtonian theory. We can use the Newtonian theory, however, to quickly obtain the size of a black hole. To escape from the surface of a spherical mass M of radius R requires a launch speed $v_{\mathrm{esc}} \equiv \sqrt{2GM/R}$. By setting $v_{\mathrm{esc}} = c$, one has, within the confines of classical physics, the size of an object $R = 2GM/c^2$ from which not even light can escape. Remarkably, the same result emerges from general relativity. The solution to Einstein's field equation[2] exterior to a spherical mass M is the *Schwarzschild spacetime separation*[3]

$$(\mathrm{d}s)^2 = -\left(1 - \frac{r_S}{r}\right)(c\mathrm{d}t)^2 + \left(1 - \frac{r_S}{r}\right)^{-1}(\mathrm{d}r)^2 + r^2\left[(\mathrm{d}\theta)^2 + \sin^2\theta(\mathrm{d}\phi)^2\right] , \qquad (13.1)$$

where t is the time, r, θ, ϕ are spherical coordinates, and the Schwarzschild radius $r_S \equiv 2GM/c^2$.

---

[1] A basic idea of the theory of relativity is that space and time do not exist independently of each other, but rather are aspects of a single entity, *spacetime*. Spacetime as a fixed backdrop for the events of physics is the province of the special theory of relativity. The general theory of relativity describes how spacetime is affected by the presence of matter-energy-momentum in that spacetime, and how spacetime affects the motion of particles.

[2] The Einstein field equation $G_{\mu\nu} = \left(8\pi G/c^4\right)T_{\mu\nu}$ is a relation between second-rank tensor fields in four-dimensional spacetime, where $\mu, \nu = 0, 1, 2, 3$, with the value 0 reserved for the time coordinate. $G_{\mu\nu}$ is the spacetime curvature tensor and $T_{\mu\nu}$ is the energy-momentum tensor that describes the density and flux of energy-momentum in spacetime. The field equation says that the local curvature of spacetime is determined by the local density of energy-momentum in that spacetime. Exterior to a well-localized mass, such as a star, the density of matter is presumed sufficiently small that we can approximate $T_{\mu\nu} = 0$, leaving us with the *vacuum equation* $G_{\mu\nu} = 0$. The curvature tensor is a complicated expression involving derivatives of the spacetime metric tensor field, $g_{\mu\nu}$, where $(\mathrm{d}s)^2 = \sum_{\mu=0}^{3}\sum_{\nu=0}^{3} g_{\mu\nu}\mathrm{d}x^\mu\mathrm{d}x^\nu$, with $x^0 \equiv ct$. Equation (13.1) is one of the few exact solutions of the Einstein vacuum equation.

[3] Equation (13.1) specifies the spacetime *separation* $\mathrm{d}s$, the "distance" between neighboring points in spacetime. Semantically, what would you call a *distance* between points in spacetime, when it's not necessarily spatial, nor is it necessarily a time interval, it's both—the word separation is apt. Note from Eq. (13.1) that $(\mathrm{d}s)^2$ can be either positive, negative, or zero. The geometry of spacetime is not Euclidean.

Example. The Schwarzschild radius associated with the mass of the earth is $r_S \approx 1$ cm, while that associated with the sun's mass is ≈ 3 km. A black hole has its mass entirely contained within r_S.

We see in Eq. (13.1) that $(\mathrm{d}s)^2$ is singular at $r = r_S$ (and at $r = 0$), and thus we expect something significant is associated with the Schwarzschild radius. The singularity at $r = r_S$ is a *coordinate singularity*: The coordinate system (t, r, θ, ϕ) *breaks down* at $r = r_S$, i.e., it fails to provide a unique coordinate for every point in spacetime. Other coordinate systems have been discovered (e.g, Kruskal-Szekeres coordinates) such that the expression for $(\mathrm{d}s)^2$ is free of a singularity at $r = r_S$. Even in these coordinate systems, however, the Schwarzschild radius has a physical significance. For $r < r_S$, the spacetime trajectories (worldlines) of particles and photons all connect with $r = 0$, which is a true singularity of the geometry; a particle interior to r_S can *only* fall into $r = 0$. For that reason, the surface $r = r_S$ is called the *event horizon*—signals (*information*) emitted in the interior of r_S cannot reach points exterior to r_S. *Black holes can be characterized as objects possessing an event horizon*. Particles and radiation can cross the event horizon from the outside, but they can't get back out.[4] At $r = 0$, general relativity breaks down; it's hoped that a quantum theory of gravity would ameliorate that conclusion.

Are black holes amenable to a thermodynamic description?

Is it obvious that black holes should be subject to thermodynamic descriptions? Thermodynamics pertains to macroscopic systems in equilibrium. Thus, are black holes macroscopic objects, and are they in equilibrium, knowing that equilibrium implies equilibrium with an environment?

Are black holes macroscopic objects?

The Schwarzschild solution pertains to a *non-rotating* mass M. Other solutions of the vacuum equation are known for objects having angular momentum J and charge Q. After particles fall into a black hole, nothing is known about them; their identity is lost. Regardless of how a black hole forms, such as in the collapse of a star, any details of the collapsing structure are lost; the only quantities describing black holes[5] are M, J, and Q, a result summarized by saying that black holes "have no hair"—no details, all we can know about a black hole are (M, J, Q). Black holes are thus the *ultimate* macroscopic object! It begs the question, however, from the perspective of thermodynamics of what are the *microscopic* degrees of freedom of a black hole that are unaccounted for in the macroscopic description provided by M, J, and Q?

Are black holes in equilibrium?

Are black holes in states of thermodynamic equilibrium?[6] That begs the question of how a black hole interacts with its environment. The classic picture is there is "nothing" exterior to black holes— anything exterior to the horizon falls in, never to return. Black holes thus resemble *black bodies* which absorb radiation without reflection. Black bodies, however, *emit* photons with a spectral distribution characteristic of its temperature[7] (Planck function). Could the same be true of black holes? In 1974, Stephen Hawking announced the remarkable discovery that black holes emit particles,[8] and

---

[4]The event horizon thus "sorts" particles, and in that sense resembles a Maxwell demon.

[5]The *size* of a black hole isn't included in this list, which is determined by M, J, and Q.

[6]Is a star in equilibrium? Chapter 14 presents non-equilibrium thermodynamics, an extension of thermodynamics to systems not globally in equilibrium, but, like a star, are such that each small volume of the system can be treated as effectively homogeneous.

[7]The temperature of a black body is a *given*; we didn't ask in Chapter 5 where its temperature comes from.

[8]Particles don't flow black out through the event horizon; rather, at the horizon virtual particle pairs are disrupted, giving the appearance to an outside observer that particles flow away from the black hole.

moreover that the spectral distribution of emitted bosons is *thermal*—that of the Planck function—characterized by a well defined temperature.[90] *Black holes are thus black bodies.* Black body radiation is a quantum phenomenon (Section 5.2); black hole radiation is quantum in origin as well.

Are black holes demons?

Consider lowering a box of atoms into a black hole, past the event horizon. If, per the no-hair theorem nothing is known to an exterior observer about those atoms, other than they contribute to M, J, and Q, what becomes of their *entropy*? Unless black holes can be said to have entropy, the entropy of the universe visible to an observer would decrease. Is a black hole a Maxwell demon, with the event horizon being the trap door?[9] We'll see that black holes do indeed have entropy.

13.2 HAWKING RADIATION

Hawking showed that the *environment* of a black hole (in the sense we use the word in thermodynamics) is the *quantum vacuum* in the vicinity of the event horizon. The "vacuum" of quantum field theory is anything but quiescent. Virtual particle-antiparticle pairs are incessantly being created, only to annihilate. By the energy-time uncertainty relation of quantum mechanics, $\Delta E \Delta t \gtrsim \hbar$, energy $\Delta E = 2mc^2$ can be "borrowed" to create virtual particles, if in time $\Delta t \sim \hbar/\Delta E$ they annihilate, returning the energy. This picture of *vacuum fluctuations* is supported experimentally by the *Lamb shift*—energy differences between the states of the hydrogen atom that are not accounted for by the Dirac equation.[10] The vacuum, nominally conceived of as *nothingness*, is crackling with energy. How to incorporate quantum effects within general relativity is an active field of research.

Example. An electron-positron pair created in a vacuum fluctuation can exist for a time

$$\Delta t \sim \frac{\hbar}{2m_e c^2} = 6 \times 10^{-22} \text{ s},$$

where m_e is the electron mass. In that time, it can travel a distance $\approx 2 \times 10^{-13}$ m, less than the Compton wavelength of an electron.

What Hawking showed is that, of the virtual particle pairs created in the vicinity of the event horizon, one of them can cross the horizon, leaving the other to propagate away from the black hole, what's termed *Hawking radiation*. He found that the spectral distribution of the radiated energy (from a non-rotating mass M) is characterized by the temperature

$$T_H \equiv \frac{\hbar c^3}{8\pi G M k_B} = 6 \times 10^{-8} \left(\frac{M_\odot}{M}\right) \text{ K},\tag{13.2}$$

where $M_\odot$ denotes the solar mass. If a star can be considered an energy conversion device, converting gravitational energy into light at the expense of the nuclear binding energy in the star, a black hole converts vacuum fluctuations into radiation, at the expense of the mass energy of the black hole. Antiparticles that cross the event horizon (which are in negative energy states) *decrease* the energy (and hence the mass) of the black hole—black holes evaporate (see below), just as stars eventually consume themselves. Note that we can let $\hbar \to 0$ in Eq. (13.2); classically, black holes have zero temperature. The Hawking temperature is too small to have astrophysical relevance; it's

---

[9]The early history of black hole thermodynamics is recounted in Bekenstein,[91] where he describes throwing entropy into black holes as "Wheeler's demon."

[10]The Dirac equation governs the relativistic quantum mechanics of spin-1/2 particles; it accounts for the spin of the electron and the existence of antimatter. Vacuum fluctuations are explained by quantum electrodynamics.

mere existence however is regarded as the first step in combining general relativity with quantum effects.

There is no simple way to account for Hawking's result, which requires dexterity in the methods of quantum field theory (in curved space no less). However, now that we *know* black holes radiate thermally, we can estimate the temperature. The energy of photons emitted from black bodies are proportional to[11] $k_B T_H$. Then,

$$E_{\text{photon}} \propto k_B T_H = \frac{hc}{\lambda} \sim \frac{hc}{r_S} \implies T_H \sim \frac{hc^3}{GMk_B} \,, \tag{13.3}$$

where we've taken $\lambda = r_S$ because the Schwarzschild radius is the only length in the problem. Equation (13.3) cannot be expected to agree numerically with Eq. (13.2) (which emerged from a detailed calculation) because of the assumptions we've made, but it gets rights the dependence on the relevant variables of the problem.

Black hole evaporation

If black holes radiate like black bodies, and because $E = Mc^2$, we can derive the lifetime of black holes subject to radiative losses. Using Eq. (5.9),

$$\frac{d(Mc^2)}{dt} = -\sigma \left(4\pi r_S^2\right) T_H^4 = -\frac{\hbar c^6}{15360\pi G^2} \frac{1}{M^2} \,. \tag{13.4}$$

Equation (13.4) can be integrated to determine the lifetime τ of a black hole:

$$\tau = 5120\pi \frac{G}{\hbar c^4} M_0^3 \,, \tag{13.5}$$

where M_0 is the initial mass. If the mass of the black hole is larger than roughly that of the moon, however, it *can't* evaporate because it receives more energy from the cosmic microwave background than it loses through Hawking radiation (see Exercise 13.4).

13.3 ENTROPY AND MISSING INFORMATION

Absolute temperature and entropy go together—fraternal twins from the same mother: Carnot's theorem.[12] Now that we have the temperature T_H of a black hole, it's natural to ask about its entropy. We could *estimate* the entropy of a black hole if we knew the number of "things" inside it (see Exercise 7.16), yet that's precisely what we don't know (no-hair theorem). Let's estimate the number of "energy units" by dividing the mass energy Mc^2 by the approximate energy per degree of freedom[13] $k_B T_H$ (whatever those degrees of freedom might be):

$$N \equiv \frac{Mc^2}{k_B T_H} \implies N = \frac{8\pi GM^2}{\hbar c} = 2 \times 10^{77} \left(\frac{M}{M_\odot}\right)^2 \,. \tag{13.6}$$

Accepting this number, a black hole of solar mass would have an entropy $S_{\text{bh}} \sim 10^{77} k_B$, *far larger* than what one would estimate for the sun, $S_\odot \sim 10^{57} k_B$ (Exercise 7.16). The far-larger entropy of a black hole compared to an ordinary star (of the same mass) is in accord with the no-hair theorem: We know nothing about a non-rotating black hole other than its mass. There is far more *missing*

---

[11]See Exercise 5.7. The energy of the photon associated with maximum emissivity is $E = 4.97k_B T$; the average photon energy (found by dividing the total energy aVT^4 by the average number of photons, Eq. (5.15)) is $E = 2.7k_B T$.

[12]Absolute temperature is a consequence of Carnot's theorem—Section 2.4—and entropy emerges from the Clausius inequality—Section 3.1—itself a consequence of Carnot's theorem.

[13]We're implicitly invoking the equipartition theorem, what's not been shown in this book.

information to a black hole, compared to a star, information that's lost in the formation of a black hole, which by Landauer's principle, is associated with an increase in entropy (see Chapter 12).

Because N in Eq. (13.6) scales with M^2, it scales with r_S^2, and thus N scales with the surface area of the event horizon. For $A \equiv 4\pi r_S^2$, $N = c^3 A/(2G\hbar)$. One would then have that the entropy of a black hole scales with area,

$$S \sim \frac{c^3 k_B}{G\hbar} A \, . \tag{13.7}$$

Equation (13.7) is just a guess, however. It can be made precise using a result established in the theory of black holes:[92]

$$\mathrm{d}\left(Mc^2\right) = \frac{\kappa c^2}{8\pi G}\mathrm{d}A + \Omega \mathrm{d}J + \Phi \mathrm{d}Q \, , \tag{13.8}$$

where $\kappa \equiv GM/r_S^2$ is the surface gravity (see Exercise 13.1), Ω is the angular frequency of the event horizon, and Φ is the electrostatic potential. Equation (13.8) is a statement of energy balance—*the first law of thermodynamics applied to black holes*. One can increase the energy of the black hole, i.e., its mass, by increasing the surface area, by increasing the angular momentum, or by increasing its charge. As particles fall into the black hole, its mass increases, increasing its area, and *lowering* its temperature. The heat capacity of a black hole is therefore *negative*—the hallmark of an unstable system (see Exercise 13.5).

From the form of Eq. (13.8), $\Omega \mathrm{d}J$ and $\Phi \mathrm{d}Q$ represent work terms—deformations of the extensible quantities associated with the system (see Chapter 1) implied by the no-hair theorem (M, J, Q). That suggests the area term in Eq. (13.8) is the analog of "heat" ($\mathrm{d}(\text{heat}) = \mathrm{d}U - \mathrm{d}W$), what we see by rewriting Eq. (13.8) :

$$\frac{\kappa c^2}{8\pi G}\mathrm{d}A = \mathrm{d}\left(Mc^2\right) - \Omega \mathrm{d}J - \Phi \mathrm{d}Q \, . \tag{13.9}$$

Equation (13.9) suggests therefore the correspondence

$$\frac{\kappa c^2}{8\pi G}\mathrm{d}A \longleftrightarrow T_H \mathrm{d}S = \frac{\hbar \kappa}{2\pi k_B c}\mathrm{d}S \, ,$$

and thus $\mathrm{d}S = \left[k_B c^3/(4\hbar G)\right]\mathrm{d}A$, which can be integrated

$$S = \frac{k_B c^3}{4\hbar G} A \, . \tag{13.10}$$

If we did not know of the existence of T_H, we would not have been able to arrive at Eq. (13.10). Equation (13.10) differs from Eq. (13.7) by a factor of four, *lending credence to the approach taken in motivating* Eq. (13.6).

Note that S scales with A, and not with volume as we would expect. In fact, S scales with the *square* of the total mass M, and not simply the mass. Equation (13.10) is equivalent to

$$\frac{S}{k_B} = \frac{4\pi G}{\hbar c}M^2 = 1.05 \times 10^{77} \left(\frac{M}{M_\odot}\right)^2 \, . \tag{13.11}$$

Black holes therefore behave differently in their scaling properties from ordinary thermodynamic systems. Entropy, in standard formulations of thermodynamics, is an extensive quantity (see Eq. (3.28)), where S scales with the size of the system. The "size" of a black hole is thus its area, not its volume, as one might expect if the black hole *is* the event horizon. That observation gives a big hint of where the microscopic degrees of freedom of a black hole are located. Our interpretation of heat is energy transferred to microscopic degrees of freedom, and we've associated the increase of area in Eq. (13.8) with heat transfers. Are the degrees of freedom associated with a black hole located on its surface? That possibility has led researchers to propose the universe is *holographic*, that *information seemingly lost in black holes is encoded in their surfaces*, just as a three-dimensional image can be constructed from information in a two-dimensional piece of film (the hologram).[93]

Statistical mechanics of black holes?

That there is such a good "fit" between black hole physics and thermodynamics is perhaps surprising, yet it shows—through the existence of the Hawking temperature—that quantum mechanics is required to understand black holes. As we've noted elsewhere in this book, the consistency of thermodynamics relies on the underlying quantum nature of the physical world. And surely black holes have more to teach us about quantum mechanics. A step in that direction would be to take the connection with entropy to the next level: Given that entropy is the bridge between microscopic and macroscopic, what are the degrees of freedom of a black hole that the entropy is a measure of?[14] That is, if $S = k_B \ln W$, what degrees of freedom can we "count" as comprising W? In Chapter 7 we found that Planck's constant allowed the states of a gas to be *countable*. Is there a *new* "Planck's constant" waiting to be discovered? Perhaps $4\hbar G/c^3 \sim 10^{-69}$ m^2 is a fundamental unit of area? We learned in Chapter 12 that maximum information is proportional to the logarithm of the number N of possibilities available to a system. Consistency would require that $N \sim \exp(c^3 A/4\hbar G)$.

13.4 LAWS OF BLACK HOLE THERMODYNAMICS

Area theorem

If entropy as given by Eq. (13.10) is to be interpreted as the physical entropy of a black hole, it should behave as we expect from the second law of thermodynamics. In 1971, Hawking proved an important result, known as the *area theorem*, which states that the area A of a classical black hole can never decrease in any process:[95]

$$\Delta A \geq 0 . \tag{13.12}$$

As noted by Jacob Bekenstein,[96] the area theorem of black holes is closely analogous to the behavior of entropy, that both tend to increase irreversibly:

$$\Delta S \geq 0 . \tag{13.13}$$

The connection suggested by the similarity between Eqs. (13.13) and (13.12) cannot be substantiated unless we know of the Hawking temperature. Black hole evaporation is not in conflict with the area theorem—a proviso of the theorem is that it holds for positive energies; antimatter—the driver of Hawking radiation—is in a negative energy state.

Laws of black holes

Two of the four laws of thermodynamics are reflected in the laws of black hole physics, Eqs. (13.8) and (13.12). What about the others? The zeroth law of black hole thermodynamics (proven in [92]) is that *surface gravity is a constant at all points of the event horizon* ($T_H \propto \kappa$, Exercise 13.1). That's in accord with the zeroth law of thermodynamics (Section 1.5) that what systems in mutual equilibrium have in common is temperature. For a black hole all points of the event horizon have the same temperature (surface gravity). Can we interpret (based on the zeroth law of thermodynamics) that the surface of a black hole is a network of "systems" each occupying a certain area of the surface? The third law of black hole thermodynamics is that it is impossible by any process to reduce κ to zero.[15] These relations are summarized in Table 13.1.

---

[14]Considerable effort has been devoted to that question. See Susskind for a popular discussion.[94]

[15]The third law of black hole thermodynamics has a different status from the other three; see Wald.[97] The third law of thermodynamics similarly has a different status from the others—see Chapter 8.

Table 13.1 Laws of black holes and thermodynamics

Law	Black holes	Thermodynamics
Zeroth	Surface gravity κ is constant and Hawking temperature $T_H \propto \kappa$	Temperature characterizes systems in equilibrium
First	$d(Mc^2) = \dfrac{\pi c^2}{8\pi G} dA +$ "work terms"	$dU = TdS +$ "work terms"
Second	$dA \geq 0$ and $S \propto A$	$dS \geq 0$ for an isolated system
Third	$\kappa = 0$ cannot be reached	$T = 0$ cannot be reached

13.5 IS GRAVITY THERMODYNAMICS?

Black holes are a prediction of general relativity, a discipline seemingly far removed from thermodynamics. In 1995, Ted Jacobson asked:[98] "How did classical general relativity know that the horizon area would turn out to be a form of entropy, and that surface gravity is a temperature?" His answer is surprising. He proved that *Einstein's field equation can be derived from thermodynamics*, taking the primacy of $dQ = TdS$ and the proportionality between entropy and the horizon area as fundamental.[98] The field equation can then be seen as an *equation of state for spacetime*. Space does not permit us to explore this intriguing idea, which requires a deep knowledge of general relativity. Gravity has long been recognized as unlike the other forces of nature—*gravity is a property of spacetime*. Whereas the fundamental fields of physics are defined *on* spacetime, gravity *is* spacetime. Jacobson showed that gravity as a manifestation of curved spacetime is *required by thermodynamics*. In his work, dQ is the flux of an energy quantity across a horizon.[16] He assumed a proportionality between entropy and horizon area not because it holds for black holes according to Bekenstein and Hawking, but because the *entanglement* of quantum fluctuations across the horizon should lead to an entropy proportional to area. Such an entropy is dominated by the shortest wavelength modes of fluctuations. Without a way to cut off short-wavelength degrees of freedom, the entropy would be infinite. The assumption that microphysics regulates this divergence and makes entropy finite is what requires gravity. *Gravity is a manifestation of the thermodynamics of the vacuum.*

Such considerations prompt the question: *Is spacetime itself a manifestation of thermodynamics?* In 2011, Erik Verlinde put forward the idea that spacetime is an *emergent phenomenon*, which arises from entropic tendencies of truly microscopic degrees of freedom, such as exhibited in string theories.[99] Verlinde's proposal is speculative, but potentially has far-reaching consequences. Whether Verlinde's idea linking gravity to an entropic force acting on microscopic degrees of freedom will pan out ultimately is a matter of experiment to decide.

CHAPTER SUMMARY

This chapter provided a brief overview of the relationship between black holes and thermodynamics discovered in the 1970s, and which has continued as an active field of research ever since.

- Black holes are macroscopic objects described by mass, angular momentum, and charge. The connection with thermodynamics is made possible by: 1) the area theorem, Eq. (13.12), that the area of a black hole cannot decrease in any process, suggesting an analogy with entropy; and 2) the existence of the Hawking temperature T_H, Eq. (13.2), that black holes radiate like

---

[16]The boost Killing energy.

a black body at temperature T_H. The entropy of a black hole is simply related to its surface area, Eq. (13.10); S is proportional to A.

- Black hole entropies are far larger than the entropy of stars of the same mass. Black holes exemplify the *missing information* paradigm developed in Chapter 12; what we know about structures that form black holes is lost in the process, and loss of information is associated with increases in entropy.

EXERCISES

13.1 a. Show that the Hawking temperature can be written

$$T_H = \frac{\hbar c}{4\pi k_B r_S} .$$

b. A useful concept is the *surface gravity*, denoted κ, the acceleration at the event horizon,

$$\kappa \equiv \frac{GM}{r_S^2} = \frac{c^4}{4GM} = \frac{c^2}{2r_S} .$$

Show that the Hawking temperature is proportional to the surface gravity

$$T_H = \frac{\hbar}{2\pi k_B c}\kappa .$$

Thus, large black holes emit *less* radiation than small black holes.

13.2 a. Fill in the details leading to Eq. (13.4), and then derive Eq. (13.5). Show that the lifetime can be written

$$\tau = 2 \times 10^{67} \left(\frac{M_0}{M_\odot}\right)^3 \text{ yr} . \tag{P13.1}$$

The Hawking temperature is so small: Loss of mass-energy by black body radiation can take a long time. The lifetime of a solar-mass black hole would exceed the age of the universe.

b. What is the initial mass of a black hole, assumed to have been formed at the time of the big bang (14 billion years ago), that would just be finished with evaporation today? The mass of a mountain here on Earth, perhaps?

13.3 Put in the numbers in Eq. (13.6) to verify the relation shown.

13.4 a. Show that

$$r_S = 2.95 \left(\frac{M}{M_\odot}\right) \text{ km} .$$

b. What is the size (mass and radius) of a black hole with Hawking temperature equal to the temperature of the cosmic microwave background, 2.7 K? Compare the mass you calculate with the mass of the moon. Black holes larger than this size receive more energy from the microwave background than they lose through Hawking radiation.

13.5 a. Show from Eq. (13.9) that the entropy of the black hole can be written

$$S = \frac{\hbar c^5}{16\pi G k_B}\frac{1}{T_H^2} .$$

b. Show that $S \propto T^{-2}$ implies a negative heat capacity. Black holes are not in stable equilibrium.

13.6 Show that the wavelength of a photon associated with energy $k_B T_H$ is $\lambda = 8\pi^2 r_S$.

CHAPTER 14

Non-equilibrium thermodynamics

N ON-EQUILIBRIUM thermodynamics is an extension of thermodynamics whereby *state variables* become *field quantities*, functions of spatial coordinates and the time. In classical thermodynamics, equilibrium is independent of time and space, with variables like temperature having the same value throughout a system.[1] Homogeneous, time-invariant states are possible only when the system is sufficiently isolated from the environment. We now relax these assumptions and allow T and other variables to vary smoothly in space and time, $T(r, t)$. Non-equilibrium thermodynamics describes systems not strictly in equilibrium, but *slightly* out of equilibrium such that temperature and other variables are locally well defined, yet vary spatially and temporally. Such variations occur in systems that permit *flows* of heat, electricity, or particles across its boundaries. If flows occur at a steady rate, the system at any point will reach a steady condition—a time-invariant, *inhomogeneous* system. Non-equilibrium thermodynamics is more comprehensive than thermodynamics; equilibrium is a limiting case of a stationary state where fluxes from the environment approach zero.[2] This chapter is a brief introduction to non-equilibrium thermodynamics.

To set up the theory, its basic equations should be *local*, i.e., refer only to the values of quantities at a single point in space and time, such as we find in Maxwell's equations or the equations of hydrodynamics.[3] That implies establishing partial differential equations for the space and time variations of its variables. We'll be guided by the use of *balance equations*, integral equations which account for the change, in time, in the amount of a quantity in a region of space. The equations of hydrodynamics follow from balance equations for mass, energy, and momentum, as we'll show. We'll establish a balance equation for entropy, where changes in entropy are controlled by the flow of mass, energy and momentum, *and* by the *generation* of entropy by irreversible processes. One of the goals of non-equilibrium thermodynamics is to identify the role of inhomogeneities in contributing to entropy creation.

---

[1] From the zeroth law, the temperature of an equilibrium system must be spatially uniform; we can conceive a system to be composed of subsystems such as depicted in Fig. 3.5, which all must have the same temperature.

[2] Contrast this characterization of equilibrium with the statement in Chapter 1 that equilibrium is the state where "nothing is happening" macroscopically.

[3] One could have *non-local* equations, where the local response is due to forces imposed at distant locations, such as in time-domain electrodynamics where the most general, linear response of the current density $J(r, t)$ to the electric field $E(r', t')$ is given by $J(r, t) = \int \mathrm{d}^3 r' \int \mathrm{d}t' \sigma(r, r', t, t') E(r', t')$. One could also envision non-linear extensions of thermodynamics, such as in the field of non-linear optics.

14.1 NON-EQUILIBRIUM PROCESSES

Basic flow processes are described by phenomenological equations expressing a proportionality between *fluxes* and *gradients*. Familiar examples are *Fourier's law*, a proportionality between heat flux J_Q (units of W m^{-2}) and a temperature gradient, $J_Q = -\kappa \nabla T$, where κ is the thermal conductivity (W K^{-1} m^{-1}); *Fick's law*, between a particle flux J_n (m^{-2} s^{-1}) and a concentration gradient, $J_n = -D\nabla n$, where D is the diffusion coefficient (m^2 s^{-1}), and n is the particle density, number of particles per volume; *Ohm's law*, between the electric current density J_q (A m^{-2}) and the gradient of the electrostatic potential, ϕ, $J_q = -\sigma\nabla\phi$, where σ is the electrical conductivity (S m^{-1}); and *Newton's law of viscosity*, that the shear stress T_{xy} (Pa) between adjacent layers in a fluid is proportional to a gradient of the velocity field, $T_{xy} = -\nu \partial v_x / \partial y$, where ν is the viscosity (Pa s). These relations are summarized in Table 14.1.

Table 14.1 Gradient-driven fluxes

Flux	Gradient	Phenomonelogical relation	Name
Heat, J_Q	∇T	$J_Q = -\kappa\nabla T$	Fourier's law
Particle, J_n	∇n	$J_n = -D\nabla n$	Fick's law
Charge, J_q	$\nabla\phi$	$J_q = -\sigma\nabla\phi$	Ohm's law
(Momentum)$_x$ through y	$\dfrac{\partial p_x}{\partial y}$	$T_{xy} = -\nu\dfrac{\partial v_x}{\partial y}$	Newton's law

More complex phenomena arise when two or more of these basic effects are present simultaneously. In *thermoelectricity*, electrical conduction and heat conduction occur together. When a temperature gradient is set up, not only does heat flow but an electric field is created. The strength of the field is proportional to the temperature gradient, $E = P\nabla T$, where P (V K^{-1}) is called the *thermoelectric power* or simply the *thermopower* of the material.[4] Charges flow from hotter to colder regions of a material, i.e., *against* ∇T. Because the system is electrically neutral, a region depleted of carriers sets up an electric field to oppose the motion of charge. For negative charge carriers, E is directed against ∇T, and $P < 0$. The voltage difference ΔV between two points held at a temperature difference ΔT is $\Delta V = \int_{T_1}^{T_2} P(T)\mathrm{d}T$. This relation is called the *Seebeck effect*, and is written $\Delta V = P\Delta T$ for small temperature differences. In this form, $P = \Delta V/\Delta T$ is called the *Seebeck coefficient*. The thermopower of a single material is difficult to measure. However, the *difference* in thermopower of two materials is easy to measure. The voltage difference generated by *junctions* of materials (A, B), held at a temperature difference ΔT, is written $\Delta V = P_{AB}\Delta T$. Charges carry energy as well as current in response to temperature differences. Charges arriving from a region of higher temperature carry with them "excess" energy relative to the environment at the colder location. The rate of energy deposition, $\dot{Q} = \Pi I$, is proportional to the current I for small currents (the *Peltier effect*), where Π is the *Peltier coefficient* (V). The Peltier effect is said to be *conjugate* to the Seebeck effect in a way we'll describe shortly. For a junction, $\dot{Q} = \Pi_{AB}I$. The two effects (Seebeck and Peltier) describe the same process: In the Seebeck effect, charges in motion (because of ΔT) create a voltage difference ΔV; in the Peltier effect, charges in motion (because of ΔV) create heat generation. From the Peltier equation, $\Delta Q = \Pi_{AB}I\Delta t = \Pi_{AB}\Delta q$, where Δq is the charge delivered by the current in time Δt. But from the heat capacity of the junction, $\Delta Q = C\Delta T$. The temperature change is then given by $\Delta T = \Pi_{AB}\Delta q/C$. From the Seebeck

---

[4]The thermopower and the Seebeck coefficient are often denoted with the symbol S. To avoid confusion with entropy, we've used the symbol P.

effect, $\Delta T = \Delta V / P_{AB}$. Equating the two expressions for ΔT,

$$\frac{\Pi_{AB}\Delta q}{C} = \frac{\Delta V}{P_{AB}} .$$

Now, Δq and ΔV are related by the *capacitance* of the junction, $\Delta q = C_{AB}\Delta V$. The Peltier and Seebeck coefficients are thus not independent and are related by

$$\Pi_{AB}P_{AB} = \frac{C}{C_{AB}} . \tag{14.1}$$

It is in this sense that the Peltier and Seebeck effects are referred to as conjugate; they're different descriptions of the same physics. Another example of a conjugate pair of non-equilibrium effects is the coupling of diffusion and heat conduction. In the *Soret effect* a concentration gradient is established as a result of a temperature gradient. In the conjugate effect, known as the *Dufour effect*, a temperature difference arises from a concentration gradient.

14.2 ONSAGER THEORY

Gradients in temperature, concentration, electric potential, etc., are generically referred to as *thermodynamic forces* and are traditionally denoted[5] X_i, $i = 1, \cdots, n$. Such "forces" have nothing to do with Newtonian forces; in this context *forces* are *the phenomenological causes behind non-equilibrium effects*. Relations between fluxes and forces (such as in Table 14.1) are phenomenological in the sense of being verified by experiment, but are not part of a comprehensive theory of irreversible processes.

Lars Onsager developed, in 1931,[100][101] a method for treating non-equilibrium processes based on the rate of *entropy production*.[6] Onsager assumed generally that any flux can result from any force, and he took as a starting point the assumption that

$$\boldsymbol{J}_i = \sum_{k=1}^{n} L_{ik}\boldsymbol{X}_k , \tag{14.2}$$

where the quantities L_{ik} are phenomenological coefficients. The diagonal terms L_{ii} are related to the basic transport coefficients (thermal conductivity, diffusion coefficient, etc.). The off-diagonal terms L_{ik}, $k \neq i$, are connected to the "interference" phenomena discussed above between, for example, heat conduction and electrical conduction. Onsager proved, provided a proper choice is made for the forces and fluxes, that the "cross terms" have a fundamental symmetry

$$L_{ik} = L_{ki} . \tag{14.3}$$

These identities are known as the *reciprocal relations*. We explain below what's meant by a "proper choice" of forces and fluxes. A *proof* of Eq. (14.3) requires the methods of statistical mechanics. At the level of macroscopic theory, the reciprocal relations can be seen as *another law of thermodynamics*, whose justification is the same as the second law of thermodynamics: through experience. The consequences of the reciprocal relations have been experimentally verified.[102] What the reciprocal relations indicate is a kind of Newton's third law, that if flow $\boldsymbol{J}_i$ is influenced by force $\boldsymbol{X}_k$, then flow $\boldsymbol{J}_k$ is influenced by force $\boldsymbol{X}_i$ with the same coefficients $L_{ik} = L_{ki}$.

Let state variables be denoted[7] x_i, so that $S = S(x_1, \cdots, x_n)$, with the equilibrium values denoted x_i^0. Define the *fluctuation* in each variable[8] by $\Delta x_i \equiv x_i - x_i^0$. Because entropy is a

---

[5]In Section 1.9, X_i was used to denote extensive quantities, with $\mathrm{d}X_i$ called generalized displacements and intensive quantities Y_i generalized forces, such that $Y_i\mathrm{d}X_i$ has the dimension of energy. You can handle the change in notation.

[6]Onsager received the 1968 Nobel Prize in Chemistry for this work.

[7]Change of notation again—you can handle it. Denote thermodynamic variables as x_i, fluxes as X_i. S is entropy here.

[8]The notational scheme implies that the theory applies to systems slightly out of equilibrium; we speak of the non-equilibrium values of variables which strictly are defined only in equilibrium.

maximum in equilibrium, fluctuations in S are such that (see Section 3.10)

$$\Delta S = -\frac{1}{2} \sum_{jk} g_{jk} \Delta x_j \Delta x_k , \tag{14.4}$$

where $\Delta S \equiv S - S^0$, and the "metric" g_{jk} is symmetric and positive definite,

$$g_{jk} \equiv \left(\frac{\partial^2 S}{\partial x_j \partial x_k} \right)^0 .$$

In Onsager's theory, fluxes are provisionally defined as the time rate of change of fluctuations,

$$J_i \equiv \frac{\mathrm{d}}{\mathrm{d}t} (\Delta x_i) , \tag{14.5}$$

and we'll explain what's meant by *provisional*, and forces are defined provisionally as

$$X_i \equiv \frac{\partial (\Delta S)}{\partial (\Delta x_i)} = -\sum_j g_{ij} \Delta x_j . \tag{14.6}$$

Together, these definitions *imply* a time rate of change[9] of S,

$$\sum_k J_k X_k \Rightarrow \frac{\mathrm{d}S}{\mathrm{d}t} \equiv \int_V \sigma \mathrm{d}V , \tag{14.7}$$

where σ is a *local* rate of entropy production per volume. There's a bit of *legerdemain* here.[10] Fluxes as defined by Eq. (14.5) are not in the form one expects—*local* rates of flow per area (the kind of quantities listed in Table 14.1), and forces as defined by Eq. (14.6) are not in the form of spatial gradients. Instead, Eqs. (14.5) and (14.6) refer to thermodynamic fluctuations, which in this theory are independent of position. The "bridge" between global thermodynamic quantities and a local property of the system occurs on the right side of Eq. (14.7) with the introduction of the quantity σ, the local rate of entropy production per volume. If one considers a *vector* of flux quantities $\boldsymbol{J} = (J_1, J_2, \cdots)$ and a vector of forces $\boldsymbol{X} = (X_1, X_2, \cdots)$, the left side of Eq. (14.7) is in the form of an inner product, $\boldsymbol{J} \cdot \boldsymbol{X}$, and inner products are invariant under suitable transformations. That permits latitude in defining forces and fluxes as long as $\boldsymbol{J} \cdot \boldsymbol{X}$ is maintained.[11] For fluxes conventionally defined as the rate at which a quantity ψ flows per surface area, and forces as related to spatial gradients, we have through dimensional analysis (square brackets denote dimension, T denotes time and L length)

$$[J_\psi \times X_\psi] = \left[\frac{\psi}{TL^2} \times \nabla \left(\frac{\partial S}{\partial \psi} \right) \right] = \left[\frac{S}{TL^3} \right] ,$$

the same dimension as σ. For Onsager's theory to work, an expression for σ must be found that occurs in the form $\sigma = \sum_i \boldsymbol{J}_i \cdot \boldsymbol{X}_i$ where the product $\boldsymbol{J}_i \cdot \boldsymbol{X}_i$ is a *density* (per volume quantity). Finding σ *requires a detour through fluid mechanics*, as shown in the next section. The reader uninterested in the details should skip to Eq. (14.45).

---

[9]Equation (14.7) does not emerge from an underlying dynamical theory; it's a way of specifying the rate at which entropy is produced by irreversible processes. Equation (14.7) follows by differentiating Eq. (14.4) with respect to time, using Eqs. (14.5) and (14.6) and the symmetry of g_{jk}.

[10]As there must be in going from a global theory (thermodynamics) to a local theory (non-equilibrium thermodynamics).

[11]In analytic mechanics one has the flexibility to choose generalized coordinates q_i and momenta p_i in such a way that $[p_i] \times [q_i] = $ [energy].

14.3   ENTROPY BALANCE EQUATION

The goal of this section is to derive the balance equation for entropy—a lengthy calculation—which provides for the rate of entropy production per volume, σ_S. To start, express thermodynamic variables as *specific*, per mass, quantities. Write $S = sM$, $U = uM$, $V = vM$, where M is the total mass of all the chemical components of the system; s, u, v are the entropy, internal energy, and volume per mass. For multicomponent systems, we need to specify particle numbers on a per-mass basis. The total mass of chemical species k, M_k, has $N_k = M_k/m_k$ particles, where m_k is the *molecular mass* of species k. We can then write $\mu_k \mathrm{d}N_k = (\mu_k/m_k)\,\mathrm{d}M_k \equiv \tilde{\mu}_k \mathrm{d}M_k$, where $\tilde{\mu}_k$ is the chemical potential per molecular mass. The first law of thermodynamics can then be written (divide Eq. (3.16) by M)

$$ \mathrm{d}u = T\mathrm{d}s - P\mathrm{d}v + \sum_{k=1}^{n} \tilde{\mu}_k \mathrm{d}c_k , \tag{14.8} $$

where there are n chemical species, each with a *mass fraction* $c_k = M_k/M$ ($\sum_{k=1}^{n} c_k = 1$). From Eq. (14.8), we have an "equation of motion" for entropy: (divide Eq. (14.8) by $\mathrm{d}t$)

$$ T\frac{\mathrm{d}s}{\mathrm{d}t} \equiv \frac{\mathrm{d}u}{\mathrm{d}t} + P\frac{\mathrm{d}v}{\mathrm{d}t} - \sum_{k=1}^{n} \tilde{\mu}_k \frac{\mathrm{d}c_k}{\mathrm{d}t} . \tag{14.9} $$

Thus, we're *defining* the time variation of entropy to be that effected by the time variations of energy, volume, and mass fractions, connected through the first law of thermodynamics. In the rest of this section, we'll find equations of motion for u, v, and c_k (based on their balance equations), which will then be combined in Eq. (14.9).

Balance equations

Balance equations (for any quantity ψ) have the form

$$ \frac{\mathrm{d}}{\mathrm{d}t} \int_V \rho_\psi \mathrm{d}V = -\oint_S \boldsymbol{J}_\psi \cdot \mathrm{d}\boldsymbol{S} + \int_V \sigma_\psi \mathrm{d}V . \tag{14.10} $$

Here V is a volume bounded by surface S having outward-pointing surface element $\mathrm{d}\boldsymbol{S}$, ρ_ψ is the density of ψ ($[\psi]$ m^{-3}), $\boldsymbol{J}_\psi$ is the flux of ψ through S ($[\psi]$ m^{-2} s^{-1}), and σ_ψ is a *source term* representing the rate per volume at which ψ is created or destroyed in V, ($[\psi]$ m^{-3} s^{-1}). Equation (14.10) specifies that the rate of change of the amount of ψ in V is accounted for by flows through S, and by the net rate of creation of ψ in V by means other than flow; *there are no other possibilities.*[12] Note that balance equations do not presume the system to be in thermal equilibrium. When the transport of ψ is carried by a fluid, $\boldsymbol{J}_\psi = \rho_\psi \boldsymbol{v}$, where $\boldsymbol{v}$ is the fluid velocity.[13] By applying the divergence theorem to Eq. (14.10), we arrive at the *local* form of a balance equation,

$$ \frac{\partial \rho_\psi}{\partial t} + \boldsymbol{\nabla} \cdot \boldsymbol{J}_\psi = \sigma_\psi . \tag{14.11} $$

If $\sigma_\psi = 0$, ψ is *conserved*; in that case the only way ψ in V can change is to flow through S.

---

[12]An example of a source term occurs in semiconductor physics, where additional charge carriers are generated by the interaction of electromagnetic radiation with the semiconductor.

[13]For example, the magnitude of the Poynting vector S, the flux of electromagnetic energy, is given by $S = uc$ with u is the density of electromagnetic energy and c the speed of light (see Section 5.2).

Mass balance

The balance equation for the mass of chemical species k in volume V is

$$\frac{\mathrm{d}}{\mathrm{d}t} \int_V \rho_k \mathrm{d}V = -\oint_S \rho_k \boldsymbol{v}_k \cdot \mathrm{d}\boldsymbol{S} + \sum_{j=1}^r \int_V \nu_{kj} J_j \mathrm{d}V \,, \qquad (14.12)$$

where $\rho_k = M_k/V$ is the mass density of species k, $\boldsymbol{v}_k$ is its velocity, and $\nu_{kj} J_j$ is a source term, the rate at which the mass of species k is produced per volume from the jth chemical reaction. The quantity ν_{kj} (which can be positive or negative) is the *density-weighted* stoichiometric coefficient, not what we introduced in Section 6.4: $\nu_{kj}/\rho_k = \nu_{kj}^0$, the usual stoichiometric coefficient with which k appears in chemical reaction j. The quantity J_j is the *rate* at which reaction j occurs. Because mass is conserved in each reaction,[14]

$$\sum_{k=1}^n \nu_{kj} = 0 \,. \qquad (j = 1, \cdots, r) \qquad (14.13)$$

We thus have, for each species, the local mass balance equation

$$\frac{\partial \rho_k}{\partial t} + \boldsymbol{\nabla} \cdot \rho_k \boldsymbol{v}_k = \sum_{j=1}^r \nu_{kj} J_j \,. \qquad (k = 1, \cdots, n) \qquad (14.14)$$

Summing Eq. (14.14) over k and making use of Eq. (14.13), we have the continuity equation, which expresses *total mass conservation*:

$$\frac{\partial \rho}{\partial t} + \boldsymbol{\nabla} \cdot \rho \boldsymbol{v} = 0 \,, \qquad (14.15)$$

where $\rho \equiv \sum_k \rho_k$ is the total density, and $\boldsymbol{v} \equiv \sum_k \rho_k \boldsymbol{v}_k/\rho$ is the velocity of the center of mass.

The formulas of fluid mechanics often simplify if we write the total time-derivative operator (nominally $\mathrm{d}/\mathrm{d}t$), as the *convective derivative*:[15]

$$\frac{\mathrm{D}_w}{\mathrm{D}t} \equiv \frac{\partial}{\partial t} + \boldsymbol{w} \cdot \boldsymbol{\nabla} \,. \qquad (14.16)$$

In the "lab frame," a vector field $\boldsymbol{A}(\boldsymbol{r}, t)$ changes in time and space. For small $\mathrm{d}t$ and $\mathrm{d}\boldsymbol{r}$, $\boldsymbol{A}(\boldsymbol{r} + \mathrm{d}\boldsymbol{r}, t + \mathrm{d}t) \approx \boldsymbol{A}(\boldsymbol{r}, t) + \mathrm{d}t\,(\partial \boldsymbol{A}/\partial t) + (\mathrm{d}\boldsymbol{r} \cdot \boldsymbol{\nabla})\,\boldsymbol{A}$; dividing by $\mathrm{d}t$ and taking the limit $\mathrm{d}t \to 0$, we have the total time derivative $\mathrm{D}_w \boldsymbol{A}/\mathrm{D}t = \partial \boldsymbol{A}/\partial t + (\boldsymbol{w} \cdot \boldsymbol{\nabla})\,\boldsymbol{A}$, where $\boldsymbol{w} \equiv \mathrm{d}\boldsymbol{r}/\mathrm{d}t$. In a reference frame in which the fluid is at rest, $\mathrm{D}/\mathrm{D}t \to \partial/\partial t$; the convective derivative measures the time rate of change of quantities that are convected by the fluid.[16] In terms of the convective derivative, the continuity equation is expressed (from Eq. (14.15))

$$\frac{\mathrm{D}_v}{\mathrm{D}t} \rho = -\rho \boldsymbol{\nabla} \cdot \boldsymbol{v} \,. \qquad (14.17)$$

If the density changes in time, the fluid is being compressed or expanded as it flows. If the fluid is *incompressible*, $\mathrm{D}_v \rho/\mathrm{D}t = 0$, implying $\boldsymbol{\nabla} \cdot \boldsymbol{v} = 0$.

The *diffusion flow* of species k is defined as the mass flux relative to the center of mass velocity,

$$\boldsymbol{J}_k^{\mathrm{d}} \equiv \rho_k \left(\boldsymbol{v}_k - \boldsymbol{v} \right) \,. \qquad (14.18)$$

---

[14]Changes in mass due to energies released in chemical reactions ($E = mc^2$) are negligibly small.

[15]The convective derivative has many other names, e.g., the material derivative or the substantial derivative.

[16]In the lab frame, the convective derivative D_w is the directional derivative in the direction of $\boldsymbol{w}$.

Note that $\sum_k J_k^d = 0$, the total mass flux in the center of mass frame is zero. It can be shown that

$$\frac{D_v}{Dt}\rho_k = -\boldsymbol{\nabla} \cdot \boldsymbol{J}_k^d - \rho_k \boldsymbol{\nabla} \cdot \boldsymbol{v} + \sum_{j=1}^{r} \nu_{kj} J_j \,, \qquad (k=1,\cdots,n) \qquad (14.19)$$

Summing Eq. (14.19) over k results in Eq. (14.17). Equation (14.19) simplifies if we use the mass fraction $\rho_k = \rho c_k$:

$$\rho \frac{D_v}{Dt} c_k = -\boldsymbol{\nabla} \cdot \boldsymbol{J}_k^d + \sum_{j=1}^{r} \nu_{kj} J_j \,. \qquad (k=1,\cdots,n) \qquad (14.20)$$

Equation (14.20) is one of the terms we require in Eq. (14.9).

The volume term in Eq. (14.9) is easy. Because $\rho = M/V$, $v = V/M = 1/\rho$. Hence,

$$\frac{D_v}{Dt}\left(\frac{1}{\rho}\right) = -\frac{1}{\rho^2}\frac{D_v\rho}{Dt} = \frac{1}{\rho}\boldsymbol{\nabla} \cdot \boldsymbol{v} \,, \qquad (14.21)$$

where we've used Eq. (14.17). We now have two of the components of Eq. (14.9).

Momentum balance

To obtain an energy balance equation (required in Eq. (14.9)), we must start with an accounting of momentum: The time rate of change of momentum is a force, and from a force we can obtain the mechanical energy by the work done by forces.

The balance equation for the total momentum in volume V is

$$\int_V \frac{\partial \rho \boldsymbol{v}}{\partial t} dV = -\oint_S \rho \boldsymbol{v}\boldsymbol{v} \cdot d\boldsymbol{S} + \oint_S \mathbf{T} \cdot d\boldsymbol{S} + \sum_{k=1}^{n} \int_V \rho_k \boldsymbol{F}_k dV \,. \qquad (14.22)$$

The first integral on the right of Eq. (14.22) represents the net flux of momentum through the surface. The second involves the *stress tensor* **T**, which keeps track of the short-range internal forces per area acting on the surface. In the third integral, $\boldsymbol{F}_k$ is an external force (per mass) that couples to species k, a source term for creating momentum (Newton's second law). Equation (14.22) written in terms of vector components is:

$$\int_V \frac{\partial \rho v_i}{\partial t} dV = -\oint_S \rho v_i \boldsymbol{v} \cdot d\boldsymbol{S} + \sum_{j=1}^{3} \oint_S T_{ij} dS_j + \sum_{k=1}^{n} \int_V \rho_k F_{k,i} dV \,.$$

The first integral on the right gives the transport of the ith component of the momentum density, ρv_i, through S. The tensor element T_{ij} is the ith component of the force acting on the jth component of the surface area, $T_{ij} = F_i/(\Delta S_j)$. The form of the stress tensor is determined by the nature of the short-range forces acting at the surface.[17] The normal component of the surface force per area is provided by the pressure, P. The components of the stress tensor associated with normal forces are simple: $T_{ij} = -P\delta_{ij}$. This type of stress tensor ignores internal friction due to viscosity. To include viscous effects, the elements of the stress tensor are modified, with $T_{ij} = -P\delta_{ij} + \Pi_{ij}$, where Π_{ij} is an element of the *viscous stress tensor*,[18] **Π**.

We then have the local equation of momentum balance from Eq. (14.22)

$$\frac{\partial}{\partial t}(\rho \boldsymbol{v}) + \boldsymbol{\nabla} \cdot (\rho \boldsymbol{v}\boldsymbol{v}) = \sum_{k=1}^{n} \rho_k \boldsymbol{F}_k + \boldsymbol{\nabla} \cdot \mathbf{T} \,, \qquad (14.23)$$

---

[17]The stress tensor is symmetric in its indices, $T_{ij} = T_{ji}$, to conserve angular momentum.

[18]Explicit expressions for the components of the viscous stress tensor are given in books on fluid mechanics, e.g., Landau and Lifshitz.[103] The momentum convection term $\rho \boldsymbol{v}\boldsymbol{v}$ is often included in the definition of the stress tensor.

where the divergence of a second-rank tensor is a vector,

$$[\boldsymbol{\nabla} \cdot \mathbf{T}]_i \equiv \sum_j \frac{\partial}{\partial x_j} T_{ij} \, ,$$

and where we are using *dyadic notation* with $(\boldsymbol{vv})_{ij} \equiv v_i v_j$. The $\rho \boldsymbol{vv}$ term in Eq. (14.23) represents the *convection* of momentum out of the volume due to the macroscopic motion of the fluid. The term $\boldsymbol{\nabla} \cdot \mathbf{T}$ represents the creation of momentum density through internal forces (the divergence of the stress tensor is a force density). It can be shown that Eq. (14.23) is equivalent to:

$$\rho \frac{D_v}{Dt} \boldsymbol{v} = \sum_{k=1}^{n} \rho_k \boldsymbol{F}_k + \boldsymbol{\nabla} \cdot \mathbf{T} \, . \tag{14.24}$$

Equation (14.24) is the *force equation*, Newton's second law for fluids. If we ignore viscous forces ($\boldsymbol{\Pi} = 0$), Eq. (14.24) becomes the *Euler equation*. If we retain viscous forces, and treat the fluid as incompressible ($\boldsymbol{\nabla} \cdot \boldsymbol{v} = 0$), Eq. (14.24) would become the *Navier-Stokes equation*.

Energy balance

To derive the local balance equation for energy, we develop separately balance equations for kinetic and potential energies.

Kinetic energy

First we make use of the result, readily shown, that

$$\rho \frac{D_v}{Dt} v^2 = \frac{\partial}{\partial t}(\rho v^2) + \boldsymbol{\nabla} \cdot (\rho v^2 \boldsymbol{v}) \, . \tag{14.25}$$

Using the fact that $D_v v^2/Dt = 2\boldsymbol{v} \cdot D_v \boldsymbol{v}/Dt$, we find, by taking the inner product of $\boldsymbol{v}$ with Eq. (14.24), and comparing with Eq. (14.25):

$$\frac{1}{2}\frac{\partial}{\partial t}(\rho v^2) + \frac{1}{2}\boldsymbol{\nabla} \cdot (\rho v^2 \boldsymbol{v}) = \sum_{k=1}^{n} \rho_k \boldsymbol{F}_k \cdot \boldsymbol{v} + \boldsymbol{v} \cdot [\boldsymbol{\nabla} \cdot \mathbf{T}] \, . \tag{14.26}$$

Here

$$\boldsymbol{v} \cdot [\boldsymbol{\nabla} \cdot \mathbf{T}] \equiv \sum_{ij=1}^{3} v_i \frac{\partial T_{ij}}{\partial x_j} = \sum_{ij=1}^{3} \frac{\partial}{\partial x_j}(v_i T_{ij}) - \sum_{ij=1}^{3} T_{ij} \frac{\partial v_i}{\partial x_j} \equiv \boldsymbol{\nabla} \cdot [\boldsymbol{v} \cdot \mathbf{T}] - \mathbf{T} : \boldsymbol{\nabla}\mathbf{v} \, .$$

Both of these terms are scalars. Combining with Eq. (14.26), we have the work-energy theorem for fluids:

$$\frac{\partial}{\partial t}\left(\tfrac{1}{2}\rho v^2\right) = -\boldsymbol{\nabla} \cdot \left(\tfrac{1}{2}\rho v^2 \boldsymbol{v} - \boldsymbol{v} \cdot \mathbf{T}\right) + \sum_{k=1}^{n} \rho_k \boldsymbol{F}_k \cdot \boldsymbol{v} - \mathbf{T} : \boldsymbol{\nabla}\mathbf{v} \, . \tag{14.27}$$

The $\frac{1}{2}\rho v^2 \boldsymbol{v}$ term represents the convection of kinetic energy; the $\boldsymbol{v} \cdot \mathbf{T}$ term represents a *conduction* of kinetic energy, a transfer of kinetic energy through internal forces. The sources of kinetic energy involve *power*, the work done per unit time by the external forces ($\sum_k \rho_k \boldsymbol{F}_k \cdot \boldsymbol{v}$) and by the internal forces through the compression represented by $\mathbf{T} : \boldsymbol{\nabla}\mathbf{v}$.

Potential energy

Assume that $\boldsymbol{F}_k$ is derivable from a potential function, ψ_k, with $\boldsymbol{F}_k = -\boldsymbol{\nabla}\psi_k$, so that $\rho_k\psi_k$ is an energy density. Assume further that the ψ_k are time-independent, $\partial\psi_k/\partial t = 0$. The total potential energy density $\rho\psi$ is given by

$$\rho\psi \equiv \sum_{k=1}^{n} \rho_k\psi_k \ . \tag{14.28}$$

Thus,

$$\frac{\partial}{\partial t}(\rho\psi) = \sum_{k=1}^{n} \psi_k \frac{\partial\rho_k}{\partial t} = \sum_{k=1}^{n} \psi_k \left(-\boldsymbol{\nabla}\cdot(\rho_k\boldsymbol{v}_k) + \sum_{j=1}^{r} \nu_{kj} J_j \right) , \tag{14.29}$$

where we've used Eq. (14.14). If potential energy is conserved in chemical reactions, then $\sum_{k=1}^{n} \psi_k\nu_{kj} = 0$, $j = 1,\cdots,r$. The last term in Eq. (14.29) is zero for most systems. Noting that $\boldsymbol{\nabla}\cdot(\psi_k\rho_k\boldsymbol{v}_k) = \psi_k\boldsymbol{\nabla}\cdot(\rho_k\boldsymbol{v}_k) + \rho_k\boldsymbol{v}_k\cdot\boldsymbol{\nabla}\psi_k$, and $\boldsymbol{F}_k = -\boldsymbol{\nabla}\psi_k$, Eq. (14.29) can be written

$$\frac{\partial}{\partial t}(\rho\psi) = -\boldsymbol{\nabla}\cdot\left(\sum_{k=1}^{n}\psi_k\rho_k\boldsymbol{v}_k\right) - \sum_{k=1}^{n} \boldsymbol{J}_k^{\mathrm{d}}\cdot\boldsymbol{F}_k - \boldsymbol{v}\cdot\sum_{k=1}^{n}\rho_k\boldsymbol{F}_k \ . \tag{14.30}$$

Potential energy is lost by the work done per unit time by the diffusive term, $\sum_k \boldsymbol{J}_k^{\mathrm{d}}\cdot\boldsymbol{F}_k$, which is transferred into the internal energy of the system. The other term represents a conversion of potential into kinetic energy—there's an analogous term in Eq. (14.27) that it will cancel.

Total energy

Adding Eqs. (14.27) and (14.30), we have a local equation for the mechanical energy

$$\frac{\partial}{\partial t}\left(\tfrac{1}{2}\rho v^2 + \rho\psi\right) = -\boldsymbol{\nabla}\cdot\left[\tfrac{1}{2}\rho v^2 \boldsymbol{v} + \sum_{k=1}^{n}\psi_k\rho_k\boldsymbol{v}_k - \boldsymbol{v}\cdot\mathbf{T}\right] - \sum_{k=1}^{n}\boldsymbol{J}_k^{\mathrm{d}}\cdot\boldsymbol{F}_k - \mathbf{T}:\boldsymbol{\nabla}\boldsymbol{v}$$

$$\equiv -\boldsymbol{\nabla}\cdot\boldsymbol{J}_{\mathrm{mech}} + \sigma_{\mathrm{mech}} , \tag{14.31}$$

where the flux of mechanical energy and the source term are given by

$$\boldsymbol{J}_{\mathrm{mech}} = \tfrac{1}{2}\rho v^2 \boldsymbol{v} + \sum_{k=1}^{n}\psi_k\rho_k\boldsymbol{v}_k - \boldsymbol{v}\cdot\mathbf{T}$$

$$\sigma_{\mathrm{mech}} = -\sum_{k=1}^{n}\boldsymbol{J}_k^{\mathrm{d}}\cdot\boldsymbol{F}_k - \mathbf{T}:\boldsymbol{\nabla}\boldsymbol{v} \ . \tag{14.32}$$

Mechanical energy $\tfrac{1}{2}\rho v^2 + \rho\psi$ is not conserved because $\sigma_{\mathrm{mech}} \neq 0$. To achieve energy conservation, we must include the "other" potential energy, the internal energy of thermodynamics,[19] ρu (an energy density). To that end, define the total energy density ρe as the sum of the mechanical and internal energy densities,

$$\rho e \equiv \rho\left(\tfrac{1}{2}v^2 + \psi + u\right) \ . \tag{14.33}$$

We "want" the total energy to be conserved, i.e., ρe should satisfy a continuity equation

$$\frac{\partial\rho e}{\partial t} + \boldsymbol{\nabla}\cdot\boldsymbol{J}_E = 0 , \tag{14.34}$$

---

[19]See discussion on page 9 where we noted that the internal energy pertains to mechanically non-conservative systems.

where J_E is the total energy flux vector, to be determined. Because mechanical energy is not conserved, neither is the internal energy, but in such a way that the sum of mechanical and internal energy *is* conserved. Thus there will be a balance equation for internal energy:

$$\frac{\partial \rho u}{\partial t} + \nabla \cdot J_U = \sigma_U , \tag{14.35}$$

where J_U is the internal energy flux vector and σ_U is the source term. Combining these definitions:

$$\begin{aligned}
\frac{\partial \rho e}{\partial t} + \nabla \cdot J_E &= \frac{\partial \rho u}{\partial t} + \frac{\partial}{\partial t} \rho \left(\tfrac{1}{2} v^2 + \psi \right) + \nabla \cdot J_E \\
&= \sigma_U - \nabla \cdot J_U - \nabla \cdot J_{\text{mech}} + \sigma_{\text{mech}} + \nabla \cdot J_E \\
&= (\sigma_U + \sigma_{\text{mech}}) + \nabla \cdot (J_E - J_{\text{mech}} - J_U) .
\end{aligned} \tag{14.36}$$

We've introduced three symbols, J_E, J_U, and σ_U, which at this point *are unknown*. We can guarantee that Eq. (14.34) is satisfied by choosing:

$$\begin{aligned}
\sigma_U &= -\sigma_{\text{mech}} \\
J_E &= J_{\text{mech}} + J_U .
\end{aligned} \tag{14.37}$$

That leaves J_U undetermined. There should be a convection term in J_U. Thus, we take

$$J_U \equiv \rho u v + J_Q , \tag{14.38}$$

which passes the buck to J_Q, the unknown *heat flux* vector.[20] Combining Eqs. (14.38), (14.34), (14.37), and (14.32)

$$\frac{\partial (\rho u)}{\partial t} + \nabla \cdot (\rho u v + J_Q) = \sum_{k=1}^{n} J_k^{\text{d}} \cdot F_k + \mathbf{T} : \nabla v . \tag{14.39}$$

Equation (14.39) *defines* J_Q. It can be shown that Eq. (14.39) simplifies:

$$\rho \frac{\mathrm{D}_v}{\mathrm{D}t} u + \nabla \cdot J_Q = \sum_{k=1}^{n} J_k^{\text{d}} \cdot F_k + \mathbf{T} : \nabla v . \tag{14.40}$$

At this point, we use the stress tensor in the form $\mathbf{T} = -P\mathbf{I} + \mathbf{\Pi}$ in Eq. (14.40), where $\mathbf{I}$ is the unit tensor (see discussion on page 201). We have that[21]

$$\rho \frac{\mathrm{D}_v}{\mathrm{D}t} u + \nabla \cdot J_Q = \sum_{k=1}^{n} J_k^{\text{d}} \cdot F_k - P \nabla \cdot v + \mathbf{\Pi} : \nabla v . \tag{14.41}$$

Entropy balance—putting it all together

We can now assemble the pieces in Eq. (14.9). Using Eqs. (14.41), (14.21), and (14.20), we find

$$T\rho \frac{\mathrm{D}_v}{\mathrm{D}t} s = -\nabla \cdot J_Q + \sum_{k=1}^{n} \tilde{\mu}_k \nabla \cdot J_k^{\text{d}} + \sum_{k=1}^{n} J_k^{\text{d}} \cdot F_k - \sum_{j=1}^{r} A_j J_j + \mathbf{\Pi} : \nabla v , \tag{14.42}$$

---

[20] The logic of the approach here should be appreciated. The first law of thermodynamics, Eq. (14.9), is already "taken"—we're using it to define the time variation of the entropy; we can't go back to that well. We've defined a local balance equation for internal energy so that the total energy is conserved, Eq. (14.34). After pushing around so many symbols, it's perhaps not surprising that we have no independent way of defining a heat flux vector J_Q, which must be inferred from Eq. (14.39), or simply stipulated on phenomenological grounds. Heat is the transfer of energy to microscopic degrees of freedom, which we have no analytic way of characterizing other than giving it a name.

[21] $\mathbf{I} : \nabla v = \nabla \cdot v$.

where A_j, the chemical affinity,[22] drives chemical reactions (see Section 6.4)

$$A_j \equiv \sum_{k=1}^{n} \tilde{\mu}_k \nu_{kj} \ . \qquad (j = 1, \cdots, r) \ .$$

Note that the pressure P has dropped out of Eq. (14.42)—only viscous stresses contribute to entropy production.

The physical content of Eq. (14.42) can be brought out if we rewrite it as a balance equation. Using $\rho D_v s/Dt = \partial(\rho s)/\partial t + \boldsymbol{\nabla} \cdot (\rho s \boldsymbol{v})$, and the identities

$$\boldsymbol{\nabla} \cdot \left(\frac{\boldsymbol{J}_Q}{T} \right) = \frac{1}{T} \boldsymbol{\nabla} \cdot \boldsymbol{J}_Q + \boldsymbol{J}_Q \cdot \boldsymbol{\nabla} \left(\frac{1}{T} \right)$$

$$\boldsymbol{\nabla} \cdot \left(\tilde{\mu}_k \frac{\boldsymbol{J}_k^{\mathrm{d}}}{T} \right) = \frac{\tilde{\mu}_k}{T} \boldsymbol{\nabla} \cdot \boldsymbol{J}_k^{\mathrm{d}} + \boldsymbol{J}_k^{\mathrm{d}} \cdot \boldsymbol{\nabla} \left(\frac{\tilde{\mu}_k}{T} \right) \ ,$$

Eq. (14.42) can be written

$$\frac{\partial (\rho s)}{\partial t} + \boldsymbol{\nabla} \cdot \boldsymbol{J}_S = \sigma_S \ , \qquad (14.43)$$

where the entropy flux $\boldsymbol{J}_S$ and entropy source σ_S are given by

$$\boldsymbol{J}_S \equiv \rho s \boldsymbol{v} + \frac{1}{T} \left(\boldsymbol{J}_Q - \sum_{k=1}^{n} \tilde{\mu}_k \boldsymbol{J}_k^{\mathrm{d}} \right) \qquad (14.44)$$

$$\sigma_S \equiv \boldsymbol{J}_Q \cdot \boldsymbol{\nabla} \left(\frac{1}{T} \right) + \sum_{k=1}^{n} \boldsymbol{J}_k^{\mathrm{d}} \cdot \left(\frac{\boldsymbol{F}_k}{T} - \boldsymbol{\nabla} \left(\frac{\tilde{\mu}_k}{T} \right) \right) - \frac{1}{T} \sum_{j=1}^{r} A_j J_j + \frac{1}{T} \boldsymbol{\Pi} : \boldsymbol{\nabla} \mathbf{v} \ . \qquad (14.45)$$

Equations (14.43)–(14.45) are the main results of this section. Equation (14.45) is in precisely the form needed by Onsager's theory. Note, for heat flows obeying Fourier's law, with $\boldsymbol{J}_Q = -\kappa \boldsymbol{\nabla} T$, that $\boldsymbol{J}_Q \cdot \boldsymbol{\nabla}(1/T) = \kappa (\nabla T)^2 / T^2 \geq 0$.

14.4 ENTROPY FLOW AND ENTROPY CREATION

The entropy balance equation is consistent with the classification of entropy into two types, $\mathrm{d}S = \mathrm{d}S_e + \mathrm{d}S_i$ (see Eq. (3.7)), where $\mathrm{d}S_e$, the external contribution to the entropy change of a system, is what results from the transport of entropy across a boundary, and $\mathrm{d}S_i$, the internal entropy change, what's associated with irreversible processes (see Fig. 3.4). The integral form of Eq. (14.43) is:

$$\frac{\mathrm{d}}{\mathrm{d}t} \int_V \rho s \mathrm{d}V = - \oint_S \boldsymbol{J}_S \cdot \mathrm{d}\boldsymbol{S} + \int_V \sigma_S \mathrm{d}V \ . \qquad (14.46)$$

The identifications

$$S = \int_V \rho s \mathrm{d}V \qquad \frac{\mathrm{d}S_e}{\mathrm{d}t} \equiv - \oint_S \boldsymbol{J}_S \cdot \mathrm{d}\boldsymbol{S} \qquad \frac{\mathrm{d}S_i}{\mathrm{d}t} \equiv \int_V \sigma_S \mathrm{d}V$$

provide a natural way of placing Eq. (14.46) into agreement with Eq. (3.7). Equation (3.8), $\mathrm{d}S_i \geq 0$, implies that σ_S is non-negative,

$$\sigma_S \geq 0 \ . \qquad (14.47)$$

Thus, through Eq. (14.45) we have *related irreversibility to non-equilibrium phenomena*. We have, from Eqs. (14.46) and (14.47),

$$\frac{\mathrm{d}S}{\mathrm{d}t} = - \oint_S \boldsymbol{J}_S \cdot \mathrm{d}\boldsymbol{S} + \int_V \sigma_S \mathrm{d}V \geq - \oint_S \boldsymbol{J}_S \cdot \mathrm{d}\boldsymbol{S} \ , \qquad (14.48)$$

---

[22] Actually, an affinity density.

with equality holding in equilibrium. Equation (14.48) is a generalization of the Clausius inequality to open systems to allow for entropy flow.

The entropy flux, $\boldsymbol{J}_S$ (Eq. (14.44)), consists of entropy convection, $\rho s v$, in addition to a contribution from the heat flux, $T^{-1}\boldsymbol{J}_Q$, as one would expect. There is also, however, a contribution to entropy flux that's associated with the flow of particles in *open systems*: $-T^{-1}\sum_k \widetilde{\mu}_k \boldsymbol{J}_k^{\text{d}}$. For Eq. (14.48) to be consistent with the Clausius inequality, Eq. (3.6), which we can write as

$$\frac{\mathrm{d}S}{\mathrm{d}t} \geq -\oint_S \frac{\boldsymbol{J}_Q}{T} \cdot \mathrm{d}\boldsymbol{S} \,,$$

we require that on the boundary of closed systems $v = 0$ and $\boldsymbol{J}_k^{\text{d}} = 0$.

14.5   THERMOELECTRICITY

As an illustration of the Onsager theory, we consider the thermoelectric effect. A temperature difference ΔT across a thermocouple produces not only a heat current, but an electric current as well, which in turn establishes a potential difference ΔV across the thermocouple.

From Eq. (14.45), we can write σ_S as

$$\sigma_S = \boldsymbol{J}_Q \cdot \boldsymbol{\nabla}\left(\frac{1}{T}\right) + \boldsymbol{J}_E \cdot \left(\frac{\boldsymbol{F}}{T} - \boldsymbol{\nabla}\left(\frac{\widetilde{\mu}}{T}\right)\right) \equiv \boldsymbol{J}_Q \cdot \boldsymbol{X}_Q + \boldsymbol{J}_E \cdot \boldsymbol{X}_E \,, \tag{14.49}$$

where $\boldsymbol{J}_E$ is a "provisional" electric current density. We need to be mindful of units here. The units of $\boldsymbol{J}_Q$ are W m^{-2}. However, the units of what we've called the "electric" current, $\boldsymbol{J}_E$, are those of mass flux, kg m^{-2} s^{-1}. The units of $\boldsymbol{X}_E$ in Eq. (14.49) are force per mass per Kelvin, N kg^{-1} K^{-1} ($\boldsymbol{F}$ is the Coulomb force per mass). Thus, $\boldsymbol{J}_E \cdot \boldsymbol{X}_E$ has units of entropy per volume per time, what we want. To remove confusion over units, however, redefine $\boldsymbol{J}_E$ and $\boldsymbol{X}_E$ such that $\boldsymbol{J}_E \cdot \boldsymbol{X}_E = \widetilde{\boldsymbol{J}}_E \cdot \widetilde{\boldsymbol{X}}_E$, where we multiply and divide by the unit of charge so that $\widetilde{\boldsymbol{J}}_E$ has the usual units of A m^{-2} and

$$\widetilde{\boldsymbol{X}}_E = \frac{\boldsymbol{E}}{T} - \frac{1}{e}\boldsymbol{\nabla}\left(\frac{\mu}{T}\right) \equiv \frac{1}{T}\boldsymbol{\mathcal{E}} \tag{14.50}$$

has units of V m^{-1} K^{-1}, where $\boldsymbol{E}$ is the electric field, e is the magnitude of the electron charge, and μ is the usual chemical potential in Joules. The quantity $\boldsymbol{\mathcal{E}}$ in Eq. (14.50) is an effective electric field vector.

Following the Onsager theory we write, as in Eq. (14.2) (noting that $\boldsymbol{\nabla}T^{-1} = -T^{-2}\boldsymbol{\nabla}T$)

$$\widetilde{\boldsymbol{J}}_E = L_{11}\widetilde{\boldsymbol{X}}_E + L_{12}\boldsymbol{X}_Q = \frac{L_{11}}{T}\boldsymbol{\mathcal{E}} - \frac{L_{12}}{T^2}\boldsymbol{\nabla}T \tag{14.51}$$

$$\boldsymbol{J}_Q = L_{21}\widetilde{\boldsymbol{X}}_E + L_{22}\boldsymbol{X}_Q = \frac{L_{21}}{T}\boldsymbol{\mathcal{E}} - \frac{L_{22}}{T^2}\boldsymbol{\nabla}T \,, \tag{14.52}$$

where the phenomenological coefficients are such that $L_{12} = L_{21}$ (Eq. (14.3)). The quantity L_{11} has units of S K m^{-1}, L_{12} and L_{21} have units of A K m^{-1}, and L_{22} has units of W K m^{-1}.

For $\boldsymbol{\nabla}T = 0$ in Eq. (14.51), we identify L_{11} in terms of the conductivity $L_{11} = T\sigma$. Also for $\boldsymbol{\nabla}T = 0$, we have the ratio of the magnitudes of the currents,

$$\frac{J_Q}{\widetilde{J}_E} = \frac{L_{21}}{L_{11}} = \Pi \,, \tag{14.53}$$

where $\dot{Q} = \Pi I$ is the basic Pelter effect, with Π the Peltier coefficient.

For a temperature difference across the thermocouple, no current flows in the open circuit. Setting $\tilde{J}_E = 0$ in Eq. (14.51), we have $\mathcal{E} = (L_{12}/(TL_{11}))\, \nabla T$. Substituting this result in Eq. (14.52),

$$J_Q = -\frac{1}{T^2}\left[L_{22} - \frac{L_{21}L_{12}}{L_{11}}\right]\nabla T \equiv -\kappa \nabla T \, .$$

Thus, we have another connection between the Onsager coefficients and a transport coefficient,

$$\kappa = \frac{1}{T^2}\left[L_{22} - \frac{L_{21}L_{12}}{L_{11}}\right] \, .$$

Hence we have three transport coefficients (Π, σ, κ) involving the three independent Onsager coefficients. The voltage ΔV across the junction is obtained by integrating $\mathcal{E}$ around the circuit, with the result that

$$\Delta V = \int_1^2 \mathcal{E} \cdot d\boldsymbol{r} = \frac{L_{12}}{TL_{11}}\int_1^2 \nabla T \cdot d\boldsymbol{r} = \frac{L_{12}}{TL_{11}}\Delta T \, .$$

The linear relation between ΔV and ΔT is the Seebeck effect. Thus, we have

$$P = \frac{1}{T}\frac{L_{12}}{L_{11}} = \frac{1}{T}\Pi \, , \tag{14.54}$$

where we have used Eq. (14.53). Equation (14.54) is a testable prediction of the theory. We have used the Onsager phenomenological coefficients as a "bootstrap" to make a prediction that's independent of the L_{ij}.

SUMMARY

- The rate of entropy production can be given as a sum of the products of fluxes J_i and thermodynamic forces, X_i, $\dot{S} = \sum_i J_i X_i$, where there is a linear connection between forces and fluxes, $J_i = \sum_j L_{ij} X_j$, where the L_{ij} are phenomenological coefficients. Onsager's theorem is that $L_{ij} = L_{ji}$, the Onsager reciprocity relations.

- Equation (14.45) connects irreversible entropy production with non-equilibrium processes in a fluid.

- As an illustration we showed that a simple relation between the Seebeck coefficient and the Peltier coefficient is predicted by the Onsager theory, $P = \Pi/T$.

EXERCISES

14.1 Verify that Eq. (14.1) is dimensionally correct. C is the heat capacity; C_{AB} is the electrical capacitance of the junction.

14.2 a. Show for scalar quantities A and B that the convective derivative satisfies the product rule of calculus

$$\frac{D}{Dt}(AB) = A\frac{DB}{Dt} + B\frac{DA}{Dt} \, .$$

b. Show that, for any scalar function ϕ,

$$\rho\frac{D_v}{Dt}\phi = \frac{\partial}{\partial t}(\rho\phi) + \nabla \cdot (\phi\rho\boldsymbol{v}) \, .$$

Use the continuity equation. The quantity ϕ could be a scalar, or the component of a vector or a tensor.

14.3 Derive Eq. (14.19).

14.4 Show that $[\nabla \cdot (\rho vv)]_i = \rho(v \cdot \nabla)v_i + v_i \nabla \cdot \rho v$. See page 202 for the meaning of dyadic notation and the divergence of a second-rank tensor.

14.5 Show that

$$\frac{\partial (\rho v_i)}{\partial t} + [\nabla \cdot (\rho vv)]_i = \rho \frac{D_v}{Dt} v_i \ .$$

Use the continuity equation.

14.6 Show that if viscous forces are ignored, Eq. (14.24) becomes the Euler equation of fluid mechanics—the ideal classical fluid,

$$\rho \frac{D_v}{Dt} v = -\nabla P + \sum_k \rho_k F_k \ .$$

Hint: Show that $\nabla \cdot [P\mathbf{I}] = \nabla P$, where $\mathbf{I}$ is the unit tensor.

14.7 *Isentropic* flow is one for which entropy is conserved, i.e., from Eq. (14.43), $\sigma_S = 0$.

   a. Using Eq. (14.45), argue that isentropic flow implies no heat flow, and hence no temperature gradients, no diffusive flows, no chemical reactions, and neglect of viscous forces. There is only convection of entropy, as per Eq. (14.44).

   b. Show that for isentropic flow

$$\frac{D_v}{Dt} \rho s = -\rho s \nabla \cdot v \ .$$

   Argue that isentropic flow (the entropy of a small volume of fluid doesn't change as it flows) is also incompressible flow.

   c. Enthalpy was defined in Chapter 4, $H = U + PV$, the heat added at constant pressure. Define a local enthalpy, per mass, $h \equiv H/M$, so that $h = u + P/\rho$.
      i. Show that the first law of thermodynamics can be written $dh = Tds + (1/\rho)dP$. For isentropic flow, $\nabla h = \nabla(P/\rho)$.
      ii. For isentropic flows, show that the Euler equation can be written (assuming $F_k = -\nabla \psi_k$)

$$\frac{D_v}{Dt} v = -\nabla \left(h + (1/\rho) \sum_k \rho_k \psi_k \right) \ . \tag{P14.1}$$

   d. Derive, or otherwise verify, the vector identity

$$v \times \nabla \times v = \frac{1}{2}\nabla v^2 - (v \cdot \nabla) v \ .$$

   e. Show in this special case (isentropic flow) that

$$\frac{\partial v}{\partial t} = -\nabla \left(h + (1/\rho) \sum_k \rho_k \psi_k + \frac{1}{2}v^2 \right) + v \times \nabla \times v \ . \tag{P14.2}$$

   f. Finally, from Eq. (P14.2), show that

$$\frac{\partial}{\partial t} (\nabla \times v) = \nabla \times [v \times (\nabla \times v)] \ .$$

   Thus, if $\nabla \times v = 0$ at an instant of time, the velocity field remains curl-free for all times.

CHAPTER 15

Superconductors and superfluids

I N this the last chapter, we briefly introduce superconductivity and superfluidity, phenomena which are macroscopic and hence amenable to thermodynamics, yet which are quantum in nature—*macroscopic quantum phenomena.*[1] Superconductors—materials that are conductors (elements, alloys, and ceramics) and hence which possess electrical resistivity—suddenly *lose* their resistivity at low temperatures, the *critical temperature* T_c, which is different for each superconductor. Superfluids have the ability to flow without resistance. The primary example, the first to be discovered, is ^{4}He which remains in the gaseous state until temperature $T = 4.2$ K, at which point it liquefies and is known as He I. At $T_\lambda \equiv 2.17$ K (the λ-point), its properties change suddenly to that of a superfluid, what's known as He II.

15.1 LONDON THEORY

Meissner effect

The essential feature of superconductors, first observed in 1911, is that there is a precipitous loss of electrical resistance at $T = T_c$. The resistance of superconductors isn't just small for $T < T_c$, it's *zero*. It was initially thought that zero resistivity is achieved by the conductivity σ becoming infinite. Such a theory, however, leads to predictions not in accord with experimental facts. From Ohm's law, $\boldsymbol{J} = \sigma \boldsymbol{E}$, we have the simple result $\boldsymbol{E} = \boldsymbol{J}/\sigma$. Letting $\sigma \to \infty$ implies that $\boldsymbol{E} \to 0$ even though $\boldsymbol{J} \neq 0$, what's known as a *persistent current*, which *do* occur in superconductors. So far, so good. Consider Faraday's law with $\boldsymbol{E} = \boldsymbol{J}/\sigma$:

$$\boldsymbol{\nabla} \times \left(\frac{\boldsymbol{J}}{\sigma} \right) = -\frac{\partial \boldsymbol{B}}{\partial t} \ .$$

Let $\sigma \to \infty$: We would have that Faraday's law for superconductors is $\dot{\boldsymbol{B}} = 0$. If $\sigma \to \infty$ captured the correct physics, it would lead to a curious state of affairs. Start with $\boldsymbol{B} = 0$, $T > T_c$ and cool to a temperature $T < T_c$. Now turn on a magnetic field, so that $\boldsymbol{B} \neq 0$. To maintain $\dot{\boldsymbol{B}} = 0$ in the superconductor, *the magnetic field would be excluded from the superconductor.* Repeat this experiment in a different order. Start with $T > T_c$, $\boldsymbol{B} \neq 0$ and cool to $T < T_c$. In this case to preserve $\dot{\boldsymbol{B}} = 0$ we would have $\boldsymbol{B} \neq 0$ in the superconductor. Thus, two experiments that "arrive" at

---

[1]Within recent memory (that of the author), the Nobel Prize in Physics has been awarded for macroscopic quantum phenomena in 1985, 1987, 1996, 1998, 2001, and 2003. As examples we mention the Josephson effect (1973 Nobel Prize in Physics), superfluid ^{3}He, and the quantum Hall effect. While there are many device applications involving superconductivity, there are fewer applications of superfluidity, yet the field is advancing.[104]

the same set of environmental variables,[2] $B \neq 0$ and $T < T_c$ have different states for the B-field in the superconductor. If $\sigma \to \infty$ describes the physics, *it would imply that the superconducting state is not one of thermodynamic equilibrium*, because equilibrium depends only on the *state of the system* and not on how it's produced (see Chapter 1). It was shown in 1933 that in the second experiment (cool below T_c with $B \neq 0$), *the B-field is expelled from a superconductor*, a phenomenon known as the *Meissner effect* (see Fig. 15.1). The Meissner effect shows that superconductors are described not by $\dot{B} = 0$ but rather by $B = 0$, and hence that the superconducting state *is* one of thermodynamic equilibrium. Thus, *a macroscopic description of superconductors should be possible.*

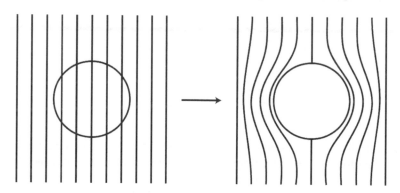

Figure 15.1   Cool a superconductor in a B-field (vertical lines) from $T > T_c$ to $T < T_c$: The superconductor expels the B-field (Meissner effect)

London equations

A successful microscopic theory of superconductivity emerged in the 1950s that makes extensive use of quantum field theory[105]—outside the scope of this book.[3] In 1935 a macroscopic theory was developed by Fritz and Heinz London that accounts for the Meissner effect[106] and which introduces many of the concepts that must be explained by the microscopic theory.

The London theory starts from the phenomenological assumption that superconductors have *two kinds of electrons*—normal and *super* electrons—with the total current density J the sum of the supercurrent J_s and the normal current, J_n: $J = J_s + J_n$. The normal current obeys Ohm's law as usual, $J_n = \sigma E$, but the super current J_s is determined by a proposed new relation with the E and B fields. The *London equations* are

$$\Lambda \nabla \times J_s = -B \tag{15.1}$$

$$\Lambda \frac{\partial J_s}{\partial t} = E , \tag{15.2}$$

where Λ is a constant characteristic of the superconductor. The dimension of Λ is $[L^2 \mu_0]$. The combination $\lambda_L \equiv \sqrt{\Lambda/\mu_0}$ is called the *London penetration depth* for reasons that we'll see. Typical values of λ_L range from 50 to 500 nm.

The London equations have an internal consistency: Equating the time derivative of Eq. (15.1) and the curl of Eq. (15.2) implies Faraday's law of induction. By introducing J_s, the London equations seem to be a way of "peeling apart" Faraday's law. We see from Eq. (15.2) that a *steady* supercurrent ($\partial J_s/\partial t = 0$) implies $E = 0$. (Equation (15.2) is a kind of Newton's first law of

---

[2]Recall that the values of intensive variables are "set" by the values of such variables in the environment, Section 3.10.
[3]The 1972 Nobel Prize in Physics was awarded for the theory developed in [105].

superconductivity—no $\boldsymbol{E}$, no change in the current.) Thus, we can have $\boldsymbol{J} \neq 0$ for $\boldsymbol{E} = 0$ (persistent currents) *without the need for infinite conductivity*. The London theory is phenomenological; Eqs. (15.1) and (15.2) are *assumptions*. Their validity derives from the experimental confirmation of the consequences they predict. We can write the supercurrent density as $\boldsymbol{J}_s = -n_s e^* \boldsymbol{v}_s$, where $\boldsymbol{v}_s$ is the velocity field of the supercurrent, n_s is the density of super electrons, and e^* is the magnitude of the charge of a super electron. Normal electrons develop a *drift velocity* when exposed to an electric field, $\boldsymbol{v}_n = -\mu \boldsymbol{E}$, where μ is the electron mobility; by Eq. (15.2), however, super electrons *accelerate* in an electric field. We would expect that $m^* d\boldsymbol{v}_s/dt = -e^* \boldsymbol{E}$, where m^* is the super-electron mass. Equation (15.2) thus implies that $\Lambda = m^*/((e^*)^2 n_s)$. In the quantum theory of superconductivity, it's found that[4] $m^* = 2m$, $e^* = 2e$, and $n_s = n/2$.

Using Faraday's law and Ohm's law for $\boldsymbol{J}_n$, we find from Eqs. (15.1) and (15.2) the generalization of the London equations for the total current $\boldsymbol{J}$:

$$\nabla \times \boldsymbol{J} = -\frac{\boldsymbol{B}}{\Lambda} - \sigma \frac{\partial \boldsymbol{B}}{\partial t} \tag{15.3}$$

$$\frac{\partial \boldsymbol{J}}{\partial t} = \frac{\boldsymbol{E}}{\Lambda} + \sigma \frac{\partial \boldsymbol{E}}{\partial t} . \tag{15.4}$$

Written this way we see that Ohm's law ($\boldsymbol{J} = \sigma \boldsymbol{E}$) is recovered in the limit $\Lambda \to \infty$.

By differentiating the continuity equation for charge ($\partial \rho/\partial t + \nabla \cdot \boldsymbol{J} = 0$) with respect to time, using Eq. (15.4), and making use of Gauss's law ($\nabla \cdot \boldsymbol{E} = \rho/\epsilon_0$) we have a second-order differential equation for the time dependence of ρ:

$$\frac{\partial^2 \rho}{\partial t^2} + \frac{\sigma}{\epsilon_0} \frac{\partial \rho}{\partial t} + \frac{1}{\Lambda \epsilon_0} \rho = 0 . \tag{15.5}$$

Try as a solution to Eq. (15.5), $\rho(t) = A_+ \exp(-\gamma_+ t) + A_- \exp(-\gamma_- t)$. We find

$$\gamma_\pm = \frac{\sigma}{2\epsilon_0} \left(1 \pm \sqrt{1 - \frac{4\epsilon_0}{\Lambda \sigma^2}} \right)$$

are the roots of $\gamma^2 - \gamma\sigma/\epsilon_0 + 1/(\Lambda\epsilon_0) = 0$. For $\Lambda \gg 4\epsilon_0/\sigma^2$, or equivalently when $\lambda_L = \sqrt{\Lambda/\mu_0} \gg 2/(c\mu_0\sigma) \approx (60\pi\sigma)^{-1}$, $\gamma_+ \approx \sigma/\epsilon_0$ and $\gamma_- \approx 1/(\Lambda\sigma)$. For good conductors $\sigma \approx 10^7$ S/m, and the inequality $\lambda_L \gg (60\pi\sigma)^{-1} \approx 0.5$ nm is easily satisfied. The quantity γ_+ is the usual rate for charge to be expelled from the interior of a conductor. Putting in numbers we find $\gamma_+ \approx 10^{18}$ s^{-1} and $\gamma_- \approx 10^{13}$ s^{-1}. The relaxation of charge from a superconductor is thus controlled by the smaller rate, $(\Lambda\sigma)^{-1}$. For frequencies $f \ll (\Lambda\sigma)^{-1} \approx 10^{13}$ Hz we conclude that $\rho = 0$ and hence that $\nabla \cdot \boldsymbol{J} = 0$.

Using Eqs. (15.3) and (15.4) together with Maxwell's equations, it's straightforward to show that in this theory $\boldsymbol{B}$, $\boldsymbol{E}$ and $\boldsymbol{J}$ each satisfy the same type of damped wave equation:

$$-\nabla \times \nabla \times \begin{Bmatrix} \boldsymbol{B} \\ \boldsymbol{E} \\ \boldsymbol{J} \end{Bmatrix} = \frac{\mu_0}{\Lambda} \begin{Bmatrix} \boldsymbol{B} \\ \boldsymbol{E} \\ \boldsymbol{J} \end{Bmatrix} + \mu_0\sigma \frac{\partial}{\partial t} \begin{Bmatrix} \boldsymbol{B} \\ \boldsymbol{E} \\ \boldsymbol{J} \end{Bmatrix} + \frac{1}{c^2} \frac{\partial^2}{\partial t^2} \begin{Bmatrix} \boldsymbol{B} \\ \boldsymbol{E} \\ \boldsymbol{J} \end{Bmatrix} . \tag{15.6}$$

The three terms on the right of Eq. (15.6) represent the contributions of the supercurrent $\boldsymbol{J}_s$, the normal current $\boldsymbol{J}_n$, and the displacement current. From Eq. (15.6) we conclude that if the inequalities $f \ll (\Lambda\sigma)^{-1} \ll \sigma/\epsilon_0$ are satisfied (the *quasistationary condition*), the *normal current and the displacement current are negligible in comparison with the supercurrent*. For $f \ll 10^{13}$ Hz, we have the equations for the electromagnetics of a superconductor

$$-\nabla \times \nabla \times \begin{Bmatrix} \boldsymbol{B} \\ \boldsymbol{E} \\ \boldsymbol{J} \end{Bmatrix} = \frac{\mu_0}{\Lambda} \begin{Bmatrix} \boldsymbol{B} \\ \boldsymbol{E} \\ \boldsymbol{J} \end{Bmatrix} . \tag{15.7}$$

---

[4]The factor of two is because it's *Cooper pairs* of electrons that are the effective super electrons of the London theory.

From Eq. (15.7), we have for the B field in particular $\nabla \times \nabla \times B = -B/\lambda_L^2$. The London theory explains the Meissner effect in that B is excluded from the *bulk* of the superconductor; however, B is nonzero within a distance λ_L of the surface. Thus, B does not vanish abruptly at the surface of a superconductor; rather it decays exponentially over the characteristic length scale λ_L. Moreover, J_s is confined to within a short distance of the surface, over the same length scale λ_L (from Eq. (15.1), the curl of J_s is largest where B is largest). Just as normal conductors screen out E-fields by moving charges to the surface, superconductors *set up surface currents that screen the interior from B-fields*.

Spherical superconductor

As an illustration, consider a spherical superconductor of radius R in a uniform external magnetic field, $B = B_0 \hat{z}$. In spherical coordinates $\hat{z} = \cos\theta \hat{r} - \sin\theta \hat{\theta}$. For $r > R$, $\nabla \cdot B = 0$ and $\nabla \times B = 0$. These equations can be solved by taking $B = \nabla \Phi_m$, where Φ_m satisfies the Laplace equation, $\nabla^2 \Phi_m = 0$. We can therefore add to the applied field the gradient of any solution of the Laplace equation, which in spherical coordinates is given as an expansion in Legendre polynomials. We need include only the $l = 1$ term; this will be matched up with B for $r < R$ through the boundary conditions at $r = R$. For $r > R$, the components of B are

$$B_r = \left(B_0 + \frac{2\mu_0 M}{r^3}\right)\cos\theta \qquad B_\theta = \left(-B_0 + \frac{\mu_0 M}{r^3}\right)\sin\theta \qquad B_\phi = 0 \,, \qquad (15.8)$$

where M is the induced magnetic moment of the sphere, and μ_0 is thrown in to get the units right. For $r < R$, we first solve for the supercurrent using Eq. (15.7), and then use Eq. (15.1) to get B. If we want the curl of J_s to have only $\hat{r}$, $\hat{\theta}$ components, it suffices to take $J_s = J_\phi(r,\theta)\hat{\phi}$. From Eq. (15.7), we require that $-\nabla \times \nabla \times J_s = J_s/\lambda^2$. An analysis of this equation shows that $J_\phi(r,\theta) = f(r)\sin\theta$ is a solution, where f satisfies the differential equation

$$f'' + \frac{2}{r}f' - \left(\frac{2}{r^2} + \frac{1}{\lambda^2}\right)f = 0 \,. \qquad (15.9)$$

The solution of Eq. (15.9) finite at $r = 0$ is

$$f(r) = C\left[\frac{1}{r^2}\sinh(r/\lambda) - \frac{1}{\lambda r}\cosh(r/\lambda)\right] \,, \qquad (15.10)$$

where C is a constant. Taking the curl of J_s, we find from Eq. (15.1) (using $\Lambda = \mu_0 \lambda^2$), that for $r < R$,

$$B_r = -2\mu_0 \lambda^2 \frac{\cos\theta}{r}f(r) \qquad B_\theta = \mu_0 \lambda^2 \frac{\sin\theta}{r}\frac{\partial}{\partial r}[rf(r)] \,. \qquad (15.11)$$

The two constants C and M are obtained by imposing the boundary condition that B_r and B_θ are continuous at $r = R$. We find that $C = 3B_0 R/(2\mu_0 \sinh(R/\lambda))$ and

$$M = -\frac{B_0 R^3}{2\mu_0}\left[1 - 3\frac{\lambda}{R}\coth(R/\lambda) + 3\frac{\lambda^2}{R^2}\right] \,. \qquad (15.12)$$

For $R \gg \lambda$, $M \approx -B_0 R^3/(2\mu_0)$.

15.2   ROTATING SUPERCONDUCTOR, LONDON MOMENT

What would a *rotating* superconductor do in the *absence* of an applied field? As we have just seen, a stationary superconductor in an external field develops a circulating supercurrent that expels the field from the interior of the superconductor. Would the rotation of super electrons *produce* a field? This is indeed what happens.

Consider a spherical superconductor of radius R rotating with constant angular velocity $\boldsymbol{\omega} = \omega\hat{z}$. Is the velocity field $\boldsymbol{v}_s$ equal to the local velocity of the superconductor at $\boldsymbol{r}$, $\boldsymbol{v}_0 \equiv \boldsymbol{\omega} \times \boldsymbol{r}$? If so, the current density of the super electrons is equal and opposite to that of the positive charge of the lattice and normal electrons. If not, there is a net current density $\boldsymbol{J} = n_s e^*(\boldsymbol{v}_0 - \boldsymbol{v}_s)$. As we show, the quantity $\boldsymbol{v}_0 - \boldsymbol{v}_s$ decays to zero over the distance λ_L. This implies that near the surface, where $\boldsymbol{v}_s \neq \boldsymbol{v}_0$, there is a net current that gives rise to a magnetic moment known as the *London moment*.

In the quasistatic approximation, Ampere's law is

$$\nabla \times \boldsymbol{B} = \mu_0 n_s e^* (\boldsymbol{v}_0 - \boldsymbol{v}_s) \ . \tag{15.13}$$

Equation (15.1) can then be written, adding $\nabla \times \boldsymbol{v}_0 = 2\boldsymbol{\omega}$ to both sides,

$$\nabla \times (\boldsymbol{v}_0 - \boldsymbol{v}_s) = -\frac{\boldsymbol{B}}{n_s e^* \Lambda} + 2\boldsymbol{\omega} \ . \tag{15.14}$$

By taking the curl of Eq. (15.14), using Eq. (15.13) and $\nabla \times \boldsymbol{\omega} = 0$, we have

$$\nabla \times \nabla \times (\boldsymbol{v}_0 - \boldsymbol{v}_s) = -\frac{1}{\lambda_L^2} (\boldsymbol{v}_0 - \boldsymbol{v}_s) \ . \tag{15.15}$$

The difference in velocities $\boldsymbol{v}_0 - \boldsymbol{v}_s$ therefore decays exponentially over a distance λ_L. The magnetic field is obtained from Eq. (15.14),

$$\boldsymbol{B} = \frac{m^*}{e^*} [2\boldsymbol{\omega} - \nabla \times (\boldsymbol{v}_0 - \boldsymbol{v}_s)] \ . \tag{15.16}$$

For a rotating superconductor there is then a finite magnetic field at the center of the sphere, $\boldsymbol{B} \approx 2m^*\boldsymbol{\omega}/e^*$. This is a kind of "inverse Meissner effect."

The solution to Eq. (15.15) is $\boldsymbol{v}_0 - \boldsymbol{v}_s = f(r)\sin\theta\hat{\phi}$, where $f(r)$ is given by Eq. (15.10). We find from Eq. (15.16) that for $r < R$,

$$B_r = \frac{m^*}{e^*} \left[2\omega - \frac{2f(r)}{r}\right]\cos\theta \qquad B_\theta = \frac{m^*}{e^*}\left[-2\omega + \frac{1}{r}\frac{\partial}{\partial r}(rf(r))\right]\sin\theta \ . \tag{15.17}$$

We impose continuity at $r = R$ between Eq. (15.17) and the form of the magnetic field for $r > R$, which is given by Eq. (15.8) with $B_0 = 0$. This gives us two conditions for the constants C and M. We find $C = -3\omega R\lambda^2/\sinh(R/\lambda)$ and

$$\mu_0 M = \omega R^3 \frac{m^*}{e^*}\left[1 - 3\frac{\lambda}{R}\coth(R/\lambda) + 3\frac{\lambda^2}{R^2}\right] \ . \tag{15.18}$$

For $R \gg \lambda$, the London moment $\mu_0 M \approx m^* R^3 \omega/e^*$. The London moment has been measured experimentally, confirming the theoretical prediction.[107][108][109] The Gravity Probe B experiment, which tested predictions of the general theory of relativity, made use of the London moment in gyroscopes made of superconducting materials.

15.3 TWO-FLUID MODEL

A macroscopic theory of superfluidity was developed by Lev Landau in 1941, the *two-fluid model*.[5][103, p507] Like the London theory, the basic idea is that He II behaves *as if* it were a

---

[5]Landau was awarded the 1962 Nobel Prize in Physics for this work.

mixture of two liquids, a *superfluid* of density ρ_s, which moves without viscosity, and a *normal fluid* of density ρ_n, which exhibits normal flow behavior. The two fluids are assumed to move without friction between them in their relative motion, i.e., no momentum is transferred from one to the other.[6] As noted by Landau and Lifshitz, "...regarding the liquid as a mixture of normal and superfluid parts is no more than a convenient description of the phenomena which occur in a fluid where quantum effects are important. Like any description of quantum phenomena in classical terms, it falls short of adequacy."[103, p507] Many properties of He II can be accounted for by the two-fluid model, without having to delve into a fully quantum-based theory.

While the model refers to a superfluid component, and a normal component, it should not be construed that He II can be separated into normal and superfluids. The model asserts that He II can execute two motions at once, one having a mass density ρ_s, moving with velocity v_s, and the other having mass density ρ_n moving with velocity v_n. At each point in the fluid, *there are two independent velocity fields,*[7] v_s and v_n. The total mass density ρ is the sum of the two densities:

$$\rho = \rho_s + \rho_n . \tag{15.19}$$

The fraction ρ_n/ρ has been measured as a function of temperature,[8] and the data fit the empirical relation[110, p66]

$$\frac{\rho_n}{\rho} = \begin{cases} (T/T_\lambda)^{5.6} & T < T_\lambda \\ 1 & T > T_\lambda . \end{cases}$$

The ratio $\rho_s/\rho = 1 - \rho_n/\rho$ decreases from unity[9] at $T = 0$, to zero at T_λ.

The total mass flux is the sum of the fluxes of the normal and superfluid components,

$$J = \rho v = \rho_s v_s + \rho_n v_n . \tag{15.20}$$

Total mass is conserved (see Eq. (14.15)):

$$\frac{\partial \rho}{\partial t} + \nabla \cdot \rho v = 0 .$$

A key idea of the two-fluid model is that entropy of He II is carried by the normal component only. Entropy is taken to be conserved, in the sense of no dissipation, i.e., the flow of the superfluid (taken as a whole) is *reversible.* This idea is expressed in a continuity equation for entropy (see Eq. (14.43))

$$\frac{\partial \rho s}{\partial t} + \nabla \cdot \rho s v_n = 0 , \tag{15.21}$$

where s is the entropy per mass (see Section 14.3). Overall momentum balance is the same as described by Eq. (14.23), except that we must include the convection of momenta for the super and normal components:

$$\frac{\partial \rho v}{\partial t} + \nabla \cdot (\rho_s v_s v_s + \rho_n v_n v_n) = -\nabla P , \tag{15.22}$$

where we ignore viscous forces, we ignore body forces, e.g., the force of gravity, and ρv is given in Eq. (15.20). The two-fluid model is thus comprised of eight scalar quantities ρ_s, ρ_n, v_s, and v_n described by (so far) five equations: the mass continuity equation, Eq. (14.15), the entropy continuity equation, Eq. (15.21), and the force equation, Eq. (15.22). We need another equation of motion.

---

[6]Dissipative effects occur in superfluids and superconductors when flow rates exceed a critical value.
[7]Just as at any point in space there can be an E-field and a B-field.
[8]The Andronikashvili, or rotating disk, experiment is the most direct method of measuring ρ_n.
[9]The same idea occurs in the theory of Bose-Einstein condensation (which requires the machinery of statistical mechanics), of two kinds of particles, those in the ground state and those in excited states.

Landau argued that the force driving the superfluid component is the gradient of the chemical potential:

$$\frac{D_{v_s}}{Dt} \boldsymbol{v}_s = -\boldsymbol{\nabla}\widetilde{\mu} = s\boldsymbol{\nabla}T - \frac{1}{\rho}\boldsymbol{\nabla}P \,, \tag{15.23}$$

where $\widetilde{\mu}$ is the chemical potential of He II per molecular mass (Section 14.3) and we've used Eq. (3.32). We now have eight equations in eight unknowns. Note that as a consequence of Eq. (15.23),

$$\frac{D_{v_s}}{Dt}\left(\boldsymbol{\nabla}\times\boldsymbol{v}_s\right) = 0 \,.$$

If the superfluid velocity field is initially curl-free, it remains that way. Landau argued that $\boldsymbol{v}_s$ must be curl-free on general grounds.

15.4 FOUNTAIN EFFECT

The equations just presented constitute the two-fluid model. It successfully accounts for many properties of He II, what would be outside the scope of this book for us to explore.[10] The *fountain effect*[11] is an experiment that demonstrates Eq. (15.23), what's schematically illustrated in Fig. 15.2. An in-

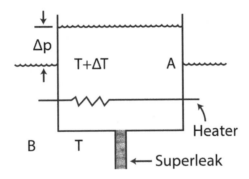

Figure 15.2 Fountain effect

ner vessel, A, is immersed in vessel B that's filled with He II at $T < T_\lambda$. Connected to A is a *superleak*, a narrow capillary packed with fine powder sufficiently dense that no normal fluid can flow through it. The heat caused by a current passed through a resistor raises the temperature of the He II in A to $T + \Delta T$. As a result of the temperature rise, some of the superfluid component is converted to normal fluid, lowering the concentration of superfluid in A relative to that in B at temperature T, and increasing the concentration of normal fluid. Superfluid flows from B to A to equalize the concentrations of superfluids, causing the level of the He II in A to rise. Only the superfluid component can flow through the capillary, however, which has zero viscosity; the normal fluid cannot flow through the capillary in an attempt to equalize the concentrations of normal fluid between A and B. The capillary therefore acts as a semi-permeable membrane. After a steady state is reached (when the heating rate to the resistor is balanced by heat losses to the walls of the vessel), the left side of Eq. (15.23) is zero, predicting a pressure difference ΔP associated with ΔT:

$$\Delta P = \rho s\Delta T \,. \tag{15.24}$$

Equation (15.24) has been well tested experimentally. If the level of He II increases by Δz, then $\Delta P = \rho g\Delta z$, where g is the gravitational acceleration, implying that $g\Delta z = s\Delta T$. The effect is quite large—see Exercise 15.3.

---

[10]The books by Wilks[44] and Putterman[111] contain many applications of the two-fluid model.
[11]The 1978 Nobel Prize in Physics was award to Pyotr Kaptisa for the fountain effect.

Heat flows from hot to cold, as does entropy (Section 3.11). In the fountain effect, however, the superfluid passing through the capillary moves *in the direction of the temperature gradient,* from cold to hot. There's no violation of the second law *if the superfluid carries no entropy.* Under this assumption, the specific entropy in B goes up (same entropy distributed over less mass), while that in A goes down (same entropy, more mass). Through the motion of the superfluid, entropy has effectively been transferred from high temperature to low. As the level of the He II rises, its rate of rise can be measured, and, knowing the cross-sectional area of the vessel, the rate of volume flow $\dot{V}$ through the superleak can be inferred. It's found experimentally that the heating rate (Joule heating from the resistor) is far higher than that required to raise by ΔT the temperature of the He II already in A, plus the heat losses to the walls. The excess heat can only go to heating up the incoming fluid. From $dQ = TdS = Ts\rho dV$, $\dot{Q} = Ts\rho\dot{V}$, and thus

$$\frac{\dot{Q}}{\rho T\dot{V}} = s \,. \tag{15.25}$$

We can calculate $\dot{Q}$ (excess heating rate after accounting for heat required to warm the He II by ΔT and the losses to the walls), we can measure ρ, we can calculate $\dot{V}$, and we can measure T. All quantities on the left of Eq. (15.25) are known. When such measurements are made, it's found that $s = s_A$, the specific entropy of the fluid[12] already in A! *The arriving superfluid through the capillary carries no entropy,* providing experimental confirmation of the assumptions of the two-fluid model.

We don't have to rely on Eq. (15.23) to arrive at Eq. (15.24). In steady state we can consider that the He II in the two vessels have reached equilibrium, implying equality of their chemical potentials,[13] $\mu(P_A, T_A) = \mu(P_B, T_B)$. A simple Taylor expansion

$$\mu(P_A, T_A) \approx \mu(P_B, T_B) + \left(\frac{\partial\mu}{\partial P}\right)\Delta P + \left(\frac{\partial\mu}{\partial T}\right)\Delta T \,,$$

implies that in steady state

$$\Delta P = -\frac{(\partial\mu/\partial T)}{(\partial\mu/\partial P)}\Delta T = \frac{S}{V}\Delta T = \rho s\Delta T \,,$$

the same as Eq. (15.24), where we've used Eq. (3.33).

EXERCISES

15.1 Derive Eq. (15.6).

15.2 Show for the spherical superconductor of radius R in an external magnetic field of strength B_0, that at the center of the sphere,

$$\lim_{r\to 0}\sqrt{B_r^2 + B_\theta^2} = B_0\frac{R/\lambda}{\sinh(R/\lambda)} \,.$$

For $R \gg \lambda$, the field strength is exponentially small at the center of the sphere.

15.3 The specific entropy of He II (inferred from heat capacity measurements) at $T = 2$ K is 0.94 J g^{-1} K^{-1}.[44, p666]. What change in height Δz is predicted to occur in the fountain effect produced by $\Delta T = 10^{-3}$ K? A: 9.6 cm.

---

[12]Entropy is calculated from heat capacity measurements.
[13]We know from Eq. (3.32) that chemical potential is a function of T and P.

Epilogue: Where to now?

W E'VE cut a wide swath through thermodynamics. The basics were presented in the first eight chapters, with the last six devoted to modern developments. What *didn't* we cover, and what are directions for further study?

Omissions

- An obvious omission is phase transitions. There is much that thermodynamics has to say about the subject. But there is even more to be said using statistical mechanics, and it seems appropriate to defer a treatment of phase transitions until statistical mechanics is developed. One of the most significant achievements in statistical mechanics in modern times was the development in the 1970s of the *renormalization group* method, in its ability to handle critical fluctuations associated with second-order phase transitions.[1]

- A topic in macroscopic physics not included in this book, but we would have wanted to include, is the theory of elasticity, where thermodynamic considerations play a vital role. Length restrictions precluded a treatment of elasticity in this book.

Limitations

- A limitation of thermodynamics is *fluctuations*, which do not naturally fit into its framework. Thermodynamics is concerned with the state of equilibrium, where its state variables have fixed values. Statistical physics is naturally able to treat fluctuations through its stronger reliance on probability methods than are used in thermodynamics.

- The state of thermodynamic equilibrium is time invariant. We did in Chapter 14 introduce non-equilibrium thermodynamics, an extension of classical thermodynamics to systems slightly out of equilibrium, where the traditional variables of thermodynamics are taken to be well defined locally, yet varying in space and time. *Kinetic theory* is a branch of statistical mechanics that can treat systems strongly out of equilibrium.

- Thermodynamics cannot incorporate interactions between atoms, except phenomenologically through equations of state such as the virial expansion. Thermodynamics is best suited for systems having *independent* constituents (ideal gas, ideal paramagnet, photon gas). Statistical mechanics naturally incorporates interactions between atoms, either in the guise of classical or quantum mechanics.

$\implies$ Fluctuations, time dependence, and interactions: Time to proceed on to statistical mechanics.

---

[1]The 1982 Nobel Prize in Physics was awarded to Kenneth Wilson for his work on the renormalization group.

Bibliography

[1] H.B. Callen. *Thermodynamics*. John Wiley, 1960.

[2] J.C. Maxwell. *The Scientific Papers of James Clerk Maxwell: Volume 2*. Dover, 1965.

[3] J.W. Gibbs. *The Scientific Papers of J. Willard Gibbs: Thermodynamics*. Ox Bow Press, 1993.

[4] K. De Raedt, H. Michielsen, H. De Raedt, et al. Massively parallel quantum computer simulator. *Computer Physics Communications*, 176:121–136, 2007.

[5] W. Kohn. Electronic structure of matter-wave functions and density functionals. *Rev. Mod. Phys.*, 71:1253–1266, 1999.

[6] R.P. Feynman. *Statistical Mechanics*. W.A. Benjamin, 1972.

[7] T.L. Hill. *Thermodynamics of Small Systems, Part I*. Dover, 1994.

[8] I.N. Sneddon. *Elements of Partial Differential Equations*. McGraw-Hill, 1957.

[9] R. Courant. *Differential and Integral Calculus, Volume II*. Interscience Publishers, 1953.

[10] R. Fowler and E.A. Guggenheim. *Statistical Thermodynamics*. Cambridge University Press, 1939.

[11] M. Planck. *Treatise on Thermodynamics*. Dover, 1945.

[12] M.W. Zemansky. *Heat and Thermodynamics*. McGraw-Hill, 1968.

[13] G.-J. Su. Modified law of corresponding states for real gases. *Industrial and Engineering Chemistry*, 38:803–806, 1946.

[14] E.A. Coddington and N. Levinson. *Theory of Ordinary Differential Equations*. McGraw-Hill, 1955.

[15] R. Clausius. *The Mechanical Theory of Heat*. John van Voorst, 1867.

[16] A.S. Eddington. *The Nature of the Physical World*. Macmillan, 1929.

[17] E.T. Jaynes. The Gibbs Paradox. In *Maximum Entropy and Bayesian Methods*, pages 1–22. Springer, 1992.

[18] I.M. Gelfand. *Lectures on Linear Algebra*. Interscience Publishers, 1961.

[19] G.F. Mazenko. *Equilibrium Statistical Mechanics*. John Wiley, 2000.

[20] L.D. Landau and E.M. Lifshitz. *Statistical Physics*. Pergamon Press, 1978.

[21] V.I. Arnold. *Mathematical Methods of Classical Mechanics*. Springer, 1989.

[22] H. Goldstein, C. Poole, and J. Safko. *Classical Mechanics*. Addison Wesley, 2002.

[23] W. Thomson. *Popular Lectures and Addresses, Vol. 1*. Cambridge University Press, 2011.

[24] S.G. Brush. *Statistical Physics and the Atomic Theory of Matter, from Boyle and Newton to Landau and Onsager*. Princeton University Press, 1983.

[25] B. Russell. *Why I Am Not a Christian*. Simon and Schuster, 1957.

[26] N.W. Ashcroft and N.D. Mermin. *Solid State Physics*. Saunders College Publishing, 1976.

[27] M. Planck. *The Theory of Heat Radiation*. Dover, 1991.

[28] M. Planck. *The Theory of Heat*. Macmillan, 1932.

[29] G.S. Rushbrooke. *Introduction to Statistical Mechanics*. Oxford University Press, 1949.

[30] D.D. Nolte. The tangled tale of phase space. *Physics Today*, 63:33–38, 2010.

[31] W.M. Haynes. *CRC Handbook of Chemistry and Physics, 93rd Edition*. CRC Press, 2012.

[32] J.D. Cox, D.D. Wagman, and V.A. Medvedev. *CODATA Key Values for Thermodynamics*. Hemisphere Publishing Corporation, 1989.

[33] R.P. Feynman. *The Character of Physical Law*. M.I.T. Press, 1967.

[34] J.W. Gibbs. *Elementary Principles in Statistical Mechanics*. Dover, 2014.

[35] E. Schrödinger. *Statistical Thermodynamics*. Cambridge University Press, 1952.

[36] R.P. Feynman. *The Feynman Lectures on Physics, Volume III*. Addison-Wesley, 1965.

[37] M. Gell-Mann and C. Tsallis. *Nonextensive Entropy*. Oxford University Press, 2004.

[38] E.T. Jaynes. Where Do We Stand on Maximum Entropy. In *The Maximum Entropy Formalism*, pages 15–118. The MIT Press, 1979.

[39] K.G. Denbigh and J.S. Denbigh. *Entropy in Relation to Incomplete Knowledge*. Cambridge University Press, 1985.

[40] G.K. White and J.G. Collins. The thermal expansion of alkali halides at low temperature. *Proc. R. Soc. Lond. A*, 333:237–259, 1973.

[41] J.O. Clayton and W.F. Giauque. The heat capacity and entropy of carbon monoxide. *J. Am. Chem. Soc.*, 54:2610–2626, 1932.

[42] F.E. Simon. On the third law of thermodynamics. *Physica*, 4:1089, 1937.

[43] H. Nishimori. *Statistical Physics of Spin Glasses and Information Processing*. Oxford University Press, 2001.

[44] J. Wilks. *The Properties of Liquid and Solid Helium*. Oxford University Press, 1967.

[45] L. Pauling. The structure and entropy of ice and other crystals with some randomness of atomic arrangement. *J. Am. Chem. Soc.*, 57:2680, 1935.

[46] W.F. Giauque and J.W. Stout. The entropy of water and the third law of thermodynamics. *J. Am. Chem. Soc.*, 58:1144, 1936.

[47] P.W. Bridgman. *The Nature of Thermodynamics*. Harvard University Press, 1941.

[48] W. Nernst. *The New Heat Theorem*. Dover, 1969.

[49] I. Prigogine and R. Defay. *Chemical Thermodynamics*. Longman Greens, 1954.

[50] E.E. Daub. Maxwell's demon. *Studies Hist. & Phil. Sci.*, 1:213–227, 1970.

[51] A. Einstein. *Albert Einstein: Philosopher Scientist, Vol. 1*. Open Court, 1949.

[52] M. Born. *Natural Philosophy of Cause and Chance*. Oxford University Press, 1949.

[53] J.B. Boyling. Carathéodory's principle and the existence of global integrating factors. *Communications in Mathematical Physics*, 10:52–68, 1968.

[54] U.M. Titulaer and N.G. van Kampen. On the deduction of Carathéodory's axiom from Kelvin's principle. *Physica*, 31:1029–1032, 1965.

[55] J. Dunning-Davies. Carathéodory's principle and the Kelvin statement of the second law. *Nature*, 208:576–577, 1965.

[56] E.M. Purcell and R.V. Pound. A nuclear spin system at negative temperature. *Physical Review*, 81:279–280, 1951.

[57] S. Braun, J.P. Ronzheimer, M. Schreiber, et al. Negative absolute temperature for motional degrees of freedom. *Science*, 339:52–55, 2013.

[58] N.F. Ramsey. Thermodynamics and statistical mechanics at negative absolute temperatures. *Phys. Rev.*, 103:20–28, 1956.

[59] P.J. Hakonen, K.K. Nummila, R.T. Vuorinen, et al. Observation of nuclear ferromagnetic ordering in silver at negative nanokelvin temperatures. *Phys. Rev. Lett.*, 68:365–368, 1992.

[60] A. Abragam. *The Principles of Nuclear Magnetism*. Oxford University Press, 1961.

[61] H.S. Leff and A.F. Rex. *Maxwell's Demon: Entropy, Information, Computing*. Princeton University Press, 1990.

[62] J.C. Maxwell. *Theory of Heat*. Dover, 2001.

[63] E. Lutz and S. Ciliberto. Information: From Maxwell's demon to Landauer's eraser. *Physics Today*, 68:30–35, 2015.

[64] L. Rosenfeld, R.S. Cohen, and J.J. Stachel. *Selected papers of Léon Rosenfeld*. Synthese library. D. Reidel Pub. Co., 1979.

[65] P. A. Skordos and W. H. Zurek. Maxwell's demon, rectifiers, and the second law: Computer simulation of Smoluchowski's trapdoor. *American Journal of Physics*, 60:876–882, 1992.

[66] L. Brillouin. Maxwell's demon cannot operate: Information and entropy. I. *Journal of Applied Physics*, 22:334–337, 1951.

[67] D. Gabor. Light and Information. In *Progress in Optics, Volume 1*, pages 111–153. North Holland Publishing, 1961.

[68] M.G. Raizen. Comprehensive control of atomic motion. *Science*, 324:1403–1406, 2009.

[69] M. D. Vidrighin, O. Dahlsten, M. Barbieri, et al. Photonic Maxwell's demon. *Phys. Rev. Lett.*, 116:050401, Feb 2016.

[70] J.A. Wheeler and W.H. Zurek. *Quantum Theory and Measurement*. Princeton University Press, 1983.

[71] L. Brillouin. *Science and Information Theory*. Dover, 2004.

[72] C.H. Bennett. Demons, engines, and the second law. *Scientific American*, 257(5):108–116, 1987.

[73] S. Toyabe, T. Sagawa, M. Ueda, et al. Experimental demonstration of information-to-energy conversion and validation of the generalized Jarzynski equality. *Nature Phys.*, 6:988–992, 2010.

[74] J.V. Koski, V.F. Maisi, J.P. Pekola, et al. Experimental realization of a Szilard engine with a single electron. *PNAS*, 111:13786–13789, 2014.

[75] C. Shannon. A mathematical theory of communication. *Bell System Technical Journal*, 27:379–423,623–656, 1948.

[76] C. Shannon and W. Weaver. *Mathematical Theory of Communication*. University of Illinois Press, 1949.

[77] E.T. Jaynes. Information theory and statistical mechanics. *Physical Review*, 106:620, 1957.

[78] A.I. Khinchin. *Mathematical Foundations of Information Theory*. Dover Publications, 1957.

[79] R. Landauer. Information is a physical entity. *Physica A*, 263:63–67, 1999.

[80] R. Landauer. Irreversibility and heat generation in the computing process. *IBM Journal of Research and Development*, 5:183–191, 1961.

[81] A. Berut, A. Arakelyan, A. Petrosyan, et al. Experimental verification of Landauer's principle linking information and thermodynamics. *Nature*, 483:187–189, 2012.

[82] J. Hong, B. Lambson, S. Dhuey, et al. Experimental test of Landauer's principle in single-bit operations on nanomagnetic memory bits. *Sci. Adv.*, 2:e1501492, 2016.

[83] Y.Jun, M. Gavrilov, and J. Bechhoefer. High-precision test of Landauer's principle in a feedback trap. *Physical Review Letters*, 113:190601, 2014.

[84] C.H. Bennett. Logical reversibility of computation. *IBM Journal of Research and Development*, 17:525–532, 1973.

[85] C.H. Bennett. The thermodynamics of computation—a review. *Int. J. Theor. Phys.*, 21:905–940, 1982.

[86] R.P. Feynman. *Feynman Lectures on Computation*. Addison-Wesley, 1996.

[87] M.A. Nielsen and I.L. Chuang. *Quantum Computation and Quantum Information*. Cambridge University Press, 2000.

[88] C.H. Bennett. Notes on the history of reversible computation. *IBM Journal of Research and Development*, 32:16–23, 1988.

[89] O. Penrose. *Foundations of Statistical Mechanics*. Pergamon Press, 1970.

[90] S.W. Hawking. Black hole explosions? *Nature*, 248:30–31, 1974.

[91] J.D. Bekenstein. Black-hole thermodynamics. *Physics Today*, 33:24–30, 1980.

[92] J.M. Bardeen, B. Carter, and S.W. Hawking. The four laws of black hole mechanics. *Commun. Math. Phys.*, 31:161–170, 1973.

[93] J.D. Bekenstein. Information in the holographic universe. *Scientific American*, 289:58–65, 2003.

[94] L. Susskind. *The Black Hole War*. Little, Brown and Company, 2009.

[95] S.W. Hawking. Gravitational radiation from colliding black holes. *Phys. Rev. Lett.*, 26:1344–1346, 1971.

[96] J.D. Bekenstein. Black holes and entropy. *Phys. Rev. D*, 7:2333–2346, 1973.

[97] R.M. Wald. "Nernst theorem" and black hole thermodynamics. *Phys. Rev. D*, 56:6467–6474, 1997.

[98] T. Jacobson. Thermodynamics of spacetime: The Einstein equation of state. *Phys. Rev. Lett.*, 75:1260–1263, 1995.

[99] E. Verlinde. On the origin of gravity and the laws of Newton. *Journal of High Energy Physics*, 1104:029, 2011.

[100] L. Onsager. Reciprocal relations in irreversible processes. I. *Phys. Rev.*, 37:405–426, 1931.

[101] L. Onsager. Reciprocal relations in irreversible processes. II. *Phys. Rev.*, 38:2265–2279, 1931.

[102] D.G. Miller. Thermodynamics of irreversible processes: The experimental verification of the Onsager reciprocal relations. *Chemical Review*, 60:15–37, 1960.

[103] L.D. Landau and E.M. Lifshitz. *Fluid Mechanics*. Pergamon Press, 1959.

[104] Y. Sato and R. Packard. Superfluid helium interferometers. *Physics Today*, 65:31–36, 2012.

[105] J. Bardeen, L.N. Cooper, and J.R. Schrieffer. Theory of superconductivity. *Phys. Rev.*, 108:1175–1204, 1957.

[106] F. London and H. London. The electromagnetic equations of the supraconductor. *Proc. R. Soc. Lond. A*, 149:71–88, 1935.

[107] A.F. Hildebrandt. Magnetic field of a rotating superconductor. *Phys. Rev. Lett.*, 12:190–191, 1964.

[108] J. Tate, S.B. Feich, and B. Cabrera. Determination of the Cooper-pair mass in niobium. *Phys. Rev. B*, 42:7885–7893, 1990.

[109] M.A. Sanzari, H.L Cui, and F. Karwacki. London moment for heavy-fermion superconductors. *Applied Physics Letters*, 68:3802–3804, 1996.

[110] F. London. *Superfluids, Volume II*. Dover, 1964.

[111] S.J. Putterman. *Superfluid Hydrodynamics*. North Holland, 1974.

Index

Milton Keynes UK
Ingram Content Group UK Ltd.
UKHW051951071024
449327UK00026B/2269